Histories of Technology, the Environment and Modern Britain

Histories of Technology, the Environment and Modern Britain

Edited by Jon Agar and Jacob Ward

First published in 2018 by
UCL Press
University College London
Gower Street
London WC1E 6BT

Available to download free: www.ucl.ac.uk/ucl-press

A CIP catalogue record for this book is available from The British Library.

ISBN: 978-1-911576-59-4 (Hbk.)
ISBN: 978-1-911576-58-7 (Pbk.)
ISBN: 978-1-911576-57-0 (PDF)
ISBN: 978-1-911576-60-0 (epub)
ISBN: 978-1-911576-61-7 (mobi)
ISBN: 978-1-911576-62-4 (html)
DOI: https://doi.org/10.14324/111.9781911576570

Acknowledgements

The editors would like to express their sincere thanks to all who helped see this collection to publication. In particular, the British Society for the History of Science and the UCL Department of Science and Technology Studies (STS) financially supported the initial workshop, 'Technology, Environment and Modern Britain', held in April 2016. Many thanks to all who participated in the workshop's presentations and discussions. Thank you to Sue Walsh for her help in organizing them. The editors would also like to express their appreciation to UCL Press, especially Chris Penfold, and all involved in the anonymous peer review process.

Contents

Figures

Contributors

Jon Agar is Professor of Science and Technology Studies at the Department of Science and Technology Studies (STS), UCL. He is the author of *Science in the Twentieth Century and Beyond* (2012), *Constant Touch: a Global History of the Mobile Phone* (2003, second edition 2013), and *The Government Machine* (2003). He was awarded the Wilkins-Bernal-Medawar Prize and Lecture by the Royal Society in 2016.

Dominic J. Berry is a Research Fellow on the Engineering Life project (ERC grant number 616510-ENLIFE). His research integrates the history, philosophy, and sociology of biology and biotechnology from 1900 to the present, concentrating on heredity, agriculture and biological engineering.

Tim Cole is Professor of Social History at the University of Bristol and currently Director of the Brigstow Institute. His research interests are primarily in historical landscapes, with a particular focus on geographies of the Holocaust. He is the author of *Images of the Holocaust/ Selling the Holocaust* (1999), *Holocaust City* (2003), *Traces of the Holocaust* (2011) and *Holocaust Landscapes* (2016), and co-editor of *Militarised Landscapes* (2010) and *Geographies of the Holocaust* (2014). He is currently writing an environmental history of post-war Britain.

Ralph Harrington studies at the School of Fine Art, History of Art and Cultural Studies at the University of Leeds.

Matthew Holmes is completing his PhD thesis at the University of Leeds exploring the history of plant biotechnology and its application to British agriculture since the 1950s, from the manipulation of crop plants through hybridisation and irradiation to the rise of genetic biotechnology. His other research interests include species history and the history of nineteenth-century natural history.

Matthew Kelly is Professor of Modern History in the Department of Humanities, Northumbria University. He works on modern British history, focusing on the development of environmental policy in the post-war period, the cultural history of landscape, and the history of National Parks and nature conservation. Matthew joined Northumbria University in 2016 as a professor. Previously he was Associate Professor at Southampton, following a British Academy Postdoctoral Research Fellowship at the University of Oxford. In 2012–13, he was a Fellow of the Rachel Carson Center, Ludwig Maximilian University, Munich and in 2016 Visiting Researcher at St John's College, Oxford.

David Matless is Professor of Cultural Geography at the University of Nottingham. He is the author of *Landscape and Englishness* (1998, revised edition 2016), and editor of *The Place of Music* (1998) and *Geographies of British Modernity* (2003). His 2014 book, *In the Nature of Landscape: Cultural Geography on the Norfolk Broads*, is a study in regional cultural landscape. A related 2015 volume, *The Regional Book*, develops work on geographical description through an account of the Broads region. His wider areas of publication include the life and work of ecologist/artist Marietta Pallis (with Laura Cameron), British geographical studies of the Eastern bloc (with Adam Swain and Jonathan Oldfield), geographies of sound, the landscapes of documentary film and the geography of ghosts. Current research addresses landscape and English identity since the 1960s, and the cultural geographies of the Anthropocene via the theme of the coastal 'Anthroposcenic'.

Mat Paskins is Postdoctoral Research Assistant on the AHRC–funded *Unsettling Scientific Stories* project, Aberystwyth University.

Esa Ruuskanen is an Academy Research Fellow in the Department of History, Cultures and Communication Studies at the University of Oulu, Finland. His research interests are environmental history and the environmental humanities with a focus on environmental values and environment–technology interaction. In recent years, he has focused on human–peatland relationships and the emergence of conservation ideas in the Nordic countries and Ireland from the eighteenth century onwards.

Simone Turchetti is a Lecturer at the Centre for History of Science, Technology and Medicine at the University of Manchester. His research interests span the history of twentieth-century science and technology, the history of the geosciences, the historiography of science, and science and international relations. He was Principal Investigator on the ERC-funded project *The Earth Under Surveillance* (TEUS).

Thomas Turnbull is a Visiting Postdoctoral Fellow at the Max Planck Institute for the History of Science (MPIWG) Department 1, having completed a PhD at the University of Oxford in 2017. His work for this volume was supported by the ESRC and the Institute for Electrical and Electronics Engineers (IEEE) Fellowship in the History of Electrical and Computing Technology.

Jessica van Horssen is a Senior Lecturer at the School of Cultural Studies & Humanities, Leeds Beckett University. She is the author of *A Town Called Asbestos: Environmental Contamination, Health, and Resilience in a Resource Community* (2016).

Jennifer Wallis is a Lecturer in Cultural and Intellectual History at the School of History, Queen Mary University of London. Her primary research interest is the history of psychiatry and medicine in the nineteenth and early twentieth centuries. She is particularly interested in integrating these fields with histories of the body; for example, looking at how patients interacted with medical technologies and the body as a site of scientific investigation and experimentation. She is completing her first monograph, *Investigating the Body in the Victorian Asylum*, and beginning work on a second, on the history of resuscitation from the nineteenth century to the present. She also works on film history, particularly British film and the horror/exploitation genres.

Jacob Ward is completing a PhD thesis at UCL on the history of information, control and research at the UK General Post Office and British Telecom, before taking up a position as The Bodleian's Byrne-Bussey Marconi Visiting Fellow, University of Oxford.

1

Technology, environment and modern Britain: historiography and intersections

Jon Agar

This collection explores the interconnected histories of technology and the environment in the context of modern Britain, broadly speaking from the late eighteenth century to the late twentieth. It is an extraordinarily rich subject, and one of immense potential. The histories of technology and the environment should be considered together for two compelling reasons. First, the artificial and the natural are not separate; technologies are made from materials that have been extracted and modified from environments, while nature has, to varying extents, been engineered. Technologies are typically assemblages, most often technological systems, with components that can be material or social in character, and many of the components will have been derived, ultimately, from natural sources. Likewise, organisms have 'become tools when human beings use them to serve human ends'.[1] This point can of course be extended to include not just organisms but modified, natural environments more generally. Environments, when cast as means towards ends, are technological in form.[2] Second, technologies and environmental and living systems share the feature of having often complex, functionally understood internal structure; they are types, even predominant types, of material organisation that surround and shape us. As organised entities they are at least as important as, say, our political structures for making us who we are, or enabling or limiting what we can do. They also, crucially, have an intertwined history. To understand the environment or technological systems of a lived-in place such as Britain, where there are layers upon layers of use and re-use, requires us to recognise and uncover their 'essential historical' character.[3] The historical analyses that emerge

1

are, necessarily, combinations of environmental history, history of technology, social, political and cultural histories.

In this introduction I have three aims. First, I will reflect on the historical studies of technology and environment, as applied or institutionalised in Britain. Second, I offer an eightfold categorisation of ways of intersection between environment and technology as a guide to thinking about the subject. These are: (1) environment as an input into a technological system; (2) environment as something natural made into, or a component within, a technological system; (3) environment as something changed, usually damaged, by outputs of technological process; (4) environment as something alongside an artificial world; (5) environment as something untouched by artifice; (6) environment as something represented through technology; (7) environmental knowledge as something organised by being registered with technology; and (8) environment and technologies as interconnected cultural imaginaries. Finally, I will aim to survey the relevant secondary literature and introduce the contributors' necessarily diverse chapters.

History of technology and environmental history in – and of – Britain

Even though the limitation is problematic, for reasons that will be stated, the historical understanding of the intersection of technologies and the natural environment can be begun (but certainly not finished) by considering the intersection of two specialities, the history of technology and environmental history. In the Anglophone world, a self-styled history of technology was institutionalised in the mid-twentieth century, with relevant markers being Londoner Charles Singer's edited volumes *A History of Technology* (first volume 1954, eventually reaching eight tomes) and the establishment of the Society for the History of Technology in 1958 in the United States.[4] Environmental history as a speciality organised itself a little later, growing rapidly in the United States in the 1980s. Both specialities could claim a roster of scholarly ancestors, from George Perkins Marsh to Lewis Mumford.

In the 1990s, the notion that the intersection of history of technology and environmental history was a growth point was already widely held. Jeffrey Stine and Joel Tarr, in their 1998 survey article and manifesto, began with the observation that it was 'difficult to write environmental history without paying at least passing attention to technology', before arguing that a 'review of past literature reveals numerous authors who have touched upon the interactions of technology and the environment, but few have pursued the topic directly'.[5] Areas of attention they found in the American literature included the environment in urban settings, public and occupational health, industry and pollution, the control of natural resources (notably water) and environmental policy and politics. 'Topics ripe for historical analysis' were also identified by Stine and Tarr.[6] The intersection has been revisited several times since, evidence of sustained historical interest.[7]

But the intersection in the case of Britain might, at first glance, seem to be stymied by the apparent weakness of both fields. Take environmental history. The prominence of environmental history in the United States has begged unflattering and unfair comparisons with the state of the subject in the United Kingdom. Clapp began his survey text with the statement that the book was 'a foray into environmental history, a branch of historical writing not yet widely practiced in Britain'.[8] Luckin in 2004 noted that in Britain environmental history has 'long remained at the margin of academic debate'.[9] He accounted for this marginalisation by the constriction of working within established scholarly frameworks for understanding industrialisation, and the availability of the social history of medicine (as well as demographic and public health histories) as an 'alternative focus' for urban-environmental studies. He also identified encouraging, if disparate, signs of change, including studies of pollution (Wohl,[10] Brimblecombe,[11] Hamlin,[12] Mosley,[13] to which could be added Thorsheim[14] and Winter[15]) and of nature/culture relations (Passmore,[16] Thomas,[17] Coates[18]) as well as new institutional homes and sources of research funding. Nevertheless, Luckin diagnosed a 'missed opportunity', as a result of which environmental history in Britain remained 'underdeveloped'. Tim Cooper, in an online survey article for the Institute of Historical Research broadly agreed with Luckin.[19] He noted that 'historical concerns with environmental questions have originated from different historical and disciplinary circumstances in Britain', not least geography, history of the British Empire,[20] economic history,[21] and landscape studies,[22] to which should be added the distinctive and immense contribution of Oliver Rackham.[23] Furthermore, despite landmark surveys of the subject by Simmons[24] and Sheail,[25] Cooper also identified 'an apparent reluctance among environmental historians working in Britain to address the environmental transformation of the British Isles'.[26]

The history of technology in the United Kingdom has attracted less direct commentary on its status and state,[27] and what there is has hardly displayed edification or a meeting of minds.[28] However, two points can be made. First, history of technology has not sustained and grown its institutional presence as a singular identity in the United Kingdom compared with, say, the United States or Germany.[29] Second, and this largely accounts for the first point, history of technology has been explored in an extremely diverse set of speciality frameworks, including history of science, cultural geography, industrial archaeology, economic and social history (especially of industrialisation), economics, history of human and veterinary medicine, agricultural history, history of architecture and design and the autochthonous historiography of engineers and other technical experts.

Technology, like the environment, is something that exists at many scales, and the national is not necessarily the best scale to choose as a frame of analysis. Indeed, a focus on the national has been superseded by interest in the transnational in both history of technology and environmental history.

Therefore the history of technology and environmental history in – and of – the United Kingdom is often hidden within many disciplinary specialities. This diversity is no bad thing. As Cooper writes, if 'we take the environment in both its material and cultural forms to form an important object of study regardless of disciplinary perspective, there is hope for a period of historical research that will be more holistic and integrative in approach'.[30] The same can be said of technology. But when we are surveying the intersection of environmental history and history of technology in modern Britain we are necessarily going to have to pull together and make sense of a heterogeneous collection of scholarship.

Eight types of combination

McNeill, in his environmental history of the twentieth century, organised his subject matter into spheres: a lithosphere and pedosphere of rock and soil, the atmosphere (urban, regional and global air and air pollution), the hydrosphere (water use and supply, rivers, seas, groundwater, dams, floods, wetlands and coasts), and biosphere (land use and agriculture, whaling and fishing and biodiversity), while also considering technological change specifically in three later case studies (chainsaws, automobiles and nuclear reactors).[31] Since each of the spheres also contained histories of technologies it would be possible to use this sort of classification to organise thoughts on how technology and environment have intersected in modern Britain. But such an organisation would also put the primacy on environmental categories. So, rather than divide the subject matter by spheres, I will review eight types of technology/

environment interaction, noting work done, introducing the chapters, and offering thoughts for further research.

(1) Environment as an input into a technological system

The engineer-turned-historian Thomas P. Hughes generalised his historical investigations of the electrical power-and-light networks[32] to offer an influential historiographical model of the growth of technological systems.[33] For Hughes, technological systems 'contain messy, complex problem-solving components', which can be 'physical artefacts, such as the turbogenerators, transformers and transmission lines in electric light and power systems', organisations, legislative artefacts and when 'they are socially constructed and adapted in order to function in systems, natural resources, such as coal mines'.[34] These systems were orchestrated by 'system builders', either independent inventors (such as Edison, typically beginning with radical invention) or corporations (such as General Electric, typically focusing on conservative, cumulative invention). Engineered natural resources are part of the system – see (2) – but outside the system lay a further 'environment':

> A technological system usually has an environment consisting of intractable factors not under the control of system managers, but these are not all organizational. If a factor in the environment – say, a supply of energy – should come under the control of the system, it is then an interacting part of it. Over time, technological systems manage increasingly to incorporate environment into the system, thereby eliminating sources of uncertainty.[35]

Take British industrial history. A system builder such as Isambard Kingdom Brunel sourced environmental inputs for his Great Western Railway. Components to which inventive focus is applied are, in Hughes' model, 'reverse salients'. The external natural environmental elements here would include coal and timber, while wooden sleepers would be classic reverse salients. Eventually (after prior experimentation with stone, which created an uncomfortable ride) sleepers were made from softwood spruce, fir or pine, imported from the Baltic, cut and laid heart-side down.[36] Timber as an input into British ship-building is discussed in this volume by Mat Paskins.[37] Sometimes the external environment to a technological system can be on immense scales. The ocean and even outer space had to be configured as safe spaces for telecommunications, as Jacob Ward shows in his chapter on the cables and satellite projects of the British Post Office.[38] Another exemplary technological system, Metropolitan Board of Works' chief engineer Joseph Bazalgette's London sewers of the 1850s and 1860s, took as inputs human excreta, waste water and rain.[39] Earlier, as Christopher Hamlin has shown, developing Hughes' analysis, the would-be systems builder Edwin Chadwick was confronted by anti-systems opponents across London's political landscape during the 'pipe-and-bricks sewer war'.[40] The linkage between sewage and British agriculture has been explored by other historians.[41]

My observation here is twofold. First, natural environmental inputs into technological systems can be found for all systems that make the infrastructure of modern Britain, and they have a history. Commodity history is an important source for such histories of environmental inputs. Second, we could, if we were bold, imagine an ambitious target of tracing all of these inputs through time – the result would be a substantial historical mapping of natural-technological system interfaces.

(2) Environment as something natural made into, or a component within, a technological system

Arthur McEvoy offered the generalisation that 'technology is the point of interaction between the human and the natural'.[42] But the value of drawing on Hughes' work is that it qualifies McEvoy's statement in important ways. Yes, the edge of a technological system is an interface, but engineered nature is found *within* technology as system components as well as nature lying outside as inputs, as in (1) above, or as something for which technological systems have consequences, as in (3) below.

A fine, worked example of engineered nature within a system can be found in Daniel Schneider's history of late nineteenth-century sewage treatment in England and the United States.[43] In places such as Enfield, Exeter and Davyhulme, the application of the new science of microbiology transformed traditional practices, intensifying and simplifying biological processes to form 'industrial ecosystems', hybrids of concrete, steel, organic waste, nematode worms, and bacteria, at the centre of sewage treatment systems.[44] The 1896 invention of Donald Cameron, city surveyor of Exeter, was one example: he called it the 'septic' tank to distinguish it from the anti-septic approach of others, in which putrefaction and odours were prevented by the deliberate killing of resident micro-organisms.[45]

Engineered nature is the subject of the chapters in this collection from Matthew Holmes (discussing barley) and, in a provocative and critical fashion, by Dominic Berry (discussing the potato), the start, perhaps, of 'new techno-environmental histories of Britain'.[46]

It is interesting to speculate what the agricultural or military history of Britain might look like if the industrialising organisms approach, deployed by Ann Greene in the case of the American Civil War, was applied to re-examine the horse as an organic, shaped component of a technological system of agricultural production or military logistics.[47] Another thought is that the technological systems approach becomes even more pertinent if we widen the 'natural' from the organic to the inorganic, in which case all components become engineered environment in source at least.

(3) Environment as something changed, usually damaged, by outputs of technological process

Attention to aspects of the environment that are changed by outputs of technological process has been the dominant theme in 'impact history' scholarship that has addressed technology and the environment. Furthermore, the changes analysed have been typically negative ones: the relationship being one of pollution, degradation or destruction. A search of the span (1995–2015) of the journal *Environment and History* reveals that 47 per cent of papers that took Britain as their area of inquiry focused on the consequences of pollutants, including such subjects as alkali pollution in St Helens, 'copper smoke' in Llanelli, stone decay in Oxford and Exeter, smoke pollution in Liverpool, post-war English beach pollution, and the side-effects of pesticides.[48]

There is a very large literature on industrial pollution, mostly urban and comparative in focus.[49] Brimblecombe set out to trace the ways 'our ancestors fouled the air' in his long history of London interior and exterior air pollution.[50] He has also, with Bowler, surveyed the subject for York.[51] Mosley showed how 'Manchester, once fêted as the "symbol of a new age", had come to epitomise the grimy, polluted industrial city: it was … "the chimney of the world"'; it was also a site where, by the 1840s, 'vegetation was all but banished from the city centre' and the term 'acid rain' was coined in 1872.[52] Bill Luckin has examined the politics of the polluted Thames of the nineteenth century, while Leslie Wood has described the technical means and measurement of partial twentieth-century restoration.[53] There are destructive impact histories from soil erosion[54] to workers' bodies[55] to whole landscapes.[56] Positive or neutral tones are rare.[57]

We might also include the impacts of *failures* of technological systems here, as well as the consequences of *working* technological systems noted above. The flood defences of Eastern England, for example, were most certainly a technological system, albeit an assemblage of relatively low technologies such as earthen banks and concrete walls, before the 1953 devastation that claimed the lives of 307 on land.[58] The Thames Barrier is a high-tech response, or 'technological fix' as analysed by Matthew Kelly in his chapter.[59] Likewise, Shane Ewen has argued, in his study of Sheffield's Great Flood of 1864, which claimed over 250 lives, that much more attention should be paid to the engineering politics as they intersected with other social interests in histories of municipal water supply.[60]

Some of the literature has been careful to show that the influence has been two-way, while still focusing on the downstream, negative consequences of technological change. The social and political response to negative changes has been an integral part of this literature on pollution. Frick traced the nineteenth-century smoke abatement movement.[61] Anti-noise campaigns in the twentieth century have been described by Bijsterveld[62] and Agar.[63] Indeed the literature on the rise of the conservation and environmental movements can be placed here.[64]

The 'paradox of technology, that environmental disruption is brought about by the industrial economy, but that advancement of the industrial economy has also been and will be a main route to environmental quality' reminds us that technologies were deployed in response to pollution or degradation.[65] There is a history here of water filters, smokeless fuels,[66] the separator device (introduced in 1926) that kept oil from bilge water,[67] the emergency oxygenation barges of the Thames,[68] and so on, much of it to be written.

The impacts of technologies are discussed in several chapters of this book, but form the focus of Ralph Harrington's chapter on the bulldozer, as both historical agent and metaphor, Tim Cole's chapter on the automobile, and Jessica van Horssen's account of the 'contamination chain' of asbestos from Canadian mines to Manchester council housing.[69] An interesting contrast to these destructive impacts is Jennifer Wallis' analysis of nineteenth-century 'aerotherapy', the marketing of which presented a 'harmonious relationship between modern machinery and "natural" landscapes'.[70]

(4) Environment as something alongside an artificial world

Let's do some weed theory. The notion that weeds were plants out-of-place (in Mary Douglas-esque terms) was commonplace by the nineteenth century.[71] This classification depends on both the existence of cultural boundaries (of garden or cereal field, say) but also the agency of organisms to transgress such boundaries. There are *longue durée* histories of weeds from ice age glacial moraine to opportunist colonisers of Neolithic fields and middens (such as fat-hen *Chenopodium album*). There are medium-term histories of weeds travelling as part of the Columbian exchange as a constituent in the formation of neo-Europes.[72] And there are histories of weeds that properly fall into later, modern periods, such as the arrival of Japanese knotweed in Victorian times, ironically popularised by the champion of the 'wild garden', William Robinson, in the 1880s, which became a notorious weed in the late twentieth century (costing, to take one example, an estimated £70 million to clear from the London Olympic site at Stratford).[73]

The intricate interplay of technological systems – roads, railways, buildings, canals and so on – creates a pattern of edges within which organisms grow. These weeds aren't transgressors in quite the same way that infuriates the gardener. Some are encouraged by the flow of substances through technological systems. Take, for example, the Danish scurvygrass (*Cochlearia danica*) that has spread from seaside to inland roadsides on the outwash of salt and grit applied as de-icer.[74] Or the nettles that spring up from nutrient-saturated ground, the product of fertiliser run-off from industrial arable farming and phosphates from household detergents.[75] Such an interaction of technological system and environment might better be classed under point (3) above.

But other weeds are more strictly just adaptive generalists, whose evolved strategies for propagation fit the niches of technological edgelands. Indeed, we might think through the weeds of modern Britain in a different way. The point is not that there is unexpected impact (as invasive species) or meaning (as cultural category) that would justify historical/sociological attention, but that there isn't – uncannily so. Therefore start from the observation of the uncanny unimportance of weeds. Human-built, artificial technological systems are set up *so* that nature is unimportant.[76] Great effort is expended to produce the smooth urban surfaces – hard surfaces with minimal cracks to make infrastructures resistant to weeds. Edgerton, in *The Shock of the Old*, rightly argues that historians of technology should pay far more attention to maintenance than they currently do.[77] 'That we neglect maintenance when we think and write about our technology', he writes, 'is an instance of the great gulf there is between our everyday understanding of our dealing with things and the formal understandings in … our histories.'[78] The examples he gives are all of mechanical maintenance: the repair of cars and aeroplanes. What he misses is the fact that maintenance is one of the activities most central to policing the boundaries of nature and technological system, and a proper subject for intersecting environmental history and history of technology. The places where this vigilant, systematic maintenance against nature is relaxed (and even then only partially) are significant – the selective growths of gardens and arable agriculture – and are where weeds are most powerfully meaningful.

Type (4) interactions are certainly not confined to the botanical. Indeed this is where we might place the environmental history of the commensal organisms of modern Britain, from the mammals (e.g. the brown rat and urban fox) and birds (the feral pigeon,[79] which moved from its restricted niche of sea cliffs to the artificial rockscape of city centres), to micro-organisms (consider the interactions of the *Legionella* bacterium with the technological system of cooling towers and air-conditioning,[80] or other ecologies of bacteria with the technological systems of antibiotic use in medicine and veterinary practice).

(5) Environment as something untouched by artifice

It is a cliché of the environmental history of (modern) Britain that there is no wilderness, no landscape that has not been shaped to some degree, usually profoundly, by human activities.[81] Many commentators have also noted that wilderness was a key concept for emerging environmental history in the United States, as well as being a critical site for its second wave.[82] Putting these two points together can be part of the explanation for why environmental history has not thrived in Britain.[83]

So it might seem that this category, of environment as something untouched by artifice, will have little application in our context. But, as Esa Ruuskanen shows in his chapter for this collection, Irish boglands were perceived as unspoilt frontiers, wastelands in need of commercial exploitation, by English observers in the late eighteenth century.[84] Furthermore, the very conditions that allow the cliché to exist – the widely held view that there are unspoiled parts of the land – mean that wilderness, or something like it, does have a relevant history as a cultural construct in modern Britain. The 'something like it' is, of course, 'countryside'. However much farmed, industrialised, and home to technologies (low and high) of all sorts, 'countryside' was still available to be mobilised as a category in opposition to city, urbanity, materialism and so on. The list of course is long, and has been the focus of one of the most important early types of environmental history-*manqué*: the cultural studies of Raymond Williams.[85] More recently, the deleterious effects of English rural nostalgia have been the subject of historians' debate.[86] The contrast between the Quantock hills as a past landscape of poets and as a new 'windscreen wilderness' as seen from the car are the subject of Tim Cole's analysis in this collection.[87]

Furthermore, beyond rurality and countryside, 'wilderness' itself has been occasionally mobilised, not least, as Rachel Woodward and Marianna Dudley have discussed, by military authorities in relation to land they have possessed, such as Salisbury Plain. '"Wilderness", the quality that first attracted the military to many of its training areas', notes Dudley, 'had been preserved by the military, at first by serendipity, and in more recent years by military-environmentalism' (in other words the deployment of environmental framings for military ends).[88] She follows Woodward in insisting that this wilderness aspect is not a given but a construction, a portrayal of the countryside used by the army to justify occupation.[89] A tiny creature, the fairy shrimp, a crustacean that breeds in the brief puddles that fill the ruts made by tank tracks, is the poster beast of military-environmentalism.[90]

The surprising military origins of another form of environmentalism, environmental concerns as a tool of diplomacy, are revealed in the chapter in this collection by Simone Turchetti.[91] As he demonstrates, NATO's environmental programme, shaped by British governmental advisers, was prompted by the UK's 'worst ever oil spill disaster', a pollution of the wilderness of the sea: the 1967 *Torrey Canyon* breach on rocks between Cornwall and the Isles of Scilly.

(6) Environment as something represented through technology

Media technologies, especially print, photography (from the mid-nineteenth century), recorded sound (from the 1870s), broadcast radio (from the 1920s), television (from the 1950s), the world wide web (from the 1990s) and social media (2000s), have been means of representing the environment in modern Britain. As such they form one type of intersection between environment and technology. There are histories here of production, transmission and consumption.[92]

The polymath environmentalist and civil servant Max Nicholson collaborated with Ludwig Koch to produce 78 rpm records of wild bird song in the 1930s.[93] Natural history film-making, to take an extended example, has a distinguished history of production in Britain. Born into a farming family, his father also a gamekeeper, Cherry Kearton began using the new 'scientific invention', the motion picture camera, to film birds in 1903.[94] He had previously provided photographs 'taken direct from nature' to illustrate his brother's book, *British Birds' Nests* (1895).

He took his camera abroad, filming in British East Africa (1909, 1910), India and Borneo (1911), North America and Canada (1912) and Central Africa (1913), in all senses following hunters' tracks, showing his films to British audiences. The trusted naturalist-traveller with a camera was developed further for a new medium, television, by David Attenborough, starting with *Zoo Quest* in 1953. Peter Scott's *Look* series for the BBC natural history unit at Bristol began in 1955.[95] Televisual authority – in Gouyon's terms the 'telenaturalist' – had to be carefully crafted, and distinguished from ethologists, such as Konrad Lorenz and Niko Tinbergen, whose authority came from science and the printed word.[96]

While movie cameras may be specially adapted for natural history film-making, the mediating technologies here (camera–darkroom–editing, room–transport–cinema projection for movie pictures; camera–darkroom–editing, room–radio transmission–television set for television) are largely unchanged by the fact that the mediated content was natural historical in character. They were not designed specifically to represent aspects of the environment. This feature distinguishes these mediating technologies from those in the next category.

(7) Environmental knowledge as something organised by being registered with technology

Interactions of types (6) and (7) are both forms of representation. Nature when it is mediated as in type (6) is certainly shaped – features are selected, a mediating frame is imposed, and so on. In type (7) this goes further: the technological system involved is expressly designed to register and represent aspects of the environment as data and ultimately as knowledge. Such technological systems include the central working tools of conservation and environmental management.

Historiographically, this topic falls within the intersection of history of technology, environmental history and, since it concerns systematised knowledge, history of science.[97] It would include the development of natural historical practices of classification, which in the modern period would include the acceptance, development and use of the eighteenth-century Linnaean system, the cartography of the Ordnance Survey and the Hydrographic Office, the ways of seeing and recording practices of natural historical societies (paying attention to the identities of 'amateurs' and 'professionals' as they came into focus in the later nineteenth century[98]), the role of instruments[99] and model organisms,[100] the work of museums,[101] and, vitally for the twentieth century, the government bureaucracies of environmental management as they co-developed with the campaigning work of civil society bodies. Modern Britain is a landscape overlaid with virtual classifications (7 per cent of land in England, to take one country, are Sites of Special Scientific Interest, SSSIs, while other acronyms – AONBs, NNRs, SPAs, SACs – go further[102]), many with Biodiversity Action Plans.

Indeed, noticing one of the most significant ways that the natural environment and technological systems intersect in modern Britain requires us to recognise bureaucracy as a technological system.[103] I have explored elsewhere the consequences of central government being both metaphorically presented as machine-like and being an organisation of clerical work that has itself been mechanised.[104] John Sheail, more than any other historian, has traced the development of conservation as an interplay of government and non-governmental organisations, although not from a history of technology perspective.[105] Thomas Turnbull, in his chapter, shows how computer models of environmental impact, a new form of systematised knowledge in the early 1970s, were received and criticised by British politicians and bureaucrats.[106]

It is crucial to put history of technology into the historical accounts of conservation and environmental management. A fine example that demonstrates why it matters is Jennifer Hubbard's work on the North Atlantic environment. She shows that fisheries biologists, from around 1900, first approached the marine environment in ways that were both framed by understandings of the terrestrial environment, but also decisively shaped by the technological systems of measurement at their disposal.[107] As these technologies changed, for example with scuba gear and undersea cameras in the 1950s, so were enabled different conceptions of the marine environment. Other British environments have similar, important histories to tell.

(8) Environment and technologies as interconnected cultural imaginaries

The final type of interaction is imaginative. Environmental components, especially organisms, can be sources of inspiration for new technologies. The theme here is not engineering as a mode of approach to life – although that is important elsewhere and has been explored by historians, mostly of the American life and human sciences[108] – but instead, as Peter Coates puts it: 'Rather than posing the question "can technology improve nature?" let us inquire, "can nature improve technology?"'[109] This has been popularised under the title 'biomimicry', but a historian's caution is needed. 'Various aspects of how technologies are naturalized by learning from nature require more rigorous investigation than they currently receive in biomimics' writings,' notes Coates; these 'include the character of the inspirational role that advocates of biomimicry claim for natural substances and processes; the relationship between "naturfact" and "artifact"; and the attitude toward non-human nature of the nature-inspired inventor'. He re-examined key cases: the nineteenth-century invention of barbed wire and the spines of the osage orange; the Wright Brothers and bird flight; and the Swiss electrical engineer George de Mestral and the invention of Velcro based on burdock burrs. Also, most relevant in a British context are the imaginative relationships between the giant, extraordinary *Victoria regia* leaf and the 1840s–1850s glass-house designs of Joseph Paxton.[110] We don't really know how typical such a mode of engineering imagination has been. At present, the historical writing on this theme is patchy.

But, on reflection, this category could and should be expanded. I have tended to take 'modern' in a minimal fashion, a mere period container. But the modern is also a substantial cultural construct. The simultaneous invocation of environment and technology is distinctive of imagery of modern Britain – think of the 'motoring pastoral' of Shell's County Guides.[111] It could be found, for example, in the centrepiece of the Festival of Britain of 1951, the Dome of Discovery (see also Tim Cole's chapter here[112]); it is discussed by David Matless in this volume in his chapter on the representation of environment and technologies in the Science Museum's iconic Agriculture Gallery, recently dismantled.[113] Yet the general history of the environmental and technological 'modern' cultural imaginary in Britain has yet to be written. David Nye wrote of a distinctive American 'technological sublime', in which the awe of the human-built world was grafted on to that of the natural.[114] Has there been such a thing as a British envirotechnological sublime? If not, why not?

Conclusion

In this chapter I have tried to provide an overview of the historiography of environmental history and history of technology as they intersect and as they relate broadly to modern Britain. In particular, I have offered an eightfold typology that maps the ways that technology and environment might be considered to interact, and show how some existing historical writing and arguments might fit within such a rubric.[115] The objective was to reveal some of the patterns of this historiography while also suggesting some ways forward. I am convinced that the intersection is an important and exciting focus for further work. In particular, there is the opportunity to write a new envirotechnical history of infrastructures, organisms, cities and countrysides. The chapters of this volume illustrate some ways that such history might be uncovered.

I will finish with a couple of provisos. First, it is clear that for many potential subjects a combination of types of interaction will be present – and, indeed, such promiscuity is to be expected. For example, a combined environmental and technological history of an event such as the 1953 North Sea flood or the 1976 drought would have nearly all types of interaction at play.[116] Second, there will be subjects that are important but should not be hammered into these categories like square pegs into round holes. Where, for example, does hunting with guns fit? As a frame for later film-making it might be described under (6), but a bullet hitting a woodpigeon is hardly a case of mere mediated representation! As a source of conservation and environmental politics, certainly, as argued by Sheail in his discussion of the shooting of seabirds for white feathers at Flamborough Head in the 1830s, it is allied to topics I have placed in (3) and (7).[117] As an aspect of British imperial identity the subject connects, rightly, to a broader cultural history.[118] It is also part of the politics of country and city, mentioned under (5).[119] Hunting with guns could be squeezed into (3), the hunted target understood as part of the natural environment that is damaged by the outputs of a technological process. Perhaps, in such cases, we can see the typology not as a one-size-fits-all classification but as a tool for the historiographical imagination, suggesting relations we had not previously considered.

Chapters in this collection explore the intersections of technological systems (from construction work to housing, from roads to satellites, from farms to flood barriers) and environments (from woods to cities, from boglands to outer space). Engineered nature is discussed by contributors, from the atmosphere in nineteenth-century air baths to new crop varieties of barley and potato. The representation of environments is also analysed, from agricultural land and technology in museum displays to the simulations of early and controversial computer models. But the hope is that this collection will also inspire. There is so much still to be researched about the intertwined histories of technologies and the environment in modern Britain.

Notes

1 Edmund Russell, 'Can Organisms Be Technology?', in *The Illusory Boundary: Environment and Technology in History*, ed. Martin Reuss and Stephen H. Cutcliffe (Charlottesville: University of Virginia Press, 2010), 249–62, 249.

2 Martin Heidegger, 'The Question Concerning Technology' (1949, published in German 1954), reprinted in *Philosophy of Technology: The Technological Condition: An Anthology*, ed. Robert C. Scharff and Val Dusek (Oxford: Blackwell, 2003), 252–64.

3 The phrase is taken from Paul David, 'Clio and the Economics of QWERTY', *American Economic Review* 75, no. 2 (1985): 332–7.

4 Charles Singer (ed.), *A History of Technology* (Oxford: Clarendon Press, 1954–8), 5 volumes. Vols. 6 and 7 edited by T.I. Williams. Vol. 1: *From Early Times to Fall of Ancient Empires*. Vol. 2: *The Mediterranean Civilizations and the Middle Ages, c.700 BC to c.1500 AD* Vol. 3: *From the Renaissance to the Industrial Revolution, c.1500–1750*. Vol. 4: *The Industrial Revolution, c.1750 to c.1850*. Vol. 5: *The Late Nineteenth Century, c.1850 to c.1900*. Vols. 6–7: *The Twentieth Century, c.1900 to c.1950*. Vol. 8: *Consolidated Indexes*.

5 Jeffrey K. Stine and Joel A. Tarr, 'At the Intersection of Histories: Technology and the Environment', *Technology and Culture* 39 (1998): 601–40, 601, 607.

6 These were: environmental regulation since the 1960s, the life-cycles of industries, the environmental consequences of military technology and war, the establishment of environmental standards and indicators, and the evolution of the profession of environmental engineering.

7 Edmund Russell, James Allison, Thomas Finger, John K. Brown, Brian Balogh and W. Bernard Carlson, 'The Nature of Power: Synthesizing the History of Technology and Environmental History', *Technology and Culture* 52 (April 2011): 246–59. Hugh S. Gorman and Betsy Mendelsohn, 'Where Does Nature End and Culture Begin? Converging Themes in the History of Technology and Environmental History', in Ruess and Cutcliffe, *The Illusory Boundary*, 265–90. Dolly Jørgensen, Finn Arne Jørgensen and Sara B. Pritchard (eds), *New Natures: Joining Environmental History with Science and Technology Studies* (Pittsburgh: University of Pittsburgh Press, 2013).

8 B.W. Clapp, *An Environmental History of Britain since the Industrial Revolution* (London: Taylor & Francis, 1994), xi.

9 Bill Luckin, 'At the Margin: Continuing Crisis in British Environmental History?' *Endeavour* 23, no. 3 (2004): 97–100.

10 Anthony S. Wohl, *Endangered Lives: Public Health in Victorian Britain* (London: Dent, 1983).

11 Peter Brimblecombe, *The Big Smoke: A History of Air Pollution in London since Medieval Times* (London: Methuen, 1988).

12 Luckin cites a forthcoming Hamlin text that never emerged in the form described. But earlier and later work is highly relevant: Christopher Hamlin, *A Science of Impurity: Water Analysis in Nineteenth-Century Britain* (Bristol: Adam Hilger, 1990). Christopher Hamlin, *Public Health and Social Justice in the Age of Chadwick: Britain, 1800–1854* (Cambridge: Cambridge University Press, 1998). Christopher Hamlin, *Cholera: the Biography* (Oxford: Oxford University Press, 2009).

13 Stephen Mosley, *The Chimney of the World: A History of Smoke Pollution in Victorian and Edwardian Manchester* (Cambridge: White Horse Press, 2001).

14 Peter Thorsheim, *Inventing Pollution: Coal Smoke and Culture in Britain since 1800* (Athens, OH: Ohio University Press, 2006).

15 James Winter, *Secure from Rash Assault: Sustaining the Victorian Environment* (Berkeley: University of California Press, 1999).

16 John Passmore, *Man's Responsibility for Nature: Ecological Problems and Western Traditions* (London: Duckworth, 1974).

17 Keith Thomas, *Man and the Natural World: Changing Attitudes in England 1500–1800* (Harmondsworth: Penguin, 1984).

18 Peter Coates, *Nature: Western Attitudes since Ancient Times* (Cambridge: Polity, 1998).

19 Tim Cooper, 'British Environmental History'. Last accessed 26 September 2017. www.history. ac.uk/makinghistory/resources/articles/environmental_history.html

20 Richard Grove, *Green Imperialism: Colonial Expansion, Tropical Island Edens and the Origins of Environmentalism, 1600–1860* (Cambridge: Cambridge University Press, 1994). William Beinart and Lotte Hughes, *Environment and Empire* (Oxford: Oxford University Press, 2007).

21 For example: E.A. Wrigley, 'The Supply of Raw Materials in the Industrial Revolution', *Economic History Review* 15 (1962): 1–16.

22 W.G. Hoskins, *The Making of the English Landscape* (London: Hodder & Stoughton, 1955).

23 Oliver Rackham, *The History of the Countryside* (London: J.M. Dent, 1986).

24 I.G. Simmons, *Environmental History of Great Britain from 10,000 Years Ago to the Present* (Edinburgh: Edinburgh University Press, 2001).

25 John Sheail, *An Environmental History of Twentieth-Century Britain* (London: Palgrave, 2002).

26 Cooper, 'British Environmental History'.

27 It is telling, for example, that there is no 'history of technology' survey produced under IHR auspices that would complement Cooper's survey on 'environmental history'.

28 For example, the symposium proceedings, 'The Current State of History of Technology in Britain', edited by Graham Hollister-Short, published in *History of Technology* 22 (2001). Notable papers included: David Edgerton, 'Reflections on the History of Technology in Britain'; Graeme Gooday, 'The Flourishing History of Technology in the United Kingdom: A Critique of Antiquarian Complaints of "Neglect"'; Rupert Hall, 'Where Is the History of Technology?'; and R. Angus Buchanan, 'Reflections on the Decline of the History of Technology in Britain'. See also David Cannadine, 'Engineering History, or the History of Engineering? Re-Writing the Technological Past', *Transactions of the Newcomen Society* 74 (2004): 163–80.

29 The institutional focus includes museums (not least the Science Museum) but also the Newcomen Society, which publishes *Transactions of the Newcomen Society*.

30 Cooper, 'British Environmental History'.

31 John Robert McNeill, *Something New under the Sun: An Environmental History of the Twentieth-Century World* (London: Allen Lane, 2000).

32 Thomas P. Hughes, *Networks of Power: Electrification in Western Society* (Baltimore: Johns Hopkins University Press, 1983).

33 Thomas P. Hughes, 'The Evolution of Large Technological Systems', in *The Social Construction of Large Technological Systems*, ed. Wiebe Bijker, Thomas P. Hughes and Trevor Pinch (Cambridge, MA: MIT Press, 1987), 51–82.

34 Hughes, 'The Evolution of Large Technological Systems', 51.

35 Hughes, 'The Evolution of Large Technological Systems', 52–3.

36 Andrew Dow, *The Railway: British Track since 1804* (Barnsley: Pen & Sword, 2015). Oak 'pads', 'chairs' and 'keys' were also part of the assembled system of track.

37 Mat Paskins, 'The Woods for the State', Chapter 12 this volume.

38 Jacob Ward, 'Oceanscapes and Spacescapes in North Atlantic Communications', Chapter 10 this volume.

39 As John Sheail notes in his review of the major work on Bazalgette's engineering, 'As a kind of interface, technology is affected by both environmental conditions and the attitudes and practices of society.' John Sheail, 'Review of Dale H. Porter', *The Thames Embankment: Environment, Technology and Society in Victorian London*, in *Environment and History* 4 (1998): 371–2.

40 Christopher Hamlin, 'Edwin Chadwick and the Engineers, 1842–1854: Systems and Anti-Systems in the Pipe-and-Brick Sewers War', *Technology and Culture* 33 (1992): 680–709.

41 Nicholas Goddard, 'A Mine of Wealth: the Victorians and the Agricultural Value of Sewage', *Journal of Historical Geography* 22 (1996): 274–90. John Sheail, 'Town Wastes, Agricultural Sustainability and Victorian Sewage', *Urban History* 23 (1996): 189–210

42 Arthur McEvoy, 'Working Environments: An Ecological Approach to Industrial Health and Safety', *Technology and Culture* 36 (supplement) (1995): S145–73, on S150, quoted by Stine and Tarr, 'At the Intersection of Histories', 604.

43 Daniel Schneider, *Hybrid Nature: Sewage Treatment and the Contradictions of the Industrial Ecosystem* (Cambridge, MA: MIT Press, 2011).

44 Schneider, *Hybrid Nature*, xvi, for 'industrial ecosystem'.

45 Schneider, *Hybrid Nature*, 47.

46 Matthew Holmes 'Crops in a Machine: Industrialising Barley Breeding in Twentieth-Century Britain', Chapter 8 this volume. Dominic Berry 'Plants Are Technologies', Chapter 9 this volume.

47   Ann N. Greene, 'War Horses: Equine Technology in the Civil War', in *Industrializing Organisms: Introducing Evolutionary History*, ed. Susan Schrepfer and Philip Scranton (London: Routledge, 2004), 143–65. Greene, *Horses at Work: Harnessing Power in Industrial America* (Cambridge, MA: Harvard University Press, 2008).

48   Richard Hawes, 'The Control of Alkali Pollution in St Helens, 1862–1890', *Environment and History* 1 (1995): 159–71. Edmund Newell and Simon Watts, 'The Environmental Impact of Industrialisation in South Wales in the Nineteenth Century: "Copper Smoke" and the Llanelli Copper Company', *Environment and History* 2 (1996): 309–36. S. Hipkins and S.F. Watts, 'Estimates of Air Pollution in York: 1381–1891', *Environment and History* 2 (1996): 337–45. Heather Viles, '"Unswept Stone, Besmeer'd By Sluttish Time": Air Pollution and Building Stone Decay in Oxford, 1790–1960', *Environment and History* 2 (1996): 359–72. Richard Hawes, 'The Municipal Regulation of Smoke Pollution in Liverpool, 1853–1866', *Environment and History* 4 (1998): 75–90. John Hassan, 'Were Health Resorts Bad for Your Health? Coastal Pollution Control Policy in England, 1945–76', *Environment and History* 5 (1999): 53–73. Rob Inkpen, 'Atmospheric Pollution and Stone Degradation in Nineteenth-Century Exeter', *Environment and History* 5 (1999): 209–20. Catherine Bowler and Peter Brimblecombe, 'Control of Air Pollution in Manchester Prior to the Public Health Act, 1875', *Environment and History* 6 (2000): 71–98. Stephen Mosley, '"A Network of Trust": Measuring and Monitoring Air Pollution in British Cities, 1912–1960', *Environment and History* 15 (2009): 273–302. T.C. Smout, 'Garrett Hardin, the Tragedy of the Commons and the Firth of Forth', *Environment and History* 17 (2011): 357–78. John Sheail, 'Pesticides and the British Environment: An Agricultural Perspective', *Environment and History* 19 (2013): 87–108. Historiographically the shift in focus has been from pollution to risk.

49   Christoph Bernhardt (ed.), *Environmental Problems of European Cities of the 19th and 20th Centuries* (Münster: Waxmann, 2000). Geneviève Massard-Guilbaud and Christoph Bernhardt (eds), *Le démon moderne: La pollution dans les sociétés urbaines et industrielles d'Europe* (Clermont-Ferrand: Presses Universitaires Blaise-Pascal, 2002). Joel Tarr and Gabriel Dupuy (eds), *Technology and the Rise of the Networked City in Europe and America* (Philadelphia: Temple University Press, 1988).

50   Brimblecombe, *The Big Smoke*, 1.

51   Peter Brimblecombe and Catherine Bowler, 'Air Pollution in York 1850–1900', in *The Silent Countdown: Essays in European Environmental History*, ed. Christian Pfister and Peter Brimblecombe (Berlin: Springer Verlag, 1990), 182–95.

52   Mosley, *The Chimney of the World*, 2, 5, 4. Peter Reed, however, argues that 'acid rain' was coined earlier by the alkali inspector Robert Angus Smith in 1859 in his article, 'On the Air of Towns', *Quarterly Journal of the Chemical Society* 10 (1859): 192–235, 232. See Reed, *Acid Rain and the Rise of the Environmental Chemist in Nineteenth-Century Britain: The Life and Work of Robert Angus Smith* (London: Ashgate, 2014).

53   Bill Luckin, *Pollution and Control: a Social History of the Thames in the Nineteenth Century* (Bristol: Adam Hilger, 1986). Leslie B. Wood, *The Restoration of the Tidal Thames* (Bristol: Adam Hilger, 1982).

54   A. Vogt, 'Aspects of Historical Soil Erosion in Western Europe', in *The Silent Countdown: Essays in European Environmental History*, ed. Christian Pfister and Peter Brimblecombe (Berlin: Springer Verlag, 1990), 83–91.

55   Damaged human bodies like 'pollution, occupational injury and disease are unwanted, unintended consequences of industrial development'. McEvoy, 'Working Environments', S146–7.

56   Ian D. Rotherham, *The Lost Fens: England's Greatest Ecological Disaster* (Stroud: The History Press, 2013). Ian D. Rotherham, 'Peat Cutters and Their Landscapes: Fundamental Change in a Fragile Environment', *Landscape Archaeology and Ecology* 4 (1999): 28–51.

57   Although see the contingent creation of Tay ecosystems from dropped ballast in John McManus, 'Ballast and the Tay Eider Duck Populations', *Environment and History* 5 (1999): 237–44.

58   Alexander Hall, 'The Rise of Blame and Recreancy in the United Kingdom: A Cultural, Political and Scientific Autopsy of the North Sea Flood of 1953', *Environment and History* 17 (2011): 379–408.

59   Stuart Gilbert and Ray Horner, *The Thames Barrier* (London: Thomas Telford, 1984). Matthew Kelly, 'The Thames Barrier: Climate Change, Shipping and the Transition to a New Envirotechnical Regime', Chapter 11 this volume.

60   Shane Ewen, 'Sheffield's Great Flood of 1864: Engineering Failure and the Municipalisation of Water', *Environment and History* 20 (2014): 177–207.

61 Carlos Frick, 'The Movement for Smoke Abatement in 19th-Century Britain', *Technology and Culture* 21 (1980): 29–50.

62 Karin Bijsterveld, *Mechanical Sound: Technology, Culture, and Public Problems in the Twentieth Century* (Cambridge, MA: MIT Press, 2008).

63 Jon Agar, 'Bodies, Machines, Noise', in *Bodies/Machines*, ed. Iwan Rhys Morus (Oxford: Berg, 2002), 197–220.

64 Dudley Stamp, *Nature Conservation in Britain* (London: Collins New Naturalist, 1969). John Sheail, *Nature in Trust: The History of Nature Conservation in Britain* (Glasgow: Blackie, 1976). John Sheail, *Nature Conservation in Britain: The Formative Years* (London: Stationery Office, 1998). W.M. Adams, *Future Nature: A Vision for Conservation*, revised edition (London: Earthscan, 1996/2003). David Evans, *A History of Nature Conservation in Britain*, second edition (London: Routledge, 1997). Horace Herring, 'The Conservation Society: Harbinger of the 1970s Environment Movement in the UK', *Environment and History* 7 (2001): 381–401. Stephen Bocking, 'Nature on the Home Front: British Ecologists' Advocacy for Science and Conservation', *Environment and History* 18 (2012): 261–81. Meredith Veldman, *Fantasy, the Bomb, and the Greening of Britain: Romantic Protest, 1945–1980* (Cambridge: Cambridge University Press, 1994).

65 Jesse H. Ausubel, Robert A. Frosch and Robert Herman, 'Technology and Environment: An Overview', in *Technology and Environment*, ed. Jesse H. Ausubel and Hedy E. Sladovich (Washington, DC: National Academy Press, 1989), 1–20, 2.

66 Peter Thorsheim, 'The Paradox of Smokeless Fuels: Gas, Coke and the Environment in Britain, 1813–1949', *Environment and History* 8 (2002): 381–401.

67 Sheail, *Nature in Trust*, 18.

68 Wood, *The Restoration of the Tidal Thames*, 137.

69 Ralph Harrington, 'Landscape with Bulldozer: Machines, Modernity and Environment in Post-War Britain', Chapter 3 this volume. Tim Cole, '*About Britain*: Driving the Landscape of Britain (at Speed?)', Chapter 7 this volume. Jessica van Horssen, 'Locality and Contamination along the Transnational Asbestos Commodity Chain', Chapter 4 this volume.

70 Jennifer Wallis, 'The Machine in the Garden: The Compressed Air-Bath and the Nineteenth-Century Health Resort', Chapter 5 this volume.

71 John Ruskin: a weed is a 'vegetable which has an innate disposition to get into the wrong place … It is not its being venomous, or ugly, but its being impertinent – thrusting itself where it has no business, and hinders other people's business – that makes a weed of it.' Quoted in Richard Mabey, *Weeds: The Story of Outlaw Plants* (London: Profile Books, 2012), 163–4.

72 Alfred W. Crosby, *The Columbian Exchange: Biological and Cultural Consequences of 1492* (Westport: Greenwood Press, 1972). Alfred W. Crosby, *Ecological Imperialism: The Biological Expansion of Europe, 900–1900* (Cambridge: Cambridge University Press, 1986).

73 Mabey, *Weeds*, 169, 4.

74 Botanical Society of Britain and Ireland (BSBI) interactive map. Last accessed 26 September 2017. http://bsbi.org/maps/?taxonid=2cd4p9h.yhp

75 Mabey, *Weeds*, 67.

76 Which is why urban bioblitzes feel countercultural, and we are surprised when they uncover so much flora and fauna. Oxford Bioblitz in 2012 made 1,420 records, including 712 records of plants. Last accessed 26 September 2017. http://scienceoxford.com/live/bioblitz

77 David Edgerton, *The Shock of the Old: Technology and Global History since 1900* (London: Profile, 2006), ch. 4.

78 Edgerton, *The Shock of the Old*, 77.

79 E. Simms, *The Public Life of the Street Pigeon* (London: Hutchinson Radius, 1979).

80 Last accessed 26 September 2017. www.hse.gov.uk/legionnaires/

81 Simmons: 'as will be demonstrated in the course of this book, there are few if any nooks of Great Britain which remain unaffected by human activity'. Simmons, *The Public Life of the Street Pigeon*, 8.

82 William Cronon, 'The Trouble with Wilderness; Or, Getting Back to the Wrong Nature', *Environmental History* 1 (1996): 7–28.

83 Luckin, 'At the Margin', 97.

84 Esa Ruuskanen, 'Encroaching Irish Bogland Frontiers: Science, Policy and Aspirations from the 1770s to 1840s', Chapter 2 this volume.

85 Raymond Williams, *The Country and the City* (London: Chatto & Windus, 1973). Raymond Williams, *Culture and Society 1780–1950* (London: Chatto & Windus, 1958).

86 Martin J. Wiener, *English Culture and Decline of Industrial Spirit* (Cambridge: Cambridge University Press, 1981). Alun Howkins, 'The Discovery of Rural England', in *Englishness: Politics and Culture*, ed. Robert Colls and Philip Dodd (London: Croom Helm, 1986), 62–88. Ben Knights, 'In Search of England: Travelogue and Nation between the Wars' and Stephan Kohl, 'Rural England: An Invention of the Motor Industries', both in *Landscape and Englishness*, ed. Robert Burden and Stephan Kohl (Amsterdam: Rodopi, 2006): 165–84, 185–206. Peter Mandler, 'Against Englishness: English Culture and the Limits to Rural Nostalgia, 1850–1940', *Transactions of the Royal Historical Society* 7 (1997): 155–75. David Matless, *Landscape and Englishness* (London: Reaktion, 1998).

87 Tim Cole, '*About Britain*: Driving the Landscape of Britain (at Speed?)', Chapter 7 this volume.

88 Marianna Dudley, *An Environmental History of the UK Defence Estate, 1945 to the Present* (London: Continuum, 2012), 9.

89 Rachel Woodward, *Military Geographies* (Oxford: Blackwell, 2004).

90 Marianna Dudley, 'A Fairy (Shrimp) Tale of Military Environmentalism: The "Greening" of Salisbury Plain', in *Militarized Landscapes: From Gettysburg to Salisbury Plain*, ed. Chris Pearson, Peter Coates and Tim Cole (London: Bloomsbury, 2010), 135–50.

91 Simone Turchetti, 'HMG's Environmentalism: Britain, NATO and the Origins of Environmental Diplomacy', Chapter 13 this volume.

92 For film and the United States, see Gregg Mitman, *Reel Nature: America's Romance with Wildlife on Film* (Cambridge, MA: Harvard University Press, 1999). For photography, see Jennifer Tucker, *Nature Exposed: Photography as Eyewitness in Victorian Science* (Baltimore: Johns Hopkins University Press, 2005). Steve Garlick, 'Given Time: Biology, Nature and Photographic Vision', *History of the Human Sciences* 22 (2009): 81–101. See also the striking view from above revealed in Kitty Hauser, *Bloody Old Britain: O.G.S. Crawford and the Archaeology of Modern Life* (London: Granta Books, 2008).

93 David Matless, 'Versions of Animal-Human: Broadland, c.1945–1970', in *Animal Spaces, Beastly Places*, ed. Chris Philo and Chris Wilbert (London: Routledge, 2000), 117–41, 126.

94 Jean-Baptiste Gouyon, 'From Kearton to Attenborough: Fashioning the Telenaturalist's Identity', *History of Science* 49 (2011): 25–60, 173.

95 The BBC Natural History Unit was formally established at Bristol in 1957, and Attenborough was asked to head it. He declined but remained the public face of natural history film-making for the corporation. Gouyon, 'From Kearton to Attenborough', 48. See also Jean-Baptiste Gouyon, 'The BBC Natural History Unit: Instituting Natural History Film-Making in Britain', *History of Science* 49 (2011): 425–51. Gail Davies, 'Science, Observation and Entertainment: Competing Visions of Post-War British Natural History Television, 1946–1967', *Ecumene* 7 (2000): 432–59.

96 A telenaturalist 'is a naturalist who self-confidently practices natural history on the television screen, for television, and whose claims to cognitive credibility the audience is asked to accept partly on the "television and showbiz derived basis", notably a "focus on personalities"'. Gouyon, 'From Kearton to Attenborough', 47.

97 Nicholas Jardine, James A. Secord and E.C. Spary (eds), *Cultures of Natural History* (Cambridge: Cambridge University Press, 1996).

98 David E. Allen, 'On Parallel Lines: Natural History and Biology from the Late Victorian Period', *Archives of Natural History* 25 (1998): 361–71. Anne Secord, 'Science in the Pub: Artisan Botanists in Early Nineteenth-Century Lancashire', *History of Science* 32 (1994): 269–315. Anne Secord, 'Corresponding Interests: Artisans and Gentlemen in Nineteenth-Century Natural History', *British Journal for the History of Science* 27 (1994): 383–408. Samuel J.M.M. Alberti, 'Amateurs and Professionals in One County: Biology and Natural History in Late Victorian Yorkshire', *Journal of the History of Biology* 34 (2001): 115–47.

99 Graeme Gooday, '"Nature" in the Laboratory: Domestication and Discipline with the Microscope in Victorian Life Science', *British Journal for the History of Science* 24 (1991): 307–41.

100 Jim Endersby, *A Guinea Pig's History of Biology: The Plants and Animals Who Taught Us the Facts of Life* (London: Heinemann, 2007).

101 Including 'museological' ways of knowing: John V. Pickstone, 'Museological Science? The Place of the Analytical/Comparative in 19th-Century Science, Technology and Medicine', *History of Science* 32 (1994): 111–38. See also Samuel J.M.M. Alberti, 'Objects and the Museum', *Isis* 96 (2005): 559–71. Samuel J.M.M. Alberti, 'Placing Nature: Natural History

Collections and Their Owners in Nineteenth-Century Provincial England', *British Journal for the History of Science* 35 (2002): 291–311.

102 Areas of Natural Beauty (AONB), National Nature Reserves (NNR), Special Protection Areas (SPAs), Special Areas of Conservation (SACs). Land and sea can also be classified as Ramsar Sites, Biosphere Reserves, National Parks, Local Nature Reserves, Nitrate Vulnerable Zones, Inshore and Offshore Special Protection Areas and Marine Conservation Zones. Wales, Scotland and Northern Ireland have subtly different schemes.

103 Nature writers are far more likely to reflect on literary rather than bureaucratic representations of the environment. We need more 'ecobureaucratocriticism' rather than more literary ecocriticism.

104 Jon Agar, *The Government Machine: A Revolutionary History of the Computer* (Cambridge, MA: MIT Press, 2003).

105 John Sheail, *Nature in Trust*. Sheail, *Nature Conservation in Britain: The Formative Years*. See also Stephen Bocking, *Ecologists and Environmental Politics: A History of Contemporary Ecology* (New Haven: Yale University Press, 1997).

106 Thomas Turnbull, 'Simulating the Global Environment: The British Government's Response to *The Limits to Growth*', Chapter 14 this volume.

107 Jennifer Hubbard, 'Mediating the North Atlantic Environment: Fisheries Biologists, Technology, and Marine Spaces', *Environmental History* 18 (2013): 88–100.

108 Philip J. Pauly, *Controlling Life: Jacques Loeb and the Engineering Ideal in Biology* (Oxford: Oxford University Press, 1987). Rebecca Lemov, *The World as Laboratory: Experiments with Mice, Mazes, and Men* (New York: Hill & Wang, 2005).

109 Peter Coates, 'Can Nature Improve Technology?' in Reuss and Cutcliffe, *The Illusory Boundary*, 43–65, 44

110 Joseph Paxton (1850): the giant water-lily was a 'beautiful example of natural engineering in the cantilevers which radiate from the centre', 'You will observe that nature is the engineer in this case' (quoted in Anon., 'The Grand Industrial Exhibition for 1851', *Illustrated London News* (17 August 1850), p. 141, quoted in Coates, 'Can Nature Improve Technology?' p. 54).

111 David Matless, *Landscape and Englishness*, 63–4.

112 Tim Cole, '*About Britain*: Driving the Landscape of Britain (at Speed?)', Chapter 7 this volume.

113 David Matless 'The Agriculture Gallery: Displaying Modern Farming in the Science Museum', Chapter 6 this volume.

114 David E. Nye, *American Technological Sublime* (Cambridge, MA: MIT Press, 1994).

115 There are still plenty of gaps. For environmental history it might be worth revisiting the list offered by Luckin, 'Continuing Crisis', 99: no book-length study of the Alkali Inspectorate, 'no monograph on political and social struggles over access to water in any major 19th century city', 'no authoritative account of the environmental responsibilities and achievements of the municipality in 19th- and 20th-century Britain', more on sewage from the 1850s to the First World War, the 'dustbin revolution', and recycling. Note for recycling: Peter Thorsheim, *Waste into Weapons: Recycling in Britain during the Second World War* (Cambridge: Cambridge University Press, 2015).

116 Alexander Hall, 'The Rise of Blame and Recreancy in the United Kingdom'. Vanessa Taylor, Heather Chappells, Will Medd and Frank Trentmann, 'Drought is Normal: The Socio-Technical Evolution of Drought and Water Demand in England and Wales, 1893–2006', *Journal of Historical Geography* 35 (2009): 568–91.

117 Sheail, *Nature in Trust*, 4. David E. Allen, *The Naturalist in Britain: A Social History* (London: Allen Lane, 1976).

118 John M. MacKenzie, *The Empire of Nature: Hunting, Conservation, and British Imperialism* (Manchester: Manchester University Press, 1988).

119 Michael Tichelar, '"Putting Animals into Politics": The Labour Party and Hunting in the First Half of the Twentieth Century', *Rural History* 17 (2006) 17: 213–34.

2

Encroaching Irish bogland frontiers: science, policy and aspirations from the 1770s to the 1840s

Esa Ruuskanen

Introduction

I begin this chapter with quotes by Arthur Young (1741–1820), an English social and political observer and writer, who spent years in Ireland in the 1770s and published a detailed account of his observations and a survey, extensively cited later in the nineteenth century, based on the knowledge he gathered on that journey. When he covered the topography of the island, he marvelled how 'the bogs, of which foreigners have heard so much, are very extensive in Ireland'. As a man who had devoted his life to the improvement of agriculture and the social conditions of the rural poor, he maintained that 'the means of improving them [bogs] is the most important consideration at present'. Young envisioned how a country so widely covered with 'wastelands' could possess a lush, cultivated countryside and wealthier population and serve as the granary of the industrialising England.[1]

Arthur Young's writing in a sense bespeaks the quite common notions and, moreover, visions of the future of late eighteenth- and early nineteenth-century Western and Northern European elite and scholars related to boggy, fenny and marshy areas. First of all, there were vast tracts of bogs or mires still in an almost wholly natural condition in the frontiers near or beyond the settled and almost wholly artificial areas of the then European economic or political powers – not only in Ireland but also in East Friesland, Livonia, Moscow and Nizhny Novgorod Oblasts, Norrland, Ostrobothnia and *Sápmi* (Finnish, Norwegian and Swedish Lapland) among others.[2] When we think about the most famous frontiers in the nineteenth century we usually name American prairies, the backwoods of colonial Africa and Southeast Asia, Siberian woodlands or the polar regions. These frontiers and the ambitions to map and tap them attained far more publicity in late eighteenth- and early nineteenth-century European media and popular culture than peatlands, which, nevertheless, also became the objects of economic, political and scientific meaning-making, which unavoidably began to shape the society–nature nexus in these particular areas. Bogs, fens, marshes and mires, as such, were conceived as wastelands and barriers to progress or civilisation. Increasingly commercial and technology- and science-oriented perspectives on peatlands and wetlands gathered momentum, aiming at transforming these areas into territories that were seen as being made valuable by local people or by massive intervention of state or enterprises. Consequently, quite distant and distinct environments were bound together when it comes to the appraisal of nature.

This chapter seeks to explain these meaning-making, valuation and commercialisation processes concerning bogs in Ireland from the 1770s to the 1840s. The Industrial Revolution, a crucial landmark considering the development of a new mindset regarding the set of assumptions about and notions of frontiers or wastelands, originated in Ireland in the 1770s. It was during that decade that Arthur Young also published his survey of Ireland, which, compared with earlier writings on the matter, was unique in its thoroughness and, should I say, certain imperial ethos. At the beginning of the nineteenth century the amount of publications and reports focusing on how to tame bogs increased and the first commission was charged with finding ways to improve these wastelands. The temporal end of this study, in turn, is in the 1840s before the Great Famine, which started the reassessment of past policies and priorities. Besides, late nineteenth-century plans for improving Irish bogs largely rested upon the plans and aspirations outlined already between the 1770s and 1840s.

In his pioneering work *The Making of the English Landscape*, William George Hoskins stated that his focus was on the ways in which humans, for instance, 'have reclaimed marshland, fen, and moor', 'created fields out of a wilderness' and 'made canals'.[3] My focus is rather on the question of why humans aimed at reclaiming peatlands, in this case particularly bogs, in Ireland. Therefore I do not trace how the plans and ideas expressed in the sources finally materialised, or how many acres, when and where were converted into arable and forest land or industrial-scale peat extraction sites, and how this all altered the natural flora and fauna in the drained or fragmented bog areas. This kind of study has been done excellently by Oliver Rackham on Britain's 'ordinary landscape' and how it has been made 'both by the natural world and by human activities, interacting with each other over many centuries'.[4] In this chapter I am, though, more interested in the notions, values and future visions to be traced in the texts, and, in addition, what they tell us about the changing human–peatland relationship. I do not, however, consider bog-land frontiers merely as a social construction, but equally as part of a tangible material world. Having read the sources used in this study, one can nearly sense the many-metres-thick peat layer, as well as muddy and watery soil awaiting 'industrious drainers' and 'spirited improvers' in the vast bogs of Ireland. Human–peatland relationship(s) should indeed be contemplated as an ambiguous and intricate process of interrelatedness within which both parties are active actors. To put it simply, humans also become the objects of their own agency and, therefore, the outcome of the action is difficult to predict, unlike the outcome of rule-based causal processes. Coincidence and even chaos play bigger roles than when examining natural or cultural processes as such, separated from each other.

As primary sources, I use parliamentary Committee Reports related to draining and cultivating Irish bogs and improving the inland navigation, as well as publications of, for example, Arthur Young, Thomas Newenham, Richard Griffith, James Dawson, Joseph Hume, Robert Monteath and Sir Robert Kane on the reclamation of Irish bogs for the benefit of agriculture, canal transport, silviculture and the prosperity of the country. Especially the assumptions and plans made by Arthur Young and Sir Robert Kane were etched in the general debates over the drainage issue and their conclusions became widely cited later in the nineteenth century. In addition to these works, the four detailed and practical reports from the Committee Respecting the Draining of Bogs in Ireland between 1810 and 1814 set the priorities for the ideal geographical locations of the intended drainage, as well as how the drained soil could be best utilised, and provided decision makers and landowners with the then most accurate scientific knowledge on these areas; in total, the commissioners surveyed 731,976 English acres of various bogs.[5] Altogether, it is important to contemplate whether the sources describe best practice rather than the state of affairs. In many cases they do cover the already materialised 'improvements of wastelands' up to a point, but mostly they envisage the future of agriculture and inland navigation in Ireland and Great Britain as well and how the goals could be achieved by means of the drainage.

Theoretically the issues covered in this chapter are intertwined with discussions on frontiers, environmental knowledge and enviro-technical imaginaries. Michael Redclift has outlined that 'frontiers can be seen as both material realities and as social constructions, whose ideological utility often develops slowly, without clear lines of demar-cation'. This is an apt remark when studying various historical peatland frontiers. Redclift considers migration and settlement, the manage-ment of resources and the effects of globalisation as the main processes that determine the development of the frontier.[6] In my case, those who were devoted to the drainage issue tried to fight emigration and find ways to settle the then uninhabited or sparsely populated wastelands. Their main aim was to find better ways to manage previously useless or valueless resources and convert them into valuable property. In a sense, the environment was seen as an input into a technological and commercial system.

The effects of (proto-)globalisation can be found in this case too, at least up to a point, since the Irish economy became 'inextricably bound to the rest of Britain' in the nineteenth century, as Michael Turner puts it,[7] and the British economy, in turn, can be contended to have become bound with its colonies and those European areas that were important to the empire's security of grain supply and raw materials. That question could also be turned into one that deals with the effects of British imperi-alism dominated by London. In this case, those who had adopted a British identity with its ambitions, or, paraphrasing Mary Louise Pratt, 'imperial eyes',[8] in a sense tried to execute the British civilising mission to ensure the material progress of Ireland. The civilising mission also related to nature, aiming at making it blooming and wealth-producing.

Bogland frontiers can be analysed in terms of knowledge and technological systems and how they impact the altered ecologies. I seek to disentangle how accumulating knowledge on peat soil, as well as new drainage, canal transport and peat extraction and processing tech-nologies, inspired those who devoted themselves to the civilisation of nature to reappraise peripheral areas and to promote the conversion of these areas into territories that are seen as being made valuable. As the main processes that determined the development of these frontiers I consider settlement, the management of resources and the effects of the centralised laissez-faire trading system dominated by London. The whole complex question can be placed in category (5) presented in the opening chapter by Jon Agar.[9] Later, in the twentieth century, it might be described under (3) and (7), but that is another story.

British scientists and politicians who envisaged how boglands could be transformed into valuable property took part in the building and dissemination of socio-technical imaginaries. In Jasanoffian terms, these imaginaries become enmeshed in performing diverse visions of collective good, at expanding scales of governance from communities to nation states.[10] In my case, socio-technical imaginaries articulated visions of social futures and of risk and benefit for society. It is therefore essential to consider how these imaginaries have provided underlying rationale for visions concerning drainage. Furthermore, projects that transformed the environment can be seen as the reflections of socio-technical imaginaries and changing aesthetical, economic and societal values to nature. Plans and visions – even as ideas and beliefs – meant the increasing rationalisation and commercialisation of peatlands.

When late eighteenth- and early nineteenth-century scholars and engineers talked about bogs they used a different set of definitions from the ones of today. In Ireland, peat was nearly always termed turf. Knowledge about the peat soil and the hydrology of bogs accumulated gradually and became tested in practical efforts to reclaim former peatlands. Irish bogs were divided into flat red bogs and mountain bogs, depending on their location and the composition of the peat soil. Mountain bogs – mountain blanket bogs in the current term – occur on relatively flat terrain in the mountain ranges above 200m altitude and are characterised by heavy rainfall and low evaporation, and red bogs – or raised bogs – occur throughout the midlands of Ireland and got their name because the dry peat looks brownish-red in colour. Overall, bogs have played a fundamental role in Irish culture. In some regions their reclamation for fuel peat was an important part of the local economy. However, the perceptions and the use of bogs were subject to considerable changes in the modern period.

Bogland frontiers framed and labelled

The drainage issue had been more or less on the agenda in Ireland already from the beginning of the eighteenth century. William King, Fellow of the Dublin Society, published in 1685 an article in which he labelled Irish bogs as 'a great destruction to Cattle', 'a great hindrance in passing from place to place' and 'a shelter and refuge to Tories and Thieves', and ruminated 'how they [bogs] may be remedyed, or made useful', for example, as meadows. Moreover, the Crown could consolidate its control over the remote areas and, at the same time, promote cattle farming.[11] King's arguments, however, can be viewed more as a backing to the police practices under the rubric of regimentation rather than as a particularly serious plan to steer the minds of the landowners on attacking the drainage issue. His treatise simply lacks detailed and practical survey and technical advice for entering into the work.

Some freeholders tried to advance the reclamation of Irish bogs in the eighteenth century, and the Dublin Society tried to further drainage works in principle at least. Henry Brooke, one of these drainers, even wrote a guidebook entitled *A Brief Essay on the Nature of Bogs, and the Method of Reclaiming Them* based on his experiences in 1772 and gave quite detailed descriptions of how a bog can be converted into arable land. Brooke was a writer and pamphleteer born in Rantavan, County Cavan, Ireland in 1703, who ran a farm at Rantavan in the 1770s, where he drained a lake and got a bog instead. Brooke was personally interested in transport and the commercial revolution of the eighteenth century.[12] His book, however, was targeted at those landowners living in a similar environment and having an interest in practical instructions related to the drainage. Thus, it is hard to find an imperial ethos in Brooke's plan, aiming at a comprehensive civilising mission when it comes to nature and Irish 'wastelands' as a whole. That was an angle, or even a mindset, which became apparent in Arthur Young's thorough survey *Arthur Young's Tour in Ireland, 1776–1779*.

Young was already an established agricultural and political writer when he published the survey on Ireland. He carried out many experiments on his own farms and reported on the practices of other farmers after touring the major agricultural districts of several countries in order to do a personal survey. Young also gathered information by corresponding with the leading agriculturalists of his age, both at home and abroad, including John Sinclair, the Chairman of the Board of Agriculture, and George Washington in America. Young always urged the traveller to depart from the main roads in order to survey agriculture that was less influenced by access to trade with urban or industrial areas.[13] That working method led to him travelling the vast Irish bogland areas as well, which impressed him in many ways.

'Although the proportion of waste territory is not, I apprehend, so great in Ireland as it is in England, certainly owing to the rights of commonage in the latter country, which fortunately have no existence in Ireland', Young began the description of boglands, 'yet are the tracts of desert mountains and bogs very considerable'.[14] By framing and labelling certain areas as 'waste territories' Young in a sense ignored all the land use conventions and techniques the locals had practised in these areas. He does not cover whether, for example, bogs were seen as waste territories by the rural poor. In fact, that seemed to have been quite an inessential question for the men who devoted themselves to the civilising mission.

At the turn of the nineteenth century, dichotomous ideas of wasteland as ruined or defiled nature became fully codified in Western philosophy and literature. In that context, wasteland refers to land that is as yet unmodified by civilisation. As Vittoria Di Palma puts it, 'the wasteland is defined not by what it is or what it has, but by what it lacks'.[15] In the eyes of drainers and improvers, bogs certainly lacked something, for example, permanent settlement, good road and canal connections, prosperous holdings, green cultivated land and even picturesque scenery. It remains, however, unclear whether the rural poor themselves shared that picture. Their voice is totally missing from the sources. Historically the human–bog relationship in Ireland and in almost every so-called peatland frontier area in Western and Northern Europe, however, had been quite mobile, a strategy for adapting to the environment and also resourcefully utilising the best offerings of nature rather than transforming it.[16]

In Ireland, the main function of bogs had been to provide fuel for the bulk of the rural population. Turf production by using such approved low-tech tools as sleáns (two-sided spades), wooden wheelbarrows, etc. was an important part of the work year: between cutting, drying, harvesting and drawing home the turf, a labourer's annual supply required up to one month's work. Like the potato, the bulk of the turf harvest was not marketed. Economic historian Cormac Ó Gráda even argues that 'abundant fuel was one of the factors which made life for the poor in Ireland bearable'.[17] Boglands had been turf-cutting frontiers without any wide-ranging improvement aims for the freeholders and many landlords as well, whereas in Young's visions these areas rather represented agricultural, navigation and settlement frontiers with more thorough ambitions.

Young concluded that the main reasons for the lion's share of the Irish bogs still being unreclaimed were the custom of leaseholders and, equally, the poor grip on reality of the majority of the landlords. He described how 'the minutes of the journey show that a few gentlemen have executed very meritorious works even in these [bogs]; but as they, unfortunately for the public, do not live upon any of the very extensive bogs, the inhabitants near the latter deny the application of their remarks'. Boglands had largely remained untamed and uncultivated. 'Trifling as they have been on the Irish mountains yet are the bogs still more neglected,' Young notes. Having then given a detailed account of the costs expected from the drainage, which he considered to be 'very moderate', Young encourages both parties to diligently attack the drainage issue. 'Whatever the means used', he persuaded, 'certain it is that no meadows are equal to those gained by improving a bog.'[18]

It is possible to find all the frontiers considered in my study already in Young's pioneering survey. Boglands were agricultural frontiers as they were places waiting to be converted into pasture and, consequently, opening settlement frontiers that could become more permanently and firmly inhabited. For Young, boglands represented not so much a *terra nullius* but rather areas at the edge of cultivation or beyond it. It would be possible to, paraphrasing legal terms, dismiss a case, but wouldn't that be inefficient and a pure waste of resources given that the estimated costs were moderate and the anticipated benefits manifold? In a similar vein, Young criticised the state of affairs when it comes to inland navigation, which was quite inextricably linked to the drainage issue. 'But of all public works, none have been so much favoured as inland navigations,' Young lashes out at the decision makers, 'but under the administration of this [Navigation] Board, which consists of many of the most considerable persons in the kingdom, very great attempts have been made, but I am sorry to observe, very little completed.'[19]

Navigation frontier as the opportunity for profit or efficiency from a new technology emerged once again in a proposition that came from inside the Dublin Society. The writer using his author abbreviation W.V. proposed in 1801 'that a company be formed, consisting of subscribers … to be incorporated by act of parliament, and called the Waste Land Company of Ireland'. That company would 'have power to purchase from tenants, in fee simple, red or black bogs … lying together; or strands, or marshes, usually covered with the tides' and 'also to cut canals to neighbouring towns, for the purpose of supplying them with turf and water'. The author was quite pessimistic about finding enough capital from private persons to embark on extensive drainage works. He justified his proposition by arguing that

> it is conceived the reason, why the bogs of Ireland have not been improved by individuals is, because they have neither property nor power, property to undertake a heavy work, or power to cut drains through neighbouring lands, or to get manure elsewhere than on the ground, and therefore that great bogs can never be improved but by a company, under the powers of an act of parliament.

Having once completed the drainage, 'such a company could carry on their works as well as any canal company'. The agricultural frontier aspect is also covered in the proposition. W.V. mentions that 'bogs, in the course of their improvement, are well-suited to the growth of hemp and rape'.[20]

Young's and W.V.'s plans, even as ideas, meant the increasing commercialisation, commodification and rationalisation of boglands. Within that context, frontiers are conceived, as Gordon M. Winder and Andreas

Dix formulate it, 'in terms of modernising commercial projects backed by cultural imaginaries and scientific, technical and political calculation that are set to work in environments'.[21] With the process of modernisation, boglands as objects became increasingly socio-natural and thus connected to the desires, aspirations and demands of the British civilising mission.

Bogland frontiers surveyed and assessed

In addition to the works of Arthur Young and other late eighteenth- and early nineteenth-century improvers, the Dublin Society and the Irish Parliament also tried to prop up the drainage issue. The Irish Parliament, for example, passed an Act in 1731 to 'encourage the improvement of barren and wasteland, and bogs'. The charter of the Dublin Society from 1750, in turn, recites 'that several of the nobility and gentry of Ireland, having observed vast tracts of land and bog in Ireland uncultivated, and a general want of skill and industry in the inhabitants to improve them, had formed themselves into a voluntary society for promoting husbandry and other useful arts', including 'draining bogs'. There were, however, many drainers and improvers who were not satisfied with how things had progressed in practice in the late eighteenth century and at the beginning of the nineteenth century. The government was expected to step forward to advance the issue more sensibly.

The appointment of a special committee with special power to manage the task, including expensive and extensive surveys, was pushed by, for example, Sir Arthur Wellesley, who was Chief Secretary of Ireland in 1807 and 1808, and Thomas Newenham, an Irish political writer and former MP. They both had the political assets to win support from London. Wellesley wrote a memorandum to the Home Secretary, Lord Liverpool, on the drainage of the bogs of Ireland and proposed 'the appointment of a Commission with power to survey the different bogs and ascertain their extent, the practicability of draining them and the expense of that operation'.[22] Newenham's *A View of the Natural, Political, and Commercial Circumstances of Ireland* published in 1809 followed the optimistic perceptions of Arthur Young and positioned the Irish drainage issue as significant to the success of Great Britain as well. Newenham also saw the costs following from the drainage as very moderate. 'Such an expenditure', Newenham judged, 'would unquestionably enable Ireland to supply, most amply, the growing wants of England, after satisfying those of her own rapidly increasing population.' He estimated that if one-eighth part of this [bogs of Ireland], consisting such land as, by situation, nature of soil, and abundance of natural manures, appeared most favourable circumstanced for cultivation, were reclaimed, 3,600,000 average barrels of the different sorts of grain, even with the present defective mode of husbandry, might be annually obtained.

Ireland could be easily transformed from 'waste territories' to the granary of industrialising England, whose dependency on the grain imports was substantial.[23]

Finally, Parliament appointed the Committee Respecting the Draining of Bogs in Ireland in 1809. Already active proponents of the drainage issue, for example, Charles Vallancey, Richard Griffith, John Leslie Foster and John Staunton Rochford, were nominated to the secretary of the commissioners to be appointed to that task. Vallancey, the acting chairman of the committee, was an English-born military surveyor who was sent to Ireland in the late 1760s, and Richard Griffith was Chairman of the Board of Works, an Irish geologist and mining engineer. Rochford served on the Dublin Society's chemistry committee, and Foster was an Irish barrister and nephew of John Foster, the Chancellor of the Exchequer of Ireland. The task of surveying Irish bogs by counties, to be carried out by the assigned engineers, was directed by the Board of Works of Ireland.[24] Ultimately, the appointment of the parliamentary committee can be seen as a result of a decades-long deliberation and civilising mission in terms of the 'wastelands' of Great Britain. Besides, the time was favourable for the initiative, since Britain was at war and stronger self-sufficiency was desired.

The committee authored four detailed reports on the bogs of Ireland between 1810 and 1814 and provided the House of Commons with an impressive amount of agricultural, soil chemical and topographical knowledge and, in addition, plans on how, where and at what cost to commence the drainage. The reports also contained pioneering research into peat deposits. No other such surveys were conducted in Ireland in the nineteenth century with respect to the thoroughness, geographical coverage and the amount of work.

The commission concluded that the drainage would benefit Ireland and also England in many ways and estimated that all flat red bogs 'might be converted to the general purposes of agriculture'.[25] As for mountain bogs, the commission was rather optimistic and judged 'that not only they, but a considerable portion of the mountain soil, may be improved at a small expense, so far as to afford excellent pasture and meadow'.[26]

Not only was the state of agriculture expected to be improved, but inland navigation would also be advanced. 'Whenever the bogs shall have been perfectly drained', Richard Griffith calculated, 'it will doubtless be a matter well worthy of the consideration of the proprietors to obtain navigable communications with the main land, for the purpose of procuring gravel, lime, etc., and for the more easy and cheap conveyance to market of the produce of the bogs.'[27] Lime and fine gravel were blended with peat soil to reduce its acidity. On the whole, the commission was convinced that the increasing land value would cover the expenses of the improvements that had been made to it.[28]

The committee's contribution to the mapping of Irish bogland frontiers bound these environments with complicated trading networks in a more illustrative and reasonable way. The agricultural or navigation frontier was not just an Irish frontier but a British frontier as well. Farm products played a crucial role in stoking the Industrial Revolution. In the future visions of the commission, Ireland would bear an apparent resemblance to the then much more cultivated and manicured environments in Western Europe. The cultivated and canalised low-lying lands of Ireland would see a constant flow of materials and products from inland to the seaports and to England, and vice versa. Even the most 'desolated' and 'abandoned' mountain districts would be used as pasture. Trading networks and their associated practices and interdependencies allowed boglands to be viewed through ever more calculative and speculative eyes. That desired march of progress demonstrated how new technology became embedded in the future visions. The age of canals and steam would be realised in the back garden of industrialising England and an infinitely greater bulk of goods could be carried in much greater security on barges than in wagons and at a very much smaller expenditure of horse power and labour. Paradoxically, rather low technology, for example, spades, mills and weirs, and lots of strong men and women were still needed to drain the water from boggy areas. Actually, that was the way peatlands were prepared for improvements until the late nineteenth century when machines and dynamite were tested in the drainage.

Consequently, boglands would become almost wholly artificial environments. In fact, there would hardly be intact bogs in Ireland any more except for the roughest and most desolate ones, should the plans of the drainers and improvers be fulfilled – only meadows, fields, turf extraction sites, canals and afforested mountains instead. Consequently, peatland ecologies would begin to tilt towards monoculture: natural, rare and diverse bog flora, including *Sphagnum* mosses, bog cotton, heathers, black bog rushes, bog myrtle, royal fern and sundew, would be replaced by plant species that humans valued most both economically and often aesthetically. Despite the ambitiousness and far-reachingness of the plans, the commission did not see any hazards for the locals arising from the drainage except the increasing flood risk. That, however, could be obviated

> by deepening the beds, and removing channel obstructions in these small rivers, sufficiently to render them capable of effecting the quick discharge of the waters, without flooding the country, as otherwise the injury that might be produced by the floods in the low lands would counterbalance the advantage of drainage in the bogs.[29]

Environmental learning, in this case, became a process characterised by interest in practical details, utilitarian values and the ambition to develop suitable technologies and engineering to drain the water from bogs.

Bogland frontiers revisited and reopened

The reports of the committee were ultimately not published and made open to the public. Quite soon, they were buried among the endless archive of the British parliamentary papers after the drainage issue took a backseat in the government's political agenda. Richard Griffith made an attempt to raise the issue by publishing a book in 1819 in which he largely reiterated his arguments already floated years ago in the reports. The book was dedicated to Robert Peel, former Chief Secretary for Ireland and a rising star in the Conservative party, who Griffith thought had power to promote the issue in Parliament. Griffith was, however, politically experienced enough to tie the issue to a more burning question, namely the poor law question. 'Let the legislature open to the people new sources of profitable labour, by the extension of inland navigation and by reclaiming the bogs,' Griffith writes; 'by doing this, they would afford present employment to the poor, and gradually throw three million acres into the land market; which would reduce the extravagant rate of rents; and thus meet the existing evil in a double form'. Griffith stressed that 'Ireland must now be considered as an integral part of the Empire'. The measures that to him appeared of the most immediate importance for the improvement of Ireland were 'the further extension of her Inland navigation; the draining of her Bogs, and the adoption of a liberal system of Religious and Moral

Education for her Poor'.[30] A new angle was opened in the drainage issue and socio-technical imaginations of boglands.

Robert Fraser, an Irish statistician and surveyor, wrote in *Sketches and Essays on the Present State of Ireland* in 1822 for the acting Committee of the Society how

> it is of importance to remark, that the reclaiming of these bogs is an undertaking of great national importance; inasmuch as the whole of the two millions of acres, are capable of being converted into the production of grain, and that at the same time the improvement of these bogs, would be a great source of employment, to the redundant population of this country, their full cultivation, would ensure to England, supplies of grain, at moderate prices, which might render it wholly independent of foreign countries, for the food of its manufacturing population.

Fraser wondered that 'no legislative measure has been adopted, in order to facilitate the reclaiming of those extensive tracts'.[31] Tackling the poor question in Ireland by means of promoting drainage was also addressed by a few British politicians in the 1830s. Joseph Hume, a radical MP, for example, tried to expedite the solving of the issue by publishing a scheme for 'improving bogs and other wastelands in Ireland', based on the arguments already brought forward by Arthur Young, the committee and Richard Griffith.[32]

Whether eventually applied to the poor question, agricultural improvement or inland navigation, Irish bogland frontiers became viewed ever more dominantly within the imperial context in the 1820s and 1830s. James Dawson, an entrepreneur who experimented with steam power on the Grand Canal in the early 1810s and later on also entered the mining business, argued in letters that originally appeared in *Carrick's Morning Post* in 1818 and 1819 how drainage and canalisation was the best way to 'promote the civilisation and improvement of the interior of Ireland, to supply the increasing agricultural wants of Great Britain, and provide employment for our superabundant population'.[33] Naturally that would also rejuvenate the business of canal companies, as well as mining enterprises reliant on good canal networks. Robert Monteath, a forester and silvicultural writer, in turn, promised in his book in 1829 that if

> [my] plan of improving the wastelands of Ireland, taken in connexion with my other plans of improvement on bare, rocky soils, are

set on foot, Ireland will soon be the richest and most independent of the three British nations, and the noblemen, the gentlemen, and above all, the farmer and the labourer, will, in their own sphere, be equally benefited and enriched.[34]

Monteath dedicated his book to the Duke of Northumberland, Lord Lieutenant of Ireland. For Monteath, mountain bogs represented silvicultural frontiers with splendid prospects. In this case too, frontiers connoted a perspective viewed from a centre and involved imperial power relations.

The projection of values, plans and aspirations on to the bogland frontiers remained quite unchanged until the 1840s. The perspective was dominated by a concoction of agricultural, inland navigation and settlement prospects. With the exception of Robert Monteath's work, silvicultural interests were covered only sparingly. This goes for perceiving bogs as fuel frontiers as well. Late eighteenth- and early nineteenth-century drainers and improvers almost downplayed the significance of bogs as a source of fuel. For them, cutting and drying turf by hand for heating purposes may have seemed archaic and ineffectual.

Boglands as fuel frontiers became framed later in the early 1840s by Sir Robert Kane, an Irish chemist, who focused on frontiers that extended vertically downward in terms of resources for extractive activities. Before that, bogland frontiers epitomised horizontally extensive frontiers for agricultural and transportation activities. Kane had seen how 'the employment of turf as a source of heat in industry is extending' and judged that 'there is in our bogs amassed a quantity of turf, which, if the peculiar characters of that fuel be suitably attended to, may become of eminent importance to the country'. Whereas the locals cut turf by hand and 'spoiled [it] by its mode of preparation', Kane envisioned fully mechanised turf extraction. 'It is only by operating on a great scale, and with powerful machinery, in fact,' he writes, 'only by manufacturing compressed peat largely for sale that the operation can be made to practically succeed.' And that was supposed to be just the beginning of the scientific and technological processing of peat. 'Not merely may we utilise turf in its natural condition, or compressed, or impregnated with pitchy matter,' Kane analysed, 'but we may carbonise it as we do wood, and prepare turf charcoal' for many industrial purposes, including railway and steam boat engines, ironworks, textile factories and breweries.[35]

What makes Kane's book of special interest is its divergent national overtone. Unlike Griffith or Fraser, Kane does not stress the benefits of the exploitation of peatlands for Britain but brings out how the gain resulting from new technology and activities would especially profit Ireland. He does not see, in the terms of Kenneth Pomeranz, Irish boglands as 'ghost acres'[36] for England's Industrial Revolution but as valuable resources for Ireland's underdeveloped industry and also for national self-sufficiency.

Altogether, Kane's vision was a high-tech response to the exploitation of bogs. It involved a series of then modern technologies, including, for example, hydraulic press and high pressure engines, and a series of industrial practices. To Kane's mind, the exploitation of bogs should be based on scientific-economic knowledge and calculation. Boglands were to service manifold aims and the transition to mass production of both soft and hard commodities. These aims benefited the empire and his homeland, Ireland, as well. The concept of bogland frontiers became broadened by new knowledge and technologies that repeatedly pushed the frontier into various layers of strata. Most importantly, making intact bogs valuable was no longer only to be achieved by converting them into arable land or engineered canal networks.

Later advocates of the drainage and reclamation of bogs often invoked late eighteenth-century and early nineteenth-century drainers and improvers to back up their arguments. Calculations and priorities changed, but the total aim remained the same in the late nineteenth century and early twentieth century: to transform boglands into places that are being seen to be made more valuable. Various visions put into practice had begun to alter the ecologies of untapped boglands over a wide area already in the nineteenth century, although the peak in drainage and the use of peat as a fuel occurred later in the next century.

Summary

Historically, the exploitation of peatlands has been sold to investors, policy makers, citizens and local people as the continuation of technologically driven futures with a particular notion of progress. Focusing on late eighteenth- and early nineteenth-century Irish drainage, reclamation and inland navigation plans and visions, this chapter looks at the imaginaries of technological progress and commodification of nature and how they take the form of material-discursive projects of making the future. Such visions of social futures were developed by English and Irish scientists and politicians who aimed at turning vast Irish boglands into valuable property by civilising nature.

The period under scrutiny was in many ways a tipping point in values, notions and ambitions and meant increasing calculation, commercialisation and rationalisation of bogs as natural resources. Common to these writings was the future vision they expressed for Ireland and the whole British Empire and how that vision was fundamentally based on strong faith in the advancement of science and technology to tame and refine 'wastelands', as bogs were labelled and assessed. Irish boglands became viewed through imperial eyes and tied to the British civilising mission. The aim of so-called drainers and improvers was to win support for their plans not just from Irish landowners and the economic elite but from the cabinet and Parliament.

Arthur Young's thorough survey *Arthur Young's Tour in Ireland, 1776–1779* emerged as a signpost of the new wave of scientific-technological improvers aiming at transforming boglands into territories that were seen as being made more valuable. Young's survey was rather optimistic with respect to the appraised potential of Irish bogs for pasture and tillage and the general development of the island through the improvement of wastelands. Thomas Newenham, an Irish political writer and former MP, followed the perceptions of Arthur Young and positioned the Irish drainage issue as significant to the success of Great Britain as well.

The optimistic spirit of Young and Newenham and other improvers of the time was shared by the Dublin Society and its members, envisioning an upturn in the economic and political conditions of their homeland now as an irremovable part of the empire. The appointment of a parliamentary committee in 1809 to enquire into the nature and extent of the several bogs in Ireland, and the practicability of draining and cultivating them, represented the apogee for improvers in terms of political publicity. The committee authored four detailed reports on the bogs of Ireland between 1810 and 1814 and provided the House of Commons with considerable agricultural, soil chemical and topographical knowledge and, in addition, plans on how and at what cost the drainage could be commenced. The committee's contribution to the appraisal of Irish bogland frontiers bound these environments with complicated trading networks in a more reasonable way. The agricultural or navigation frontier was not just an Irish frontier but a British frontier as well. Improvers envisaged how the cultivated and canalised low-lying lands of Ireland would see a constant flow of materials and products from inland to the seaports and to England, and vice versa. That desired march of progress demonstrated how technology, particularly steam power and barges capable of moving heavy produce, became embedded in the future visions.

The reports of the committee were ultimately not published and made open to the public. Richard Griffith, being disappointed in the outcomes as a member of the committee, tried to raise the issue by publishing a book in 1819, in which he largely reiterated the arguments already floated years ago in the reports. Griffith was, however, politically experienced enough to tie the issue to a more burning question, namely the poor law question. Griffith and Robert Fraser, an Irish statistician and surveyor who published a book on the matter in 1822, saw the drainage issue as crucial in tackling the rural poor question by providing land for settlement and extra income for rural people living in poverty.

The projection of values, plans and aspirations on to the bogland frontiers remained quite unchanged until the 1840s. The perspective was dominated by a concoction of agricultural, inland navigation and settlement prospects. Sir Robert Kane, an Irish chemist, opened a new perspective by focusing on frontiers that extended vertically downward in terms of resources of extractive activities. Kane envisioned how turf, having been cut by hand by locals for centuries, could become a vital fuel for the industrialisation of Ireland. Kane's vision was a high-tech response to the exploitation of bogs: he introduced fully mechanised turf extraction and utilisation practices and called for mass production of peat products from turf charcoal to turf coke. Making intact bogs valuable was no longer only to be achieved by converting them into arable land or engineered canal networks.

Altogether, the arguments developed by English and Irish scientists and politicians between 1770s and 1840s meant active rethinking of bogs and expectations related to them and laid the foundations for material environmental transformations. To refer to Jon Agar's *types of technology/environment interaction* (see Chapter 1), Irish boglands were perceived and also mastered as wastelands in need of commercial exploitation. That articulation conflated imaginaries of technological and social progress, capital accumulation and profits and benefit.

Notes

1   Arthur Young, *Arthur Young's Tour in Ireland (1776–1779)* (London: George Bell & Sons, 1892), 7, 101–2.
2   On Northern European future visions and notions regarding peatlands, see Esa Ruuskanen, 'Valuing Wetlands and Peatlands: Mires in the Natural Resource and Land Use Policies in the Nordic Countries from the Late Eighteenth Century to the Present Date', in *Trading Environments: Frontiers, Commercial Knowledge, and Environmental Transformation, 1750–1990*, ed. Gordon M. Winder and Andreas Dix (New York: Routledge, 2016), 118–32.
3   William George Hoskins, *The Making of the English Landscape* (Dorchester: Little Toller Books, 2014), Introduction to the first edition of 1955.

4 Oliver Rackham, *The History of the Countryside: The Classic History of Britain's Landscape, Flora and Fauna* (London: Orion Books, 1997), xiii.

5 *The Fourth Report of the Commissioners Appointed to Enquire into Nature and Extent of the Several Bogs in Ireland; and the Practicability of Draining and Cultivating Them* (London: House of Commons, 1814).

6 Michael Redclift, *Frontiers: Histories of Civil Society and Nature* (Cambridge, MA: MIT Press, 2006), 14–15, 18.

7 Michael Turner, *After the Famine: Irish Agriculture, 1850–1914* (Cambridge: Cambridge University Press, 1996), 1.

8 Mary Louise Pratt, *Imperial Eyes: Travel Writing and Transculturation* (New York: Routledge, 1992).

9 Jon Agar, 'Technology, Environment and Modern Britain: Historiography and Intersections', Chapter 1 this volume.

10 Sheila Jasanoff, 'Future Imperfect: Science, Technology, and the Imaginations of Modernity', in *Dreamscapes of Modernity: Sociotechnical Imaginaries and the Fabrication of Power*, ed. Sheila Jasanoff and Sang-Hyun Kim (Chicago: University of Chicago Press, 2015), 4–5.

11 William King, 'Of the Bogs and Loughs of Ireland by Mr William King, Fellow of the Dublin Society, as it was presented to that Society', *Philosophical Transactions* 15 (1685): 948–60.

12 Markman Ellis, *The Politics of Sensibility: Race, Gender and Commerce in the Sentimental Novel* (Cambridge: Cambridge University Press, 1996), 156–7.

13 Liam Brunt, 'Rehabilitating Arthur Young', *Economic History Review* 56, no. 2 (2003): 265–6, 281.

14 Young, *Arthur Young's Tour in Ireland*, 95.

15 Vittoria Di Palma, *Wasteland: A History* (New Haven: Yale University Press, 2014), 2–3.

16 Ruuskanen, 'Valuing Wetlands', 119–20.

17 Cormac Ó Gráda, *Ireland: A New Economic History 1780–1939* (Oxford: Oxford University Press, 1995), 15, 116–17.

18 Young, *Arthur Young's Tour in Ireland*, 100–3.

19 Young, *Arthur Young's Tour in Ireland*, 123, 142–3.

20 'Scheme for Improvement of Bogs and Other Waste Lands in Ireland' (written by W.V.), *Transactions of the Royal Dublin Society* 2 (1801): 178–82.

21 Gordon M. Winder and Andreas Dix, 'Commercial Knowledge and Environmental Transformation', in Winder and Dix, *Trading Environments*, 14.

22 R.D. Collison Black, *Economic Thought and the Irish Question 1817–1870* (Cambridge: Cambridge University Press, 1960), 182.

23 Thomas Newenham, *A View of the Natural, Political, and Commercial Circumstances of Ireland* (London: T. Cadell & W. Davies, 1809), 72, 312.

24 'Commissioners Appointed to Enquire into the Nature and Extent of the Several Bogs in Ireland', *The Journals of the House of Commons*, 64, Sess. 1809.

25 *The Fourth Report*, 16.

26 *The Fourth Report*, 8.

27 *The Second Report of the Commissioners Appointed to Enquire into Nature and Extent of the Several Bogs in Ireland; and the Practicability of Draining and Cultivating Them* (London: House of Commons, 1811), 10.

28 *The First Report of the Commissioners Appointed to Enquire into Nature and Extent of the Several Bogs in Ireland; and the Practicability of Draining and Cultivating Them* (London: House of Commons, 1810), 7.

29 *Second Report*, 11.

30 Richard Griffith, *Practical Domestic Politics: Being a Comparative and Prospective Sketch of the Agriculture and Population of Great Britain and Ireland, Including Some Suggestions on the Practicability and Expediency of Draining and Cultivating the Bogs of Ireland* (London: Sherwood, Neely & Jones, 1819), 4, 7, 18, 20.

31 Robert Fraser, *Sketches and Essays on the Present State of Ireland, Particularly Regarding the Bogs, Waste Lands & Fisheries; with Observations on the Best Means of Employing the Population in Useful and Productive Labour* (Dublin: Carrick, 1822), 6, 9.

32 Joseph Hume, *Brief Notices Respecting the Origin and Purpose of Some of the Departments of the Royal Dublin Society, Referring also to a Scheme for Improving Bogs and Other Waste Lands in Ireland* (Pamphlets, Hume Tracts, 1831).

33 James Dawson, *Canal Extensions in Ireland, Recommended to the Imperial Legislature, as the Best Means of Promoting the Agriculture–Draining the Bogs–and Employing the Poor* (Dublin: Carrick, 1819), Letters 1 and 3.

34 Robert Monteath, *A New and Easy System of Draining and Reclaiming the Bogs and Marshes of Ireland: with Plans for Improving Waste Lands in General* (Dublin: Carrick, 1829), 161.

35 Robert Kane, *The Industrial Resources of Ireland* (Dublin: Carrick, 1844), 30, 37–38, 40.

36 Kenneth Pomeranz, *The Great Divergence: China, Europe, and the Making of the Modern World Economy* (Princeton: Princeton University Press, 2000).

3

Landscape with bulldozer: machines, modernity and environment in post-war Britain

Ralph Harrington

A natural history of the bulldozer

In the landscape of post-war Britain the bulldozer was not a native species but a naturalised one. Some might even have called it invasive. The bulldozer arrived in Britain as a consequence of war and as part of an army. Massed formations of crawler tractors with bulldozer blades, manufactured in the United States, crossed the Atlantic as vital elements of the vast armoury of equipment brought to Britain by the American armed forces during the Second World War (see Figure 3.1).[1] The bulldozer itself, while in origin a civilian machine, has always had a close relationship with the technologies of mechanised conflict, and in important respects it was the Second World War that created the bulldozer as we know it today and spread its use throughout the world. Certainly it was the Second World War that played the key role in bringing the American bulldozer to Britain.[2]

The origins of the bulldozer lie in agriculture and construction in the inter-war United States, and the twentieth-century bulldozer is a distinctively American device, rooted in distinctively American circumstances: it is a machine of big spaces and big structures, of cheap land and expensive labour, of wide horizons and sweeping transformation. And the word 'bulldozer', too, was made in America. When the *Illustrated London News* introduced the bulldozer to its readers in early 1941 it explained that 'The word "bulldozer" is used in American slang phraseology to mean "to intimidate or coerce", and the coercive powers of this almost incredible machine would seem to be considerable.'[3] It is a word with a complex history, in which agriculture, mechanical engineering, conflict

Figure 3.1 A picture taken in the spring of 1944 of the US Army engineer depot at Thatcham, Berkshire, showing massed ranks of bulldozers and tractors being prepared to accompany the D-Day invasion forces.
Source: National Archives and Records Administration/US Army Signal Corps 111-SC-189366.

and politics intersect and connect, but that has consistently embodied and expressed a nexus of brute force and coercive violence. The first published uses of the word are from the United States in the 1870s and refer to organised racist violence in the politics of the post-Civil War era, particularly in the South.[4] '"Bulldozing" is the term by which all forms of this oppression are known,' explained one writer of 1879, describing 'the violent methods which have been employed to disfranchise the negroes, or compel them to vote under white dictation, in many parts of Louisiana and Mississippi'.[5] The notoriety of the 'bulldozers' of the American South evidently brought the word and its associations into more general usage during the last quarter of the nineteenth century. By circa 1900 'bulldozers' could be found in agriculture, mining and metal-working activities that involved the application of force to aspects of the physical environment that needed to be reshaped, exploited, and, in the terms used by Jon Agar in the opening chapter of the present volume to summarise the second of his technology/environment combinations, transformed into a component within a technological system.

In agriculture, the bulldozer was a vertical wooden blade used for smoothing rough ground, held in a wheeled frame and drawn by oxen, mules or by hand. These scraping implements developed in the American West and seem to have been particularly associated with the Mormon farmers of Utah, sometimes being known as 'Mormon Scrapers'.[6] In the 1880s commercially available machines such as the 'Western Bulldozer' and the 'Fresno Scraper' used wheeled frames and pivoted blades, combining some of the attributes of the later mechanised bulldozers and tractor-driven scrapers.[7] These devices were used in agriculture for clearing and preparing land and for earthmoving during construction projects such as railway building, confirming the value and adaptability of the principle that would later be employed by the mechanical bulldozer. In mining and quarrying, meanwhile, 'bulldozing' was the use of explosives to break large boulders into smaller pieces, either below ground in 'bulldozing chambers' or on the surface.[8] The direct, unmediated application of powerful force to the thing being destroyed or reshaped is the key element that unites this use of the term 'bulldozing' to its uses in the parishes of Reconstruction-era Louisiana. Similarly in metalworking the 'bulldozer press' was a machine in which plates and bars were shaped by a powerful crosshead ram moving in the horizontal plane.[9] The resemblance of this machine to the vertical blade of the bulldozer, pushing with brute force against the earth in order to reshape it, is very strong, and it is possible that the bulldozer blade, and thus the entire bulldozer vehicle, received its name from comparison with this piece of equipment, so that the derivation from the 'bulldozers' of the American South or from the Western agricultural implement may be indirect rather than direct.[10]

Precise lines of etymological descent, however, are less important than the underlying denotative and connotative content of the term 'bulldozer', which remains consistent across more than a century of usage and a broad range of applications: the shared associations of brute force and violent coercion. Since the last quarter of the nineteenth century, to *bulldoze* something has consistently meant the use of force to reshape the environment: physically, but also socially and politically. As, during the 1920s and 1930s, the bulldozer developed into an ensemble of powerful tractor running on caterpillar tracks deploying a large adjustable blade – a creation that was honed and perfected in the 1940s under the pressures of war – these associations became ever more strongly entrenched.

The modern bulldozer, as with other mechanical earthmoving equipment, was known in Britain before the Second World War but was present only in very limited numbers and was not widely used in construction or related fields: 'little was seen of this class of equipment until vast numbers were brought into the country from America during the Second World War', noted one earthmoving expert looking back on the development of British construction technology in 1964.[11] When the engineer V.W. Bone addressed the Institution of Mechanical Engineers in 1936 on 'modern developments in tractor-drawn excavator equipment' he felt it necessary to describe the bulldozer in some detail:

> My first example is the 'Bulldozer', which consists of a large rectangular plate reinforced at the bottom or cutting edge, located at the front of the tractor and attached through girders to the caterpillar track side frames. This type is especially adapted to pushing dirt, rock, debris, and other material ahead of the tractor. The blade has a vertical movement in relation to the tractor, which is under the control of the operator so that the thickness of the cut being taken by the blade on the ground or the amount of material being pushed along can be regulated.[12]

Bone went on to observe that the bulldozer was a new development that, although 'employed extensively on large excavation projects, particularly in the United States of America', had yet to achieve its full potential in Britain: 'in this country we have only touched the fringe of the possibilities of tractor-drawn excavating equipment'.[13]

It was not until the introduction of thousands of earthmovers by the Americans during the Second World War that the bulldozer became an established feature of the British landscape. These machines built ports, roads and airfields to serve the war effort. They often worked in places that had seen no such large-scale or rapid development before, and in rural districts in which mechanical earthmoving and modern construction methods of any kind were a novelty. In 1941 the bulldozer was thought by the *Yorkshire Post* to be sufficiently unfamiliar to its readers for a brief description of it to be necessary: 'American-built machines normally employed in road-making in rough virgin country'.[14] The bulldozers described in this case were, as it happens, Canadian, and were working on forestry and land clearance in Scotland, but it was their US counterparts from 1942 onwards that were to have the greatest impact on the British landscape during the war. In September 1943 an article in *Picture Post* discussed the role of the bulldozer in sustaining Allied military activity, and stressed its distinctively American identity:

> The most important new factor in our air policy is not a flying machine but a land machine – the bulldozer. This powerful caterpillar tractor, armed with a long steel blade, which digs into the ground, tears out boulders and tree stumps, and can even be turned on walls and small buildings, is the central machine in the whole

American armoury … The Americans, in fact, are teaching us how to mass-produce aerodromes.[15]

For wartime Britain the bulldozer was thus an icon of American techno-logical modernity, bringing the techniques of that quintessentially American socio-industrial technique, mass production, to the reshaping of the earth itself.

In 1944 the writer D.W. Brogan reflected on the wartime signifi-cance of the American military bulldozer in his essay 'The Bulldozer', in which he identified this machine as a key symbol of the Allied war effort and saw in it a promise of the new society that victory would ultimately bring. Brogan began by listing various impressive Allied war machines: Flying Fortress bombers, amphibious trucks, Sherman tanks. But these instruments of war, he argued, were not as socially or politic-ally significant for the Western world as what he termed 'instruments of peace turned to the uses of war':

> The bulldozer is politically mightier than the tank, for in the bull-dozer Europe has seen an instrument of power made directly by American civil society, serving indeed a military purpose, but bringing the Old World a flavour of the New, of that world of repeated mechanical novelty in which wars are not quite episodes, but are no more than great but manageable crises of American production.[16]

The bulldozer thus embodied the way the great wartime coalition had been underpinned by American technological might and industrial organisation, and had brought the strengths of the New World to the aid of the Old. Brogan saw hope for the post-war world in a continu-ation into the era of peace of this coalition created to serve the purposes of war: he saw hope in the ongoing work of the bulldozer. He ended his essay by observing that having received American help to win the war, the Europeans would certainly want 'to be left to manage their own affairs', but in clearing away the debris of conflict and building the peace 'they would like help from the country of the bulldozers, the country of humane miracles … The bulldozer has cleared away more than ruins.'[17]

Bombsites and bulldozers: rebuilding Britain

'During the Second World War', observed one construction industry engineer in 1964, 'crawler-mounted bulldozers became widely known as miracle construction equipment.'[18] The British earthmoving and construction industries were well aware of the importance of the war in bringing American mechanisation into the British landscape, and of the significance of the transformation this represented. The war 'brought to this country a far larger amount of muck shifting and other plant than we had been used to,' commented one industry insider in the *Proceedings of the Institution of Mechanical Engineers* in 1947. 'Scrapers, bulldozers, and dumpers, fathered in the USA, were of great use over here in the preparation of aerodromes, camps, temporary roads, drainage and tank traps.'[19] In his 1944 essay D.W. Brogan described the impact of the bulldozer on British industries and landscapes previously unmarked by rapid technological progress:

> First came the bulldozer; it came into parts of England which had been left outside the main stream of mechanical progress, into East Anglia, into the southern rural counties. It came in and did more to change the face of the land in a few months than had been done at the same speed since Roman times ... roadmaking and construction in general are the most backward parts of British technical practice. And the latest American devices were demonstrated in parts of England more backward than most.[20]

In other texts the impact of this new earthmoving technology on an unchanging, enduring landscape, deeply rooted in the past (or at least upon a landscape *represented* in these ways), was narrated in such a manner as to emphasise historical continuity rather than disruptive change. 'British rural roads excite admiration and wonder,' observed the American authors of *Bulldozers Come First: The Story of US War Construction in Foreign Lands* (1944). 'Constant military maneuvers, involving long convoys of trucks, tanks, and mobile guns, should rut and ravel their tar-macadam surface, but few signs of distress are visible.'[21] The endurance shown by British roads during the war, like the endurance shown by Britain as a whole, was rooted in history: 'Many British rural roads have followed their present alignments for centuries. Cromwell or even Caesar may have started the consolidation of their bases.'[22] Thus the arrival of the new construction technology could be seen as congruent with the past rather than constituting a break with it, as the new world of modern technology came to the rescue of the old but did so by building upon its achievement and its values. In that sense the American bulldozer grinding its way along the British rural road was a microcosm of the entire Anglo-American war effort.

As noted earlier, the bulldozer came to a Britain in which heavy construction and earthmoving equipment had previously found very little application.[23] This may have been partly due to the innate conservatism of an industry still dominated by craft processes and work patterns, and sceptical of technical innovation,[24] but also reflected the relatively restricted size of most construction projects compared with the huge schemes in America that were only made possible at all by extensive mechanisation. Environmental factors were also an issue restricting the use of bulldozers and similar machines in the British Isles, or were believed to be: many British engineers in the 1940s appear to have been convinced that bulldozers would not work well in wet weather and that British soils and geology were not suitable for them.[25] Yet the wartime performance of the American machines left no doubt as to the benefits they offered. As well as literally preparing the ground for new construction, their contribution to clearing rubble was invaluable, both at the fighting fronts and in bombed cities in Britain. This vital work also brought the bulldozer a high degree of public visibility. When earthmovers were working at remote (and highly secured) airfields and other military sites they were largely out of sight, but in bombed city centres they became performers on a much more public stage. Their contribution was noted as early as 1941, with newspapers introducing their readers to 'This new machine, the "Bulldozer" … now being used in London streets for clearing away bomb damage',[26] 'this almost incredible machine',[27] 'The bulldozer tractor' that 'can deal with all obstacles – debris, mud, water'.[28] Gradually, as the number of bulldozers in Britain grew and their contribution to clearing away urban war damage become increasingly well known and appreciated, the machines entered the British imagination and became icons of wartime endurance and symbols of hope for the future. There was nothing the bulldozer could not do, it seemed, in this landscape of urban rubble: the Sketch's humorous diarist even suggested in 1944 that if Britain ran out of real ruins to show the predicted influx of future tourists expecting to see the scars of war, 'all we have to do is take a bulldozer and ram a few of our surviving public buildings' to make some more.[29] American men and machines played an important and well-reported role in rebuilding bomb-damaged sites in British cities, and in constructing both temporary housing and new permanent housing estates.[30] The Daily Express reported in December 1944 on a site in London where the US Army was working on the construction of temporary housing, and drew a sharp contrast with the labour-intensive, technology-averse practices on an adjacent building site in the hands of British workers:

> American troops completely equipped with trucks and machinery, are maintaining a racing speed in clearing sites and laying bases for 600 temporary homes for Lambeth Council. On the British site yesterday, only partially cleared and levelled, a dozen workmen were unloading bricks, smoothing cement and laying drains for temporary homes. A single cement-mixer was the only machinery in action. An unused steam shovel stood at the front of the site.[31]

The *Express* reporter quoted the British foreman as telling the officer in charge of the American working party that 'One of those bulldozers in a few hours could level the ground it's going to take me weeks to clear.'[32]

The rubble of destroyed buildings that bulldozers cleared away in urban areas was often collected and reused, some being transported as 'blitz brick' to rural construction sites where, worked by yet more bulldozers, it formed an aggregate foundation for new runways and other military installations.[33] Thus the bulldozer constituted a very material link between the urban and the rural landscapes of wartime Britain. As D.W. Brogan had noted in his comments on bulldozers transforming rural areas,[34] the impact of bulldozers on the landscape of wartime Britain was not limited to bomb-damaged towns and cities: 'Bulldozers alter the face of Britain' was one newspaper's description of the way 'virgin countryside' had been 'bulldozed' for military purposes.[35] 'War has changed most things in our island home, but none more so than the countryman and the countryside,' observed the *Manchester Evening News* in October 1943: 'Chestnut-covered walks have been torn up by bulldozers and excavators, rolling out giant aerodromes.'[36] The changes wrought by the bulldozer were not limited to war construction, but reflected the increasing presence of this machine and others in the ordinary work of the land, as the countryside column of the *Western Gazette* noted: 'Ancient hedges are grubbed out by gyro-tillers; ancient banks and ditches are levelled by bulldozers.'[37] The War Agricultural Executive Committees made increasing use of bulldozers as more machines became available, and bulldozers were used by the Women's Land Army to clear and level land for farming.[38] There was concern from some observers about the potentially destructive effect such machinery would have on the British countryside: 'Because the military authorities and Government have shown us what can be done in the way of removing hedges by the aid of "bulldozers", people are wondering what will be the outcome.'[39] In general, however, the increasing involvement of modern bulldozers, scrapers, excavators and similar equipment in agriculture was welcomed as a beneficial side-effect of the war and there were efforts to expand it in order to increase the land available for agricultural exploitation. Questions were asked in Parliament about whether the government could take measures 'to make use of the "bulldozers" used on aerodromes in order to clear the land'.[40]

Beyond the work of the bulldozers in rubble clearing, civil and military engineering and agriculture within Britain, the exploits of similar earthmovers overseas at the fighting fronts of the war received much attention, further increasing the prominence of these 'new weapons of the present war ... these indispensable machines'.[41] In North Africa and the Far East, and later in Italy, France and Germany, the role of the bulldozers that were an essential element of the Allied armed forces was extensively reported. Writing in the *Daily Express* of the fighting in Sicily in the summer of 1943, Alan Moorehead commented that 'The bulldozer was the tank of this campaign ... The bulldozers pushed their ugly snouts right up to the front line.'[42] For the *Yorkshire Post* in March 1944 the bulldozer was 'becoming an essential weapon in the Far Eastern war. The Japanese are excellent at light-weight warfare in the tropics; but they cannot rival the array of heavy machines ... which the Allies are now bringing against them.'[43] For *Picture Post*, reporting on the advance of the Allied armies from their Normandy beachhead in August 1944, 'The clanking and thumping of the bulldozer is as familiar at the front as the sound of the guns.'[44] Tales of the heroism of bulldozer drivers, who were often exposed to enemy fire as they worked, were frequently reported, as in this example from the coverage of the advance of the American Fifth Army through Italy in October 1943:

> At one of our crossings of the Volturno I watched a bulldozer cutting away a newly-won river bank so that tanks could cross. Within a few minutes a German shell directed from a hill overlooking the river killed the driver. I saw another man take his seat immediately and the work proceeded until another shell destroyed him and the bulldozer. Within the next half hour two more bulldozers and their drivers were knocked out by the deadly accuracy of the enemy fire. The last one was hit just as it had pushed the last load of muddy soil aside, allowing for the passage of tanks. No fighting troops could have died more heroic deaths than these four bulldozer drivers.[45]

Bulldozer crew bravery in a lower key was reported by *The Times* in February 1945, as British soldiers advanced towards the Rhine: when a bulldozer 'came clanging along the road to Cleve on a dark night' the soldiers leading the advance advised the driver and his mate that the road ahead was under small arms fire. '"But we are armour," they protested, and continued on their way through the darkness.'[46] This account also emphasises the ambivalent position of the bulldozer, simultaneously civil and military, part earthmover and part tank. This was a continuity that worked both ways, enabling the bulldozer to take its place in the military arsenal during the war, but also allowing it to act as a conduit for the continuation of a militarised culture of coercion and clearance into the landscapes of peace, in Britain and elsewhere.[47]

The issue of bringing the wartime technologies of earthmoving to bear on the problems of the post-war British landscape was already exercising minds while the fighting was still under way. The task of reconstruction, observed the *Sphere* in December 1943, 'will require the use of every available bulldozer, muck-shifter, scraper, tipper and dumper', also noting that such machines would be 'with the exception of jeeps, almost the only vehicles made to War Office specifications which will have a commercial value in peacetime'.[48] By the autumn of 1945 the British government, keen to secure this modern technology for the economic benefit of the country, was negotiating with the Americans over the disposal of US earthmoving and construction equipment to the public works and commercial building sectors, and ministers were urging 'the building trade to accept mechanization ungrudgingly and use it to the full'.[49] Farmers too were being encouraged to adopt modern 'labour-saving machines' including crawler tractors and bulldozers, which were 'admirable for levelling, clearing sites, back-filling trenches or excavating'.[50] The 1940s and 1950s were a period in which technological modernity applied to agriculture became one of the most potent images of a modernising, progressive Britain,[51] and in farming and forestry bulldozers were adopted with enthusiasm, partly driven by the easy availability of war surplus machines, and while not all experiments in their use were successful, these adaptable machines did find increased application in clearance and ground preparation works.[52]

The use of bulldozers in British construction expanded greatly during the 1950s and 1960s, partly as a result of the huge size of new building works. Large-scale construction of building developments and new towns, roads and airports provided both extensive employment for fleets of modern bulldozers and other equipment, and an arena in which their technology and the ways in which they were used could be further

Figure 3.2   The construction of the M1 motorway through Bedfordshire in 1959 is marked by a swathe of bare earth, cleared and levelled by an army of machines such as the bulldozer prominent in the foreground. Source: photograph taken by Ben Brooksbank and reproduced by permission of the photographer.

developed. In place of the laments of civil engineers of the 1930s and 1940s that British clearance and construction projects could not offer scope for the degree of mechanisation to be found in the United States, came an eager grasping of the opportunity offered by such schemes as the 'motorway programme with its associated mammoth earthwork problem'.[53] (See Figure 3.2.)

With its power and flexibility the bulldozer emerged from the Second World War as a universally adaptable wonder-machine, and a key element in the effort of rebuilding and reshaping the post-war landscapes of Britain. Peter Merriman has noted the important role of the Caterpillar company in exploiting the transition from war to peace in Britain through the employment of their machines, which had constituted a vital part of the Anglo-American wartime workforce, on peacetime projects. The appearance of British-built Caterpillar tractors and bulldozers in particular represented a new Anglo-American collaboration:

> For the North American company Caterpillar the motorway marked the 'beginning of a new era ... of progress and opportunity' which would be aided by their reliable, efficient and economic D8 tractor. During World War Two Caterpillar tractors could be seen to be part of an army of imported machines turning Britain's waste-lands into

spaces for the practice of an efficient, modern and mechanised farming industry, but the D8 was 'now built in Great Britain', at their new Glasgow factory. This was a British tractor 'helping to build a better Britain' and helping Caterpillar to fulfil their duty to strengthen the economy, society and 'the nation'.[54]

This continuity legitimised the enterprise of physical transformation and made it part of a progressive historical process. As Scott Kirsch and Don Mitchell have written with reference to the large-scale civil engineering projects of the post-war decades, efforts to transform landscape 'were part of an ancient, and in some ways technologically cumulative, history … And the conversion of military technologies to civil uses was also a long-standing enterprise.'[55] Within this historical trajectory, the bulldozer became the symbol and the harbinger of a new world.

The dark side of the bulldozer

In Britain the bulldozer entered the public imagination in the 1940s and 1950s as a potent image of rebuilding and transformation. Even the incoming General of the Salvation Army was described in 1946 as a 'bulldozer-driver', and declared that he would 'devote everything to making the Army the bulldozer of Evangelism, seeking to drive through the ruins and desolation of our shaken civilisation, a road by which men may travel toward the Kingdom of God'.[56] Bulldozers were part of the modern mechanised transformation of a war-ravaged country, clearing away, repairing and rebuilding. The work they carried out clearing roads and rescuing people trapped by snowstorms during the severe winters of the post-war years, or clearing the damage caused by floods, also contributed to the positive image of these machines.[57] Reviewing the clear-up operations following the floods of 1953 the *Geographical Journal* specifically praised the contribution of the 'bulldozer and other war-time American machines' that had 'made a vast difference' in the recovery work.[58]

As the 1950s and 1960s went on, however, darker and more destructive images of the bulldozer began to gain more prominence in public environmental discourse. The periodical *Sport & Country* summarised the ambiguities of the bulldozer as early as 1950, observing that 'the bulldozer itself is going to be kept busy enough in connection with future agricultural enterprises' because of 'our country's desperate economic plight' but also asserting that 'in the bulldozer we have a horrid symbol of all that is inexorable in farming's mood to-day. The super-powered crawler stands for all else that is farming's "latest".'[59] The association of bulldozers with environmentally damaging extractive industries such as quarrying and open-cast mining was also an increasing matter of public concern in the 1950s.[60] 'Using bulldozers to tear away topsoil and wrench a mixture of coal and earth from seams near the surface not only disfigures more of the English landscape than we can spare but does lasting harm to the productivity of the land,' declared the *Manchester Guardian* in 1955.[61] As large-scale urban redevelopment and road building gathered pace the bulldozer became the single most potent image of the damaging schemes that threatened landscapes and communities.[62] The seemingly inexorable advance of caterpillar tracks and huge steel blades across the land gave those opposing such schemes a new sense of urgency: 'we will wake up one morning', a protester against a by-pass plan for Bangor told a reporter in 1953, 'and find the bulldozers have come' – the suggestion being that the earthmovers were effectively a hostile army of occupation moving under the cover of darkness to seize the land from its defenders.[63]

The association of bulldozers with urban rebuilding made them into an ambiguous symbol, combining destruction and regeneration in a potent, mechanised agent of transformation. Louis MacNeice captured this aspect of the bulldozer in his 1962 poem 'New Jerusalem': 'Bulldoze all memories and sanctuaries: our birthright / Means a new city, vertical, impersonal'.[64] References to bulldozers and bulldozing became staples of newspaper articles reporting on controversial and damaging development schemes and protests held against such schemes frequently focused on bulldozers as particular targets. Protesters chained themselves to the bulldozers, poured honey or sugar into their fuel tanks, blocked their paths and lay down in front of their tracks.[65] The act of throwing oneself in front of the bulldozers, or threatening to do so, became an almost ritualistic aspect of new developments,[66] to the extent that the writer Douglas Adams was able to use this scenario as the starting point of his 1978 science fiction radio series (and subsequent novel) *The Hitchhiker's Guide to the Galaxy*.[67] Some in the new protest movements were wary of the image created by such direct action: in 1973 one member of the planning board responsible for the Lake District National Park warned his fellow board members, who were divided over the possible route of a road in the district, that they would 'begin to look foolish' if they associated with those who wanted to 'lie down in front of bulldozers'.[68] The risk of appearing foolish did not appear to trouble the

Duke of Rutland, however, who declared in the summer of 1977 that 'he would lie down in front of the bulldozers' to prevent work starting on the coalfield that threatened the Vale of Belvoir and his home, Belvoir Castle.[69]

During the 1960s and 1970s in Britain the word 'bulldozer' became a shorthand for the whole process of destructive development: machine became metaphor, the bulldozer standing as the symbol of the entire vast and seemingly unstoppable machine of development and its accompaniments of heedless politicians, greedy developers and faceless bureaucracy. During these years historic streets were 'abandoned to bulldozers',[70] in towns marked for rebuilding concerned observers saw 'streets full of charm and character look sadly neglected, apparently waiting for the Great Bulldozer',[71] and archaeologists were urged to 'battle against the bulldozers'.[72] The 'bulldozing' of ancient monuments was in itself a particular focus of concern, with experts warning that 'huge archaeological treasure houses will be lost for ever before the advance of the bulldozers'.[73] Whether what was threatened was long-established communities, vulnerable historic towns, or the fragile remnants of the past, the symbolism of the bulldozer as emblematic of a mechanised society heedless of what it destroyed was ever more widely reflected in critical press coverage and protest activities.[74] Bulldozers were often anthropomorphised as destructive animals, and the sites with which they were associated were characterised in terms of war zones, battles and violent destruction: 'Bulldozers groan where the hungry children used to play,' wrote one reporter of the rebuilding of Jarrow in the 1960s, 'and the part between the centre and the Tyne is like a battlefield now.'[75] Those opposed to destructive developers saw their struggle in terms of armies: 'On the one hand stand the forces of "development", their bulldozers massed like the chariots of old, their chemical equipment as menacing as the plagues of Egypt; on the other are ranged a small government agency, the Nature Conservancy, and a growing popular movement,' wrote Bruce Campbell in the *Guardian* as he introduced the first British 'National Nature Week' in May 1963.[76]

The bulldozer was seen as the emblematic machine of a new era in landscape transformation, a slow-moving machine perhaps but one possessed of enormous brute force, momentum and transformative power. Before the 'age of the bulldozer' previous generations had been content 'to modify and improve their properties', but now small settlements were 'threatened by total demolition', warned one architect concerned at threats to old buildings and small villages in 1962.[77] The same applied to the rural landscape, in which the increased mechanisation of farming and estate management found expression in images of bulldozers uprooting trees, destroying habitats and stripping soil. An increasingly important concern of the environmental movement during the 1960s and 1970s was the threat posed to 'traditional small holders and small farmers', held to be responsible custodians of the Earth, by the 'bulldozer-farmer',[78] hungry for land to exploit through modern machinery in order to gain the greatest yield in the shortest time. The various agencies of the British state were another target for environmentalists, and these same agencies were often among the most enthusiastic early users of the bulldozer on a large scale. Particularly notable in this respect, and to environmentalists particularly notorious, was the Forestry Commission: T.C. Smout has written of the role of 'a new generation of drag-lines, bulldozers and mole ploughs' in making possible the destruction of moor, bog and fen and its replacement by regimented plantations of Sitka and lodgepole pine in Scotland during the 1950s and 1960s.[79] For people concerned to protect areas of natural beauty such as the Quantock Hills, the Forestry Commission was 'the despoiler of beauty' under whose aegis 'bulldozers will take over'.[80] As a machine equally at home in urban clearance and rural redevelopment, the use of which linked finance, economics, politics, agriculture and the environment, the bulldozer was seen as an all-encompassing symbol of the threat posed to the ecosystems of the Earth by the malign exploit- ative forces of modernity – an image summed up in *The Ecologist*'s cover illustration for July 1972 (see Figure 3.3), which showed a vast bull- dozer looming threateningly over city and countryside alike.[81]

In symbolising the destruction by faceless bureaucratic and com- mercial forces of the precious and fragile natural environment, the bull- dozer could also be seen as the emblematic expression of the use of force by the collective to crush the individual, expressing a political idea that was gaining traction during the decades of the twentieth century during which the bulldozer itself rose to ascendancy. As the historian David Schalk noted in 1971, 'The metaphor of the bulldozer or steam- roller has been frequently used to describe the condition of twentieth- century man caught up in the vast impersonality of institutional society.'[82] This situation had been conceptualised by Cyril Connolly in 1944 as presenting humanity with a dualistic choice between indi- vidual freedom which permitted growth and renewal on one side, and bureaucratised, mechanised collectivisation on the other: 'Well, which side are you on? The Corn-Goddess or the Tractor? The Womb or the Bulldozer?'[83] The year 1944 was of course also when D.W. Brogan had published his essay on the bulldozer as symbolic of the virtues of the

Figure 3.3   Front cover of *The Ecologist*, July 1972.
Source: reprinted with permission of The Resurgence Trust (www.
theecologist.org, www.resurgence.org).

New World offering hope to the war-torn lands of the Old. Brogan saw
synthesis leading to progress, whereas Connolly saw a choice between
continuation and annihilation – but the bulldozer was at the heart of the
argument for both.

On the tracks of modernity

'The green mantle of Earth is now being ravaged and pillaged in a frenzy of exploitation by a mushrooming mass of humans and bulldozers,' warned Michael Soulé in a pioneering work on conservation biology in 1980.[84] Forty years after the Second World War led to its proliferation across the globe and the acceleration of its career of brute-force clearance and transformation, the bulldozer had become an emblem of global devastation rather than global progress. This dark side of the bulldozer had always been an innate aspect of a machine whose power to create was always rooted in destruction. The work of the bulldozer is to push aside, to clear away, to bury. It is a machine that prepares the ground for the new, and in doing so it obliterates the old. It buries the evidence, levels the ground and moves on, both creator and destroyer. In the post-war world the bulldozer negotiated the passage from past to future across the rubble-strewn landscape of rebuilding. The future was highways, hospitals, power stations, shopping malls and spreading suburbia, with the bulldozer, itself becoming ever larger, more powerful and more efficient, leading the advance of urban modernity.[85]

At the beginning of the bulldozer's career in wartime and post-war Britain it represented a distinctively American technological modernity, welcomed as clearing the way for society to move towards a brighter future. Over the next half-century that promise was dimmed by uncertainties and fears, and the very Americanism of the bulldozer seemed itself an aspect of its negative characteristics as symbol of 'western or American cultural imperialism, lurching across the globe like a runaway bulldozer levelling everything in its path'.[86] The bulldozer of the 1940s and 1950s had stood as an emblem of progress, but the hope for a better future that technology had long seemed to offer was, for an increasing number of people in the era of Cold War, environmental destruction and urban expansion, far outweighed by the dangers it unleashed. A *leitmotif* of the nuclear age (hinted at in Soulé's imagery of a threatening 'mushrooming mass') was the partnering of the bulldozer with the atomic bomb to create a summary image of the all-encompassing destructive power humanity possessed in the modern technological age. In 1963 Lewis Mumford warned that 'our age will be known to the future historian as the age of the bulldozer and the exterminator' given that 'the building of a highway has about the same result upon vegetation and human structures as the passage of a tornado or the blast of an atom bomb'.[87] The architectural historian James Marston Fitch was suggesting by the early 1970s that

'man now runs the literal risk of losing *all* the past, man-made and natural – either piecemeal, to the bulldozer, or instantaneously, to nuclear
weapons',[88] and the same theme was revisited in 1984 by the philosopher and environmentalist Richard Routley to argue for the connect-
edness of environmental ethics and nuclear ethics: 'The Bomb and the
Bulldozer are out of the same technological Pandora's Box.'[89] From
being the machine that would clear away the debris of the past and
reshape the land to create a bright future, the bulldozer had become
symbol of a technologically driven apocalypse that threatened to grind
past, present and future alike into destruction beneath its inexorable
tracks and irresistible blade.

Notes

1   For the American bulldozer's role in the Second World War, see Francesca Russello Ammon,
    Bulldozer: Demolition and Clearance of the Postwar Landscape (New Haven: Yale University
    Press, 2016), 21–93.
2   Peter Merriman, '"Operation Motorway": Landscapes of Construction on England's M1
    Motorway', *Journal of Historical Geography* 31, no. 1 (January 2005): 113–33.
3   'The "Bulldozer": A Mechanical Elephant', *Illustrated London News* (8 February 1941): 187.
4   'Bull-Dose, -Doze', *Oxford English Dictionary*; Keith Haddock, *The Earthmover Encyclopedia* (St
    Paul, MN: Motorbooks, 2002), 35; Eric C. Orlemann, *Super-Duty Earthmovers* (Osceola, FL:
    Motorbooks International, 1999), 77, 84; Richard Grant White, 'Americanisms', *The Galaxy*, 24,
    no. 3 (September 1877): 383; George C. Rable, *But There Was No Peace: The Role of Violence in the
    Politics of Reconstruction* (Athens, GA: University of Georgia Press, 1984/2007), 95, 115, 176–9.
5   James B. Runnion, 'The Negro Exodus', *Atlantic Monthly*, 44, no. 262 (August 1879): 226–7.
6   William R. Haycraft, *Yellow Steel: The Story of the Earthmoving Equipment Industry* (Champaign:
    University of Illinois Press, 2000), 70; C.H. Wendel, *Encyclopedia of American Farm Implements
    and Antiques* (Iola, WI: Krause Publications, 2004), 124; Randy Leffingwell, *Caterpillar:
    Farm Tractors and Bulldozers* (Osceola, FL: Motorbooks International, 1994), 117–18; Mark
    Fiege, *Irrigated Eden: The Making of an Agricultural Landscape in the American West* (Seattle:
    University of Washington Press, 1999), 121; Joseph A. Miller, 'From Bulls to Bulldozers: A
    Memoir on the Development of Machines in the Western Woods from Letters of Ted P. Flynn',
    Forest History 7, no. 3 (Autumn 1963): 14–17, 15.
7   Miller, 'From Bulls to Bulldozers', 15; Philip L. Fradkin, *The Seven States of California: A Natural
    and Human History* (Berkeley: University of California Press, 1997), 235; Gene Rose, *The San
    Joaquin: A River Betrayed* (Clovis, CA: Quill Driver Books/Word Dancer Press, 1992/2000),
    81ff; Anon., *The Fresno Scraper, Invented in 1883: A National Historic Mechanical Engineering
    Landmark* (Fresno, CA: American Society of Mechanical Engineers, 1991).
8   'British Columbian Mines', *Economist* (25 January 1902): 112; John J. Curzon, 'Changing
    Mining Methods at the Holden Mine', *Transactions of the American Institute of Mining and
    Metallurgical Engineers* 163 (1946): 75–7; Agne Rustan, Claude Cunningham, William
    Fourney, Alex Spathis and K.R.Y. Simha (eds), *Mining and Rock Construction Technology Desk
    Reference: Rock Mechanics, Drilling and Blasting* (London: CRC Press, 2010), 11.
9   Gardiner Dexter Hiscox, *Mechanical Movements, Powers, Devices, and Appliances* (New York:
    N.W. Henley, 1904), 230.
10 Leffingwell, *Caterpillar: Farm Tractors and Bulldozers*, 117.
11 E.G. Robson, 'The Development and Testing of Earth-Moving Machinery', *Proceedings of the
    Institution of Mechanical Engineers* 179, part 3F (1964–5): 55–64; see also Charles Thomas
    Mitchell, 'Some Economical Aspects of Modern Earthmoving Equipment', *Institution of Civil
    Engineers Engineering Division Papers* 4, no. 2 (1946): 3.

12 V.W. Bone, 'Modern Developments in Tractor-Drawn Excavating Equipment', *Proceedings of the Institution of Mechanical Engineers* 137 (1937): 345–53. The original meaning of 'bulldozer' was the blade and its associated apparatus, and this is the way in which Bone is using the term, but its application quickly expanded to encompass the entire ensemble of tractor plus blade. Until after the Second World War 'bulldozers' (i.e. blades) were manufactured by different companies from those building the tractors to which they were attached.

13 Bone, 'Modern Developments in Tractor-Drawn Excavating Equipment', 348, 353.

14 'Tractors that Fill in Craters', *Yorkshire Post* (5 August 1941): 2.

15 'Bombing: A Choice of Priorities?', *Picture Post* (25 September 1943): 7.

16 D.W. Brogan, 'The Bulldozer' (1944), in *American Themes* (London: Hamish Hamilton, 1948), 202.

17 Brogan, 'The Bulldozer', 208.

18 H.A. Land, 'Crawler-Mounted Tractors and Attachments for Scraping and Dozing', *Proceedings of the Institution of Mechanical Engineers* 179, part 3F (1964–5): 89–105.

19 'Chairman's Closing Remarks at the Meeting on Handling Machinery', *Proceedings of the Institution of Mechanical Engineers* 157, part 1 (1947): 317–18, 318.

20 Brogan, 'The Bulldozer', 204.

21 Waldo G. Bowman, Harold W. Richardson, Nathan A. Bowers, Edward J. Cleary and Archie N. Carter, *Bulldozers Come First: The Story of US War Construction in Foreign Lands* (New York: McGraw-Hill, 1944), 20.

22 Bowman et al., *Bulldozers Come First*, 20.

23 Robson, 'The Development and Testing of Earth-Moving Machinery', 55.

24 Christopher Powell, *The British Building Industry Since 1800: An Economic History* (London: Routledge, 1980/2013), 134.

25 'Chairman's Closing Remarks', 318; V.W. Bone, W. Savage and R.M. Wynne-Edwards, 'Public Works Contractors' Plant', *Proceedings of the Institution of Mechanical Engineers* 157, part 1 (1947): 313–17, 313, 314.

26 *Liverpool Daily Post* (25 January 1941): 4. This text is part of the caption for a large picture of a bulldozer clearing away rubble.

27 'The "Bulldozer": A Mechanical Elephant', 187.

28 'This Will Shift Anything', *Yorkshire Evening Post* (1 February 1941): 5.

29 'Motley Notes', *The Sketch* (1 November 1944): 226.

30 'More about the English Housing Problem', *The Sphere* (23 December 1944): 379.

31 'American Troops and Bulldozers Swing into the Battle to Rebuild London', *Daily Express* (14 December 1944): 3.

32 'American Troops and Bulldozers Swing into the Battle to Rebuild London', 3.

33 Ammon, *Bulldozer*, 30.

34 Brogan, 'The Bulldozer', 204.

35 'Bulldozers Alter the Face of Britain', *Lincolnshire Echo* (9 June 1943): 2.

36 'The Plough', *Manchester Evening News* (13 October 1943): 2.

37 'The Week in Farm and Field', *Western Gazette* (24 September 1943): 5.

38 'Land Girls Reclaim Lost Acres with Bulldozers and Cub-Excavators', *Marylebone Mercury* (1 January 1944): 4.

39 'Hedges and Bulldozers', *Lichfield Mercury* (24 November 1944): 8.

40 House of Commons Debates, 'Land Cultivation', *Hansard* (29 September 1942) vol. 383, col. 732.

41 'Aspects of the World War – Mechanical and Topographical', *The Sphere* (25 December 1943): 406.

42 Alan Moorehead, 'Sicilian Summing-Up', *Daily Express* (23 August 1943): 2.

43 'Manned Bulldozer for 72 Hours Under Fire', *Yorkshire Post* (10 March 1944): 1.

44 'The Crossing of the Orne', *Picture Post* (26 August 1944): 12.

45 Noel Monks, 'The Bulldozers', *Daily Express* (16 October 1943): 1.

46 'Goch Closely Threatened', *The Times* (19 February 1945): 4.

47 Ammon, *Bulldozer*, 11.13. For analysis of another ostensibly peaceful civilian technology deeply implicated with military significance see Jacob Ward, 'Oceanscapes and Spacescapes in North Atlantic Communications', Chapter 10 this volume.

48 'When We Get Back to Road Transport', *The Sphere* (18 December 1943): 376.

49 'Mechanized Building', *The Times* (25 September 1945): 2; House of Commons Debates, Written Answers, Ministry of Supply, 29 October 1945, *Hansard*, vol. 415, col. 192W.

50 'Mechanical Aids for the Builder or Farmer', *Sport & Country* (23 November 1945): 334, 335.

51 See David Matless 'The Agriculture Gallery: Displaying Modern Farming in the Science Museum', Chapter 6 this volume, for further discussion of this point.

52 W.A. Cadman, 'Forestry and Silvicultural Developments in North Wales', *Forestry* 26, no. 2 (1953): 65–80, 69; L.C. Troup, 'The Afforestation of Chalk Downland', *Forestry* 27, no. 2 (1954): 135–44, 143; Victor Bonham-Carter, 'Reviving the Rural Industries', *The Sphere* (15 May 1954): 263; 'Woodland Reclamation', *Sport & Country* (26 May 1954): 534–6.

53 Frank H. Archer, 'Earth-Moving Machinery in Motorway Construction', *Proceedings of the Institution of Mechanical Engineers, Conference Proceedings* 179, part 3F (1964–5): 1–8, 5.

54 Peter Merriman, '"Operation Motorway"', 116–17. The quotations within this passage come from Caterpillar advertisements published to coincide with the opening of the M1 motorway on 2 November 1959.

55 Scott Kirsch and Don Mitchell, 'Earth-Moving as the "Measure of Man": Edward Teller, Geographical Engineering, and the Matter of Progress', *Social Text* 16, no. 1 (Spring 1998): 100–34, 103.

56 '"Bulldozer-Driver" Takes Over Command Today', *Daily Mirror* (21 June 1946): 8.

57 For example, see 'Snowbound Farm Relieved', *Manchester Guardian* (23 February 1951): 5; 'High Street that Vanished in the Night', *Manchester Guardian* (18 August 1952): 5; 'Sea-Wall Holds Back Tide at Lynmouth', *Manchester Guardian* (4 September 1952): 5; 'How Troops and Civilians Waged the Battle of the Floods', *The Sphere* (14 February 1953): 238.

58 J.A. Steers, 'The East Coast Floods', *Geographical Journal*, 119, no. 3 (September 1953) 280–95, 293.

59 Horace Phillips, 'Power-Farming and Game Birds', *Sport & Country* (11 January 1950): 2. The main concern of the article was the effects of increasingly mechanised farming on game-preserving and shooting.

60 'Prelude to Opencast', *Manchester Guardian* (26 May 1952): 5.

61 'Opencast Coal', *Manchester Guardian* (7 January 1955): 6.

62 'Beauty to Suffer in National Interest', *Guardian* (17 April 1969): 7; 'Bulldozers Destroying "Unique Habitat"', *Guardian* (5 February 1970): 5.

63 'Beauty and a By-Pass', *Manchester Guardian* (7 April 1953): 3.

64 'New Jerusalem' (1962) from *Collected Poems* (London: Faber, 1979): 529.

65 'Attack on the Trees', *Manchester Guardian* (19 May 1957): 19; '10 Arrested at Shell Site', *Manchester Guardian* (11 October 1958): 1; 'Before the Bulldozer', *Guardian* (20 January 1961): 9.

66 'Protest Saves Abbey Trees', *Observer* (5 November 1967): 3; 'Militancy', *The Ecologist* 2, no. 10 (October 1972): 39; 'Families Fight to Save Their Homes', *Guardian* (2 June 1978): 2.

67 *The Hitchhiker's Guide to the Galaxy*, BBC Radio Four, episode 1, broadcast 8 March 1978.

68 'Lakes Board Split Over Road', *Guardian* (23 January 1973): 5.

69 'Protagonists Marshal Forces as the Battle of Belvoir Begins', *Guardian* (19 July 1977): 2.

70 'Georgian Streets Abandoned to Bulldozers', *Guardian* (11 June 1977): 3.

71 'The Case of the Authentically Medieval River-Moated Car Parks', *Guardian* (23 May 1977): 12.

72 'Battle Against the Bulldozers', *Guardian* (8 March 1977): 6.

73 'History May Go Bust in Property Boom', *Observer* (5 March 1978): 4.

74 'Ancient Monument Destroyed', *Manchester Guardian* (28 April 1953): 1; 'Bulldozing Wat's Dyke', *Manchester Guardian* (7 December 1957): 1; 'Developers "Grave Threat" to Ancient Remains', *Guardian* (31 December 1970): 6.

75 'The Last Crusade', *Guardian* (5 October 1961): 9.

76 'A Mirror Up to Nature', *Guardian* (18 May 1963): 6.

77 'Down in the Village Something Stirs', *Guardian* (22 October 1962): 2.

78 John Seymour, 'A New Kind of Man', *Resurgence* 1, no. 12 (March/April 1968): 18–19, 18.

79 T.C. Smout, 'Bogs and People in Scotland since 1600' (1997), in *Exploring Environmental History: Selected Essays*, ed. T.C. Smout (Edinburgh: Edinburgh University Press, 2009), 104.

80 'The Future of the Quantocks: An Area of Great Natural Beauty is in Jeopardy', *The Sphere*, 3 December 1949, 347. See also Tim Cole's discussion of the significance of the Quantocks in the context of road-building: '*About Britain*: Driving the Landscape of Britain (at Speed?)', Chapter 7 this volume.

81 *The Ecologist*, 2, no. 7 (July 1972): front cover.

82 David L. Schalk, 'La trahison des clercs – 1927 and later', *French Historical Studies* 7, no. 2 (Autumn 1971): 245–63, 263, n88.

83 Cyril Connolly, *The Unquiet Grave: A Word Cycle by Palinurus* (London: Hamish Hamilton, 1944), 100.

84 Michael E. Soulé and Bruce A. Wilcox (eds), *Conservation Biology: An Evolutionary Perspective* (Sunderland: Sinauer Associates, 1980), 7.

85 On the ambiguities of technological progress, particularly in relation to landscape, see David E. Nye, 'Technologies of Landscape', in *Technologies of Landscape: From Reaping to Recycling*, ed. David E. Nye (Amherst: University of Massachusetts Press, 1999): 3–20.

86 John Boli, 'Contemporary Developments in World Culture', *International Journal of Comparative Sociology* 42, no. 5/6 (2005): 383–404, 396.

87 Lewis Mumford, *The Highway and the City* (New York: New American Library, 1963), 247.

88 James Marston Fitch, 'New Uses for Artistic Patrimony', *Journal of the Society of Architectural Historians* 30, no. 1 (March 1971): 3–16, 7.

89 Richard Routley, 'Metaphysical Fallout from the Nuclear Predicament', *Philosophy & Social Criticism* 10 (1984): 19–34, 30, n4. Richard Routley had changed his name to Richard Sylvan the year before this essay was published.

4

Locality and contamination along the transnational asbestos commodity chain

Jessica van Horssen

Natural resources often take widespread, diverse paths as they are removed from their natural habitat, shipped to factories either near or far, and processed into marketable goods available around the world. But what happens when a natural resource is toxic? How does contamination change at each stage along its global commodity chain? How have the risks toxic resources pose to human health been regulated by those in positions of power in ways that overlook those most vulnerable in society?

This chapter examines these questions by following transnational contamination along the Canadian asbestos commodity chain, which I call the 'contamination chain'. For the purpose of this chapter, I will focus on the transnational path asbestos took from Canadian mines to factories and homes in Greater Manchester during the interwar and post-war period. Canada extracted the majority of the world's asbestos in the twentieth century, and British politicians, city planners and consumers saw the fireproof mineral as being crucial to creating safe, long-lasting communities. When it came to asbestos, however, it was never as simple as supply and demand, and with this chapter, I intend to broaden our understanding of environmental and social justice through a close examination of how the people of Manchester came to experience toxicity in both the workplace and the home.

Philosopher of science Bruno Latour calls asbestos 'a perfect substance', and 'one of the last objects that can be called modernist' before the illusions of its modernity were revealed by large-scale disease rates and workers' compensation cases.[1] It is important to remember, as outlined in the opening chapter of this volume, that environmental artefacts like asbestos could only be deemed both a 'perfect substance' and a tool of contamination after going through an intense technological process. In this way, asbestos is a technological artefact, reflecting both society's ambition to protect itself from the dangers of fire and society's fear of being harmed by the very thing thought to protect.

Asbestos is a fireproof mineral formed deep within the Earth's crust during the Devonian period between 410 and 355 million years ago, as large land masses broke apart and collided. This occurred throughout the world, but in the case of Canada's 'asbestos belt', the mineral developed along with the Appalachian Mountain Range in what is now the Eastern Townships region of the French Canadian province of Quebec. The friction and heat involved in the formation of the Appalachians chemically reconstituted the serpentine rock at this particular site, and in its re-crystallisation, the chemical composition was changed and veins of asbestos fibre formed.[2] The asbestos located in this region is composed of magnesium, silicon and oxygen ($Mg_3Si_2O_5(OH)_4$) and is able to withstand temperatures in excess of 3,000°F.[3]

Although formed millions of years ago, the largescale human use of asbestos only really began in the late nineteenth century when rising population rates and the industrialisation of the Western world enabled the development of the technological systems required to process the mineral into marketable, fireproof goods. Surveyors for the Geological Survey of Canada first noted the presence of asbestos in the Eastern Townships in the 1840s and described it as a nuisance that ruined perfectly marketable slate.[4] Indeed, it wasn't until the 1870s that the fireproof qualities of the mineral became more widely known and market demand began to grow. With this came a technological revolution of the landscape, as farms and forests began to be replaced by massive open pit asbestos mines in the Eastern Townships of Quebec.

Examining the path asbestos took from local mine to global market, from Canadian mines to Manchester factories and homes, informs us of the different ways society has interpreted risk, blame and the legacies of environmental contamination and justice. Environmental contamination rarely, if ever, respects artificial borders: wind, water and species transport toxic resources and waste beyond seemingly contained sites, and result in different forms of exposure and contamination.[5] What's more, these different forms of exposure often result in different diseases, and this chapter examines the impact history of asbestos as a technological artefact impacting health and safety in a multitude of ways within different environments.

Asbestos exposure causes three main diseases: asbestosis, lung cancer and mesothelioma. Asbestosis is the hardening of the fluid lining of the lungs due to the inhalation of large amounts of asbestos fibre over a number of years, resulting in death by suffocation, when lungs can no longer expand and contract with each breath. This was particularly common in factory workers in places like Manchester, where large amounts of asbestos dust would be raised in the process of manufacturing marketable goods. Lung cancer and mesothelioma were more difficult for medical researchers to diagnose because they usually occur after fairly limited exposure to asbestos – a carcinogenic mineral – and so the causal factor in these cases was often overlooked, although mesothelioma is only caused by exposure to asbestos.

In this chapter, I analyse the different ways Canadian miners and the people of Manchester, England became exposed to asbestos at different stages along its contamination chain. Left untouched, deep within the Earth's crust, asbestos is a benign substance. It was only when humans decided to extract the mineral and apply it to modern technologies that the issue of contamination arose, but it did so differently along the commodity chain, at place-specific stages of exposure. While there can be end points for extracted natural resources, the method of processing and moving large quantities of the environment through technological systems creates legacies of contamination and disease far beyond a single resource community or end destination. Thus, this chapter examines the significance of how *locality* works to define resource toxicity along transnational commodity chains of natural resources.

Canada's asbestos culture

The first site of exposure to Canadian asbestos was in the mining communities of the Eastern Townships of Quebec, where citizens lived (and breathed) the industry on their doorsteps.[6] The residents of these communities were the French Canadian Catholic 'worker bees' and their families who were disconnected from the information and resources of their American or British counterparts due to language barriers and access issues. Foreign companies like Britain's Turner & Newall and America's Johns-Manville supplied wages, housing and medical care to these workers, and the Catholic Church supplied the union.

Asbestos was mined in these communities in large, open cast pits that allowed for maximum fibre extraction without the risk of collapsing tunnels underground. Geological engineers at the turn of the twentieth century saw open cast asbestos mining as a real advantage long before anyone was aware of the hazards the mineral posed to human health, as it allowed workers access to fresh air throughout the extraction process, which was significantly different from other mining industries like coal.[7] This was a 'healthy' way to mine.

Indeed, once the dangers of the mineral started to become known in places like Manchester, industry officials and doctors used Canadian miners as evidence to prove that asbestos was not dangerous to human health, and their open air exposure to the mineral was a key factor in this campaign. Because of the open pits in these communities, the level of hazardous dust exposure in the mines themselves was fairly low in comparison with the factory-based exposure in places like Manchester. While there were factories in the Canadian mining communities, they were mostly for light processing of the mineral before shipping it to larger-scale operations located along the transnational commodity chain.

This does not mean Canadian asbestos workers were not adversely affected by their exposure to asbestos, but rather that their diseases often differed from those of other workers at other stages in the technological process of making asbestos marketable, resulting in lung cancer and mesothelioma, which occurred around 30 years after first exposure. What's more, the companies that owned these mines kept the workers away from independent medical researchers, and much of their disease realities were obscured by a sophisticated and deliberate corporate effort to hide the hazards of asbestos.[8] As long as those exposed to the mineral in its purest form remained healthy, the companies invested in the industry could claim the diseases that occurred in Manchester factory workers, for example, were caused by other materials added along the technological process.

Company doctors did not inform these isolated Canadian workers when they contracted asbestos-related disease. The French Canadian workers themselves did not have access to literature on the hazards of the mineral written in their own language. Union leaders were primarily interested in showcasing the mild temperament of their members to make the industry appealing to foreign investors, especially in comparison to the more radical socialist movements sweeping Britain and the United States in the first decades of the twentieth century. This culminated in an intense local pride in Canadian asbestos communities, rooted in the mistaken belief that their work was helping make the world safe, and their product was not what was causing workers at other sites along the commodity chain harm. Once we follow shipments of raw asbestos out of the Canadian mines and mining communities, however, it becomes clear how site-specific exposure during the technological process of making the mineral marketable changes understandings of risk and toxicity.

Manchester's asbestos culture

While Manchester may be more widely known historically as 'Cottonopolis' and a birthplace of the Industrial Revolution,[9] it is this city's history with the cotton industry that made it an ideal place for an asbestos processing industry to develop.[10] Because it is a fibrous mineral, asbestos is broken apart during the technological process it requires to become a marketable good. In many cases, the mineral's fibres are initially carded and woven much in the same way as cotton, and the two were often blended to create asbestos cloth for a variety of goods, including firefighting and military uniforms, as well as aprons, ironing board covers and oven mitts.

While the fireproof qualities of asbestos had been known for centuries, with Charlemagne even having an asbestos table cloth for party tricks,[11] it is a natural resource particularly suited to modern, industrialised societies. The intense industrialisation process undertaken during the World Wars, combined with the fact that Britain owned several of the Canadian asbestos mines, meant that it was one of the only European countries importing this valuable material, which the American Minerals Yearbook termed 'indispensable to modern life' in 1939.[12]

Over the span of the first half of the twentieth century, asbestos became part of the culture in Manchester. Not only was the cotton industry particularly suited for an asbestos transition, but so too were the people. Manchester shared this 'asbestos culture' with those in the mining communities of Canada, although it had important, site-specific differences as well, rooted in the different forms of exposure that occurred on the factory floor, compared with the open mine.

The context in which British factory workers experienced and accessed information on industrial disease also differed from that of their Canadian counterparts. Britain had a much longer history of a strong labour movement, workers generally went to doctors who weren't paid by the companies they worked for, and the asbestos workers in Manchester spun, wove and processed the mineral in a way that created much more dust in a confined space compared with the Canadian miners who worked in an open pit.[13] These British factories were a new 'site of contamination' for Canadian asbestos and they fostered a disease awareness that was simply non-existent in Canada at this time.

Although Britain had no domestic asbestos mines, it is where the first recorded death from asbestos-related disease in the world occurred in 1907.[14] This death of a British textile worker was subject to an inquest, which discovered the victim's lungs had hardened due to the inhalation of asbestos fibres, but no action against the resource or the industry followed, and market demand for the mineral continued to grow. The fact that the first recorded asbestos-related death occurred in Britain, rather than in any of the mining communities that interacted with the mineral in its rawest form, is indicative of the impact of technological processing on the resource, bringing it into contact with humans in a way that had never been experienced before.

After the textile worker's death in 1907, the British medical community was tuned into the asbestos situation. When asbestos textile factory worker Nellie Kershaw died in 1924, her cause of death was quickly determined as being asbestos, and her family was the first to successfully sue an asbestos company for wrongful death.[15] Kershaw had worked as a spinner for British asbestos magnates Turner & Newall in Rochdale, Greater Manchester. Turner & Newall owned and operated mines in both Canada and what is now Zimbabwe, and used a mixture of asbestos fibres in their operations.[16]

Kershaw had worked for Turner & Newall from 1903, when she was 12, until 1922, when she was unable to work due to her increasingly severe disease, and obtained a National Health Insurance certificate of ill health due to 'asbestos poisoning'. She would have been exposed to large amounts of raw asbestos at work, and would have taken it home in her hair and on her clothing at the end of each shift.[17] Despite this obvious connection, Turner & Newall refused to assist the Kershaw family when she became unable to work.

Dr W.E. Cooke investigated Kerhsaw's death for the *British Medical Journal* in 1924 and discovered that her lungs were hardened beyond the ability to function due to being packed full of asbestos fibre. Two years later, American asbestos giant, the Johns-Manville Co., faced its first claim for compensation from textile workers in New Jersey, although there remained no claims from Canadian miners, again supporting the corporate lie that Canadian asbestos was benign.[18] Cooke coined the term 'asbestosis' in 1927 to describe the fatal disease that Kershaw and others had developed.[19] This was the first asbestos-related disease to be named, and textile factory workers were particularly vulnerable to it due to the high amount of fibre they were exposed to in an enclosed space.

Asbestosis resembled another industrial lung disease rooted in a natural resource that British factory workers were quite familiar with: byssinosis. Especially common in the textile factories located in and around Manchester, byssinosis occurs when lungs fill up with microscopic cotton fibre, resulting in a hardening of the lining and suffocation.[20] Thus it appeared that while a different natural fibre caused asbestosis, its symptoms and progression were similar to other industrial diseases British textile workers were all too familiar with. The British state had shown little inclination to prevent byssinosis, and the same was true for asbestosis. As long as the disease remained within the factory walls, it was not of prime concern to legislators, especially when regulating dust exposure in workers would likely result in slower production rates and lower profits.

By comparison, at this point Canadian asbestos miners had very little information about the state of their bodies, nor did they have any legal recourse if and when they did get ill: asbestos-related disease was not eligible for workers' compensation claims in Canada, and most Canadian asbestos workers were not even unionised by the time of Kershaw's death.[21] While asbestosis grew to be particularly common in British factories, only a fraction of Canadian asbestos miners were diagnosed with the disease, largely because their exposure to the mineral occurred in a completely different site of contamination, with open pit mining being the majority worker activity, rather than indoor processing.[22]

Industrial vs domestic exposure

While Cooke's identification of asbestosis made both local and national news in Britain,[23] it seemed to be simply yet another industrial disease that the working class was vulnerable to, not the general public, so there was no immediate risk prevention regulation to address this problem. Dust was simply a fact of life for many factory workers in Britain at the time,[24] and it was not seen as something that could – or should – be changed. Throughout this period, reporting on dust exposure and disease was common practice for the British press, especially the local newspapers in the north-west of England. Knowledge about this risk to workers' health was very public, yet action to prevent it was very slow. A public acknowledgement and discussion of asbestosis was almost non-existent in Canada at this time, with the first newspaper article on asbestosis only appearing in 1949.[25] There were research articles in Canadian medical journals prior to this, but these were inaccessible to French Canadian asbestos miners, and they were largely edited by the British and American companies that operated the mines.[26]

Despite a general awareness of the dangers asbestos posed to human health in Britain, the problem appeared to be in the dust that was raised while processing asbestos, not in the actual material that was becoming increasingly embedded in so many homes and businesses. This was a problem with the technological system the natural resource was processed through on its way to becoming home insulation, an additive to paint, shingles and cement or even brake pad linings for the growing automobile-reliant population. In fact, it wasn't until asbestos-related disease began appearing outside the factory walls that public awareness morphed into public fear.

Indeed, early twentieth-century newspaper advertisements from Manchester indicate how all-pervasive the mineral was in the everyday life of the region's residents. A large advertisement published in the *Manchester Guardian* in 1919 claimed that asbestos was 'the most perfect scientific building material … fire-resisting, economical, weatherproof and durable', and would allow homes to be built faster and better than wooden houses.[27] Another ad, published one year later, boasted about how 'ideal' asbestos cement was for bungalows and schools in the Manchester region.[28] Because of the manufacturing tradition of northwest England, the factories of the region were particularly adaptable to all types of asbestos processing, whether it be for fabric or building materials. Furthermore, companies like Turner & Newall, as well as Bell's Asbestos, were active in the Manchester community. Although Turner & Newall did not assist Nellie Kershaw's family when she became too ill to work due to her occupational exposure to asbestos, in 1923, they proudly donated to the hospitals and infirmaries of Manchester and Salford, helping raise two million shillings and appearing on the Roll of Honour given to the Duke of York on 6 November 1923, less than a year before Kershaw's death.[29]

Homes and hospitals were other 'sites of contamination' along the transnational asbestos commodity chain. Because the asbestos in these sites was usually contained behind walls or underneath flooring, exposure to the mineral was significantly less than occurred in factories. It took much longer for awareness of this contamination to develop because it took longer for the diseases caused by this sort of slow exposure to manifest in the families who lived in these homes, or to catch the attention of the medical professionals who worked in these hospitals. In fact, hospitals often used asbestos in the treatment of other diseases. In 1935, the mineral was used to line baths used by patients suffering from rheumatoid arthritis and was deemed a 'versatile ally of medicine', along with technologies such as the X-ray.[30] Despite rising disease rates in factory workers, it appeared as though Britain could not get enough asbestos.

As an industrial hub, Manchester was a main target during the Second World War, which brought the local population an appreciation for fireproof homes, as well as an urgent need for new ones.[31] Thousands of homes were destroyed by German bombing in Manchester, as well as other urban industrial centres like London, Liverpool and Birmingham. Children were evacuated to the countryside, factories were destroyed and entire neighbourhoods were erased. Realising this was a growing crisis, especially as the displaced population continued to increase, the British government passed the Temporary Housing Act in 1944, which saw over 100,000 prefabricated asbestos homes assembled all over Britain, including in Manchester. The Canadian Parliament discussed these prefabricated homes, as many of them were constructed with the mineral mined in Canada's asbestos belt.

After much of Manchester's city centre and industrial zone was destroyed by German bombing campaigns, the town council saw an opportunity.[32] The people who lived in these parts of Manchester were typically poor working-class families, and there were a lot of them: over-population was a serious problem. When their homes were destroyed, Manchester City Council attempted to influence or change the social standing of its displaced residents while rebuilding the town so it had a modern urban centre. At the time, even though there was a general awareness about the health risks of the mineral in the UK, the need for fireproof, durable homes was too great to omit one of the most important natural resources of the modern era.

While these prefabricated houses were purchased under the 'temporary homes' programme, many of the small prefabricated asbestos bungalows remained on Britain's urban landscape for decades following the war, and many still remain today.[33] Asbestos was a fundamental part of these prefabricated homes, as sheet after sheet of asbestos cement were used as walls, floors and ceilings to ensure the families who lived in these structures would be safe from the dangers of fire.[34] By the end of the Second World War, medical and newspaper reports of asbestos-related disease still only focused on those who developed it through factory exposure. As far as regulators were concerned, the general public was safe once housed in structures that had indoor plumbing, electricity and state of the art insulation: these homes offered modernity to an urban working class previously stuck in city centre slums.

From the factory to the home: changing sites of exposure and contamination

Medical knowledge of asbestos-related disease beyond asbestosis and beyond the factory walls took decades to develop, and even longer for this knowledge to spread further than confidential industry memos. However, the slow exposure to asbestos that occurred in domestic sites of contamination, as well as in schools, hospitals and government buildings, was eventually the mineral's undoing.

Asbestos causes cancer. Asbestos-related cancers, such as lung cancer and mesothelioma, largely occur in people who are exposed to smaller levels of asbestos fibre. This is the type of exposure that occurs in home renovations, or simply through interacting with asbestos-based products in the home, including aprons and oven gloves. It also widely occurred in Canadian asbestos miners working in an open pit, where dust levels were relatively low.

While the industry used the apparently low rates of asbestosis in Canadian asbestos workers as evidence that the mineral was safe, they also secretly studied the health of these workers because they knew this was untrue. The studies took a variety of forms, but culminated in the secret autopsy of deceased Canadian miners' lungs, which were then transported across the international border to Saranac Laboratory in upstate New York, where they were studied without any public knowledge or repercussion in the 1940s and 1950s.[35] It was at this lab in 1943 that chief researcher Dr Leroy Gardner 'unintentionally' discovered that asbestos caused cancer.[36] Saranac researchers discovered 70 cases of unreported asbestos-caused lung cancer in these lungs by 1958.[37] The families of these deceased miners were never notified, and the risk of cancer to those with low exposure to asbestos was covered up by industry leaders.

As with the case of asbestosis, it was more difficult to hide asbestos-related disease from workers and the public in Britain because of an engaged labour movement and independent medical researchers who had access to those working with the mineral. The low exposure that led to cancer developing in Canadian miners was also the type of exposure factory workers were vulnerable to once regulation had reduced the amount of fibre dust in the workplace. Manchester's asbestos workers were among the first to be publicly diagnosed with asbestos-related cancer.

In an article published in a 1960 edition of the *Lancet*, Dr E.E. Keal examined the causes of death of men and women suffering from asbestosis in British processing and manufacturing plants over a prolonged period of time. Keal found that while the majority of male subjects with asbestosis died of carcinoma of the lung, the bulk of the female asbestos-related deaths were caused by carcinoma of the ovary and breast, suggesting that the interaction between asbestos and the female body was unique.[38] While the asbestos industry did not want the dangers asbestos posed to human health to become general knowledge, it especially wanted to avoid any connection between low exposure to asbestos and cancer, which was a disease workers and the general public understood much more than they did asbestosis.

As reports on the connection between asbestos and cancer spread through British society, Liverpool dockworkers refused to unload shipments of asbestos in 1967 unless they were packaged in dust-proof containers, and in March 1968, the British government banned imports of crocidolite asbestos because of the risk it posed to both workers and the general public.[39] Britain continued to import chrysotile asbestos beyond this point, however, as the health realities of Canadian miners remained hidden, and thus the mineral was still understood to be safe. However, low exposure disease rates remained threatening to the industry because it meant that the domestic and public realm could also be vulnerable to industrial – and environmental – contamination.

Houses in Manchester were a prime location of this contamination. In 1931, Manchester purchased a neighbouring part of Cheshire, on the southern border of the city, in order to address the rising crisis of urban overpopulation. The city had been developing this land since the 1920s, and following the devastation of the Second World War the council fully developed the suburb of Wythenshawe to rehome displaced families.[40] In her examinations of the South African asbestos industry, historian Nancy Jacobs explains that, 'it is necessary to recognize that environmental and social justice are linked and that power imbalances will determine the ways men and women, rich and poor, and blacks and whites live with each other and the natural world'.[41] While asbestos was used in most post-war homes and other buildings because of its remarkable ability to prevent the spread of fire, once the threat of domestic asbestos exposure became public, this was a low-income community that was forgotten by the city when it was arranging asbestos-removal plans, despite the homes largely being owned by the council.

From its beginning, urban planners saw Wythenshawe as the ideal location to house underprivileged families from Manchester's city centre, because it was more or less a blank, green, and leafy canvas. Tree-lined streets, grassy parks and asbestos-filled homes soon came to define Wythenshawe, and it was part of 'the Garden City' movement that began in Britain at the start of the twentieth century, which emphasised the importance of green spaces.[42]

Wythenshawe quickly went from open farmland to one of the largest social housing communities in Europe. Asbestos was fundamental to this community project, as well as many others like it. This all sounded ideal, and as though asbestos and Manchester were a perfect match, not only in the region's factories, but also in its homes.

The illusions of Wythenshawe's modernity can be seen as a social experiment designed to elevate the status of the urban poor through Victorian notions of the benefits of green space. In fact, the motivation for this project was similar to the sentiment expressed in early asbestos building material advertisements relating to the speed and perfection it would add when building family homes: governments and businesses looked to the natural environment to shape the behaviours of its population.

In reality, moving poor working-class families out of the urban centre where their jobs were based, and into a distant suburb that didn't have access to reliable public transport, shops or other community infrastructure, eventually became a major problem. Parents were home late from work, the leafy streets and parks offered good cover for criminals, and the asbestos-containing materials and products in these homes posed a serious threat to the health of local underprivileged families, which has yet to be addressed.

Conclusion

Despite the corporate interest in keeping contamination in factories, asbestos was used in new housing and public buildings in Britain for much of the twentieth century, not just those inhabited by the working class. This was a domestic toxic time bomb just waiting to go off. Once the public realised the threat of non-industrial contamination, however, middle- and upper-class homeowners had the disposable income to get rid of the asbestos in their homes.

Residents of communities like Wythenshawe, however, were less able to make these changes. A pamphlet sent to residents of Wythenshawe by the council's Community Housing Group in the 2000s acknowledged the threat of asbestos in the community's homes, but emphasised that residents should not panic, stating that as long as they did not do any renovations or other activities that could disturb the fibre, they would be fine.[43] Access to safe homes and reliable knowledge about asbestos contamination remains limited for Manchester's working-class residents, and these limitations have deep historical roots.

Asbestos as a technological and environmental artefact has impacted human health in different ways depending on the locality in which people were exposed to the mineral. This varied from open air exposure in Canada's asbestos mines, intense dust exposure in Manchester's factories, and limited, yet still harmful exposure in communities like Wythenshawe. Social class has been a significant factor in the degree to which people were exposed to the mineral, as well as the speed by which government regulators managed – and continue to manage – the risk. Understanding the different ways asbestos was deemed to be toxic in different localities along its transnational commodity chain informs us of the ways in which techno-environmental artefacts are manipulated and managed, and the human repercussions of these processes.

Notes

1 Bruno Latour, *Politics of Nature* (Cambridge, MA: Harvard University Press, 2009), 23.
2 Paul D. Ryan, 'Caledonides', in *The Oxford Companion to the Earth*, ed. Paul Hancock and Brian J. Skinner (Oxford: Oxford University Press, 2000); David A. Rothery, 'Obduction', in *The Oxford Companion to the Earth*, ed. Paul Hancock and Brian J. Skinner (Oxford: Oxford University Press, 2000); David S. O'Hanley, 'The Origin of the Chrysotile Asbestos Veins in Southwestern Quebec', *Canadian Journal of Earth Sciences* 24, nos 1–3 (1987): 1–9, 8.
3 Cornelis Klein, 'Rocks, Minerals, and a Dusty World', in *Reviews in Mineralogy: Health Effects of Mineral Dusts*, ed. Brooke T. Mossman and George D. Guthrie Jr (Chelsea: Mineralogical Society of America, 1993), 17.
4 Alexander Murray, 'Report of Alexander Murray, Esq., Assistant Provincial Geologist, addressed to W.E. Logan, Esq., Provincial Geologist', *Canadian Geological Survey* (Montreal: Lovell & Gibson, 1849), 388. 'Asbestus' was an alternative spelling of 'asbestos' at this time.
5 See, for example, Greg Mitman, Michelle Murphy and Christopher Sellers (eds), 'Landscapes of Exposure: Knowledge and Illness in Modern Environments', *Osiris* 19 (2004).
6 Jessica van Horssen, *A Town Called Asbestos: Environmental Health, Contamination, and Resilience in a Resource Community* (Vancouver: University of British Columbia Press, 2016).
7 Fritz Cirkel, *Asbestos: Its Occurrence, Exploitation and Uses* (Ottawa: Mines Branch, Department of the Interior, 1905), 30.
8 For more information on the corporate obstruction of medical evidence, see van Horssen, *A Town Called Asbestos*.
9 See, for example, Michael Nevell, *Cottonopolis: An Archaeology of the Cotton Industry of North-West England* (Manchester: Manchester University Press, 2017).
10 For background on this transition, see Geoffrey Tweedale, *Magic Mineral to Toxic Dust: Turner & Newall and the Asbestos Hazard* (Oxford: Oxford University Press, 2001).
11 Rachel Maines, *Asbestos and Fire: Technological Tradeoffs and the Body at Risk* (New Brunswick: Rutgers University Press, 2005).
12 Oliver Bowles and K.G. Warner, 'Asbestos', *Minerals Yearbook*, ed. Herbert Hughes (Washington, DC: United States Government Printing Office, 1939), 1309.
13 For more on the different types of exposure in different work environments, see Tweedale, *Magic Mineral*.
14 'Asbestos Chronology', Asbestos Claims Research Facility, 4. This source was produced by Johns Manville solicitors in the 1980s and 1990s and comprises key primary documents from the company detailing corporate suppression of medical evidence.
15 'Asbestos Chronology', 4.
16 The type of asbestos found in Canada is chrysotile asbestos, whereas the type found in Zimbabwe is largely crocidolite. While there are some chemical differences between the two,

both are hazardous to human health and cause asbestosis, lung cancer and mesothelioma, among other diseases.

17 Peter Bartrip, *The Way from Dusty Death: Turner and Newall and the Regulation of Occupational Health in the British Asbestos Industry, 1890s–1970* (London: Athlone, 2001), 76.

18 'Asbestos Chronology', 4.

19 W.E. Cooke, 'Fibrosis of the Lungs due to the Inhalation of Asbestos Dust', *British Medical Journal* 2 (1924): 140–2, 147.

20 William N. Rom and Steven B. Markowitz (eds), *Environmental and Occupational Medicine* (Lippincott: Williams & Wilkins, 2007), 493.

21 Jessica van Horssen, '"À faire un peu de poussière": Environmental Health and the Asbestos Strike of 1949', *Labour/LeTravail* 70 (Fall 2012): 101–32.

22 The diagnosis of asbestosis in Canadian asbestos workers is one of controversy, as the companies that operated the mines had their own doctors tending to their employees, and they rarely informed them of their disease, instead attributing poor breathing to the fact that many of them smoked cigarettes.

23 See, for example, *Yorkshire Post and Leeds Intelligencer* (22 July 1927): 10, and *Essex Newsman* (7 December 1929): 3.

24 Jock McCulloch and Geoffrey Tweedale, *Defending the Indefensible: The Global Asbestos Industry and Its Fight for Survival* (Oxford: Oxford University Press, 2008).

25 Gérard Filion, *Le Devoir* (15 January 1949).

26 Van Horssen, *A Town Called Asbestos*.

27 Display ad 16, *Manchester Guardian* (26 November 1919): 9.

28 Display ad 46, *Manchester Guardian* (25 November 1920): 3.

29 'Lord Mayor's Appeal: Two Million Shillings', *Manchester Guardian* (20 October 1923): 14.

30 'For Diagnosis and Treatment', *Manchester Guardian* (5 February 1935): 12.

31 Graham Phythian, *Blitz Britain: Manchester and Salford* (Stroud: The History Press, 2015).

32 See, for example, Guy Hodgson, *War Torn: Manchester, Its Newspapers and the Luftwaffe's Christmas Blitz of 1940* (Chester: University of Chester Press, 2015).

33 Sonia Zhuravlyova, 'War, Prefabs and an Unlikely Friendship Between Opposing Soldiers', *Guardian* (5 June 2014).

34 See Barbara Vale, *Prefabs: The History of the UK Temporary Housing Programme* (London: Routledge, 2003).

35 Van Horssen, *A Town Called Asbestos*, 50.

36 Leroy Gardner, Saranac Laboratory, to Hektoen, JM, 15 March 1943. 'Doc. 7', 31, and Dr Leroy Gardner, 'Draft Report', 'Doc. 7', Asbestos Claims Research Facility, 29.

37 Gerrit W.H. Schepers, 'Chronology of Asbestos Cancer Discoveries: Experimental Studies of the Saranac Laboratory', *American Journal of Industrial Medicine* 27 (1995): 602–3.

38 E.E. Keal, 'Asbestosis and Abdominal Neoplasms', *Lancet* (December 1960): 1211.

39 Quebec Asbestos Mining Association Meeting Minutes, 10 August 1967, 38 and March 1968, 3, Asbestos Claims Research Facility.

40 Standish Meacham, *Regaining Paradise: Englishness and the Early Garden City Movement* (New Haven: Yale University Press, 1999), 181.

41 Nancy Jacobs, *Environment, Power, and Injustice: A South African History* (Cambridge: Cambridge University Press, 2003), 221.

42 Jacobs, *Environment, Power, and Injustice*, 221.

43 Wythenshawe Community Housing Group, 'Asbestos in the Home: Guidance for Tenants'. Wythenshawe Community Housing Group, no date.

5

A machine in the garden: the compressed air bath and the nineteenth-century health resort

Jennifer Wallis

> All the luxurious comforts of life … are as nothing, and not worthy
> to be compared with a good supply of pure air … and at last science
> has laid hold of it.[1]

In the mid-nineteenth century visitors to hydropathic establishments, 'aerotherapy' institutes and other health resorts – both in Britain and elsewhere – may have found themselves sitting inside a metal 'room' into which was pumped compressed air. This compressed air bath aimed to increase the amount of oxygen circulating in the body and – via the greater 'weight' of the air introduced – clear any obstructions of the air passages. In taking the pure air of the health resort and compressing it, bath proprietors were able to draw upon established discourses of climatic medicine – emphasising the naturally occurring health benefits of an environment – while simultaneously altering that atmosphere to deliver a hybrid form of air to their clients. As physical objects, compressed air baths were striking additions to the landscape of resorts, often positioned outdoors to capitalise upon their picturesque surroundings and fresh air. At first sight, then, they are a striking example of the unwelcome 'machine in the garden' of nineteenth-century literature[2] – machines intrusive and incongruent with their surroundings. This chapter will demonstrate, however, that in commodifying the atmosphere and delivering it to people via mechanical apparatus, compressed air baths were understood by those who marketed and used them as machines that were part of a 'mechanical pastoral' – relying upon a harmonious relationship between modern machinery and 'natural' landscape.

Nineteenth-century landscapes of technology

The compressed air bath highlights how the tendency to pitch 'environment' and 'technology' against one another as opposing forces – one 'natural' and one 'unnatural' – is a rather simplistic and reductive one. 'Environment', 'nature' and 'landscape', as well as 'technology', are historically specific. In this chapter 'landscape' is appealed to repeatedly because it was also something appealed to by contemporaries in their discussions of compressed air baths and resorts. As Nan Fairbrother writes in *New Lives, New Landscapes* (1970), 'Landscape ... is not a static background which we inhabit, but the interaction of a society and the habitat it lives in.'[3] The landscape is changeable, and it may be constructed and categorised according to prevailing social, political or scientific mores.[4] 'Landscape' may also refer to the built, as well as the rural, pastoral or 'natural' environment. William Cronon has argued that the transformation of previously 'worthless' landscapes into 'wildernesses' imbued with sacred value demonstrates that human encounters with 'nature' are often far from 'natural'.[5] Indeed, for many seekers after wilderness the obvious visual differences from more urban environments 'came to reflect the very civilization [they] sought to escape' by highlighting that civilization's absence.[6]

Language surrounding technology is equally historically contingent, changeable and contested. 'Technology', when referring to the nineteenth century, is our own useful epistemological shorthand. For the nineteenth-century commentator 'technology' was something that denoted technique or a facet of technical education. In referring to a piece of apparatus such as the compressed air bath, a commentator would likely instead have spoken in terms of 'machinery'.[7] 'Technology', however, has a certain utility, and when considering technology in the nineteenth century this chapter takes its cue from Joel D. Howell. In *Technology in the Hospital* (1995), Howell imbues medical 'technology' with three layers of meaning: a physical machine, an activity and 'what people know'.[8] The compressed air bath was undeniably a physical machine, but it was also active in structuring relationships between individuals and their environment, and could be a place of knowledge-making when serving as an observational and experimental space, as this chapter will show.

In considering technology in the nineteenth century, it is all too easy to construct a picture of contemporaries as technophobic – struggling with the expansion of industrial society and suspicious of any and all mechanical innovation.[9] Yet many early- to mid-century observers did not necessarily perceive any ideological conflict between admiration of modern engineering and enjoyment of the landscape. In his discussion of early nineteenth-century train travellers in Pennsylvania, Will B. Mackintosh writes that, rather than experiencing 'a tension between modern technology and landscape appreciation', passengers 'found that together the two created novel and pleasurable aesthetic experiences'.[10] Riding on mechanised transport could 'heighten and dramatize the beautiful and sublime aspects of the landscape', while at the same time allow the traveller to view that landscape, from afar, in 'luxurious contemplation'.[11]

The contemplative observation of new landscapes was not the sole rationale for travel in the nineteenth century, of course. Travel for health was another compelling reason for many people to journey to new places, both at home and abroad. The typical historical image of the nineteenth-century environment is one of cities shrouded in smog, slums marred by malodorous drains and rural hamlets choking on the smoke of factory chimneys. Yet this growing awareness of atmospheric pollution was accompanied by a corresponding appreciation of the potential healing properties of air, and a desire to seek out 'healthful' environments. The notion that the external atmosphere could have an impact on health was not new to the nineteenth century. In the first half of the century, though, traditional recourse to the seaside or mountainous areas in times of ill health became increasingly codified. Medical climatologists paid a great deal of attention to the specific features of natural environments, such as the salt content of waters or the moistness of soils, matching these to their patient's needs. Thus, the sea voyage might be prescribed to the nervous and withdrawn patient, who would be reinvigorated and restored by the bracing sea air, as well as the extended break from the worries of home.

Just as the nineteenth-century spa goer might visit resorts both in their own and in other countries, this chapter – while focusing primarily on British examples – draws upon work from American and European commentators. As will become clear from many of the examples used, the dialogue between doctors across continents, the supply of hydropathic appliances across national borders and the often prolific travelling of resort clients themselves, meant that knowledge about the compressed air bath travelled widely. Indeed, a global outlook was (and is) unavoidable as doctors and clients compared their knowledge of baths, resorts and climates in their quest to discover the perfect environment that would cure their particular ills.

One of the most popular means of exploiting the healing properties of the natural world in this period was hydropathy. Unlike spa treatments of the eighteenth century, hydropathy was relatively unconcerned with the specifics of the water used, such as whether it came from mineral or salt springs. Introduced to Britain in the 1840s, hydropathy quickly gained in popularity: by 1861, Malvern's hydropathic establishments were receiving around six thousand patients each year.[12] They were tightly swaddled in wet sheets, doused with buckets of water and immersed in a variety of baths. The clients of such establishments might be imagined as modern fugitives, fleeing their own immediate toxic and debilitating environments to enjoy the healing benefits of the resort. Indeed, a nineteenth-century visitor to the Ben Rhydding Hydropathic Establishment, situated on the edge of a moor just outside the Yorkshire town of Ilkley (see Figure 5.1), constructed the immediate environment as something of a utopia:

> The sun was bright, the air balmy. On the deciduous trees and shrubs the buds were sprouting, while the evergreens, rhododendron, laurel, yew, box, fir, &c., looked fresh and glistening. The enclosures everywhere echoed the melody of the birds hailing the spring, and portly, well-to-do-looking humble bees, with shining black coats, variegated by yellow stripes, plied their labours on the pink flowers of the wild currant.[13]

Figure 5.1 The Hydropathic Establishment at Ben Rhydding.
Source: Wellcome Collection, London.

It does not follow, though, that hydropathy represented a wholesale rejection of modernity and its trappings. Indeed, as this chapter will discuss, the compressed air bath is a prime example of how hydropathic establishments could successfully combine recourse to nature with mechanical novelty in their appeal to clients. In the use of powerful shower baths, for example, hydropathy was evidently not averse to incorporating mechanical appliances into its therapeutic armoury. James Baird, an enthusiastic supporter of Ben Rhydding, distinguished between 'hydrotherapeutic appliances' used in hydropathy and 'the every-day use of water and air for purposes of health and cleanliness' – an important distinction when advertising the virtues of costly hydropathic establishments to potential clients.[14] As an 'entire medical system',[15] hydropathy readily incorporated mechanical apparatus in a holistic approach to health – and not only apparatus that utilised water. 'Aerotherapy' or 'aerotherapeutics' capitalised on the healing potential of air, encompassing everything from sending patients to breathe the rarefied air of the Alps to prescribing inhalations of arsenic for the relief of asthma. At Ben Rhydding from the mid-1850s, visitors could enjoy an entirely new form of treatment that drew its healing powers not from water, but from air.

The compressed air bath and its uses

The Ben Rhydding Hydropathic Establishment had been founded in 1844 by a local man, impressed by the European spas he had visited and wishing to make such treatment available in his native Ilkley. A few years later, one William Macleod took over as manager of Ben Rhydding: an Edinburgh man who had studied medicine at St Andrews, and whose primary interests lay in the fields of hydropathy and homoeopathy. Soon after taking over as head of the establishment, Macleod introduced a Turkish bath, but also something rather more novel. This new treatment came in the form of a compressed air bath, constructed in Ben Rhydding's grounds in 1856 and allowing the establishment to claim the honour of being the first resort in Britain to have such an apparatus. Compressed air had been noted for its physiological effects for many years, but industrial expansion during the early nineteenth century had provided more opportunities to observe these effects at first hand among mine, bridge and tunnel workers. As well as negative reactions to compression such as breathlessness and headache, it seemed that in some cases the compressed environment had a positive impact on a range of conditions including hearing problems, asthma and emphysema. Emile Tabarié, a French engineer, was likely the first to design a compressed air bath for therapeutic purposes, which was put into operation in France in the 1830s under the direction of a physician.[16] The bath is testament to the often close relationship between medicine and engineering in the earlier nineteenth century, as doctors attended building sites to observe the effects of compression and engineers' reports were incorporated into medical accounts.[17] A number of medical observations were made of bridge builders, for example, with building sites such as that surrounding the Rhine bridge at Kehl converted into sites of experiment.[18]

Tabarié's bath was a circular construction consisting of iron plates riveted together 'like those of the boiler of a steam-engine', with an air-tight door and porthole windows glazed with strong glass.[19] The floor, boarded over, contained small apertures to allow the entry of the air, which was forced in via pipes connected to a steam engine. Once a patient was seated inside the bath the first half hour was spent gradually increasing the pressure, an hour spent in the pressurised atmosphere and then a final half hour to reduce the pressure back to normal. A valve in the top of the chamber could be adjusted by those inside to allow the escape of air and avoid the creation of a stagnant atmosphere. The pressure deemed appropriate for treatment was anywhere between half and two-thirds of an atmosphere (in addition to normal atmospheric pressure of $14.7 lb/in^2$, thus taking the pressure to between 22 and $25 lbs/in^2$) with a pressurisation rate of 1lb every three minutes; this is a pressure range that falls within currently accepted safe limits for hyperbaric medicine. Various modifications could be made to increase the comfort of sitters or enhance the air's therapeutic effect. As air temperature increased during compression, the pipes carrying the air were often directed through baths of iced water, especially during the summer months. Conversely, in winter, it was noted by one of Ben Rhydding's visitors that the tempera-ture inside the bath sometimes necessitated the wearing of furs and great coats.[20] Filtering air through cotton wool soaked with various substances also allowed the air being inspired to be medicated if appropriate. The most basic and important rationale of the air bath, though, was to deliver a greater proportion – or 'weight' – of oxygen into the body than that available in the normal atmosphere. By the 1880s there were over 50 baths across Europe and the United States, many of them in established health resorts and spa towns.

The immediate physical effect of the bath was usually a slight pain or discomfort in the ears, which was said to disappear after one or two sittings. S. Solis-Cohen, an American physician and advocate of aerotherapy, had visited a compressed air bath in Brussels and found it a less than enjoyable experience: 'I went into the cabinet myself, but did not remain long; the noise in the ears and the sense of fullness in the head were sufficiently unpleasant to make one wish to get out as soon as possible.'[21] For those willing to persevere beyond this initial discomfort, repeated compressed-air inhalation provided a useful course of respiratory gymnastics: respiration apparently became easier and the capacity of the lungs increased. Its application extended beyond mere respiratory conditions and it was touted as curing deafness, chronic headache caused by overwork, loss of the voice, whooping cough and menstrual complications. In stimulating the flow of saliva compressed air was also said to stimulate the digestion and increase the appetite, something of especial service when rallying convalescents. It was even touted as a cure for sterility by some of its more zealous supporters, who confidently claimed that it encouraged the 'secretion' of the generative organs.[22] Contemporary literature presented several cases in which the bath appeared to have effected wondrous cures. A baritone of the Lyons theatre, who had lost his voice to bronchitis, was said to have recovered his voice entirely after several baths at Joannis Milliet's institution in Lyons.[23]

The number of treatments needed to effect a cure varied from person to person; one physician recommended at least 24 spread over one to two months,[24] but some patients took up to 100 before being declared cured. Not all were convinced of the bath's beneficial effects, however. Responding to a letter to the *Medical Times and Gazette* in 1860, in which a reader had asked about the value of the compressed air bath in pulmonary conditions, 'Chirurgus' warded them off the idea in strong terms. A patient of his had (against his medical advice) used a bath and 'on the second day, while sitting in the little room breathing the compressed air, he suddenly felt something coming into his mouth, and, putting up his handkerchief, he found he was spitting blood'.[25] Certainly it wasn't a treatment to be employed frivolously: it was specifically advised against in patients suffering from any cardiac affliction or in the late stages of tuberculosis.

The tuberculous patient would not, in any case, have been especially welcome at Ben Rhydding; the establishment catered to the invalided and 'worried well' rather than the clearly infectious or seriously ill. Macleod had long touted the establishment as a luxurious health resort/hotel hybrid that offered hot water and vapour baths, an outdoor gymnasium and traditional amusements such as a bowling green. Like many hydropathic establishments it performed an important social function, being a noted centre for the meeting of radical liberal

Quakers and Unitarians. Most visitors to Ben Rhydding came from the professional and middling classes and it was one of many nineteenth-century arenas that acted as a marriage market for high society.[26] The visitor might find themselves in the company of officers invalided out of the army following the Indian 'Mutiny', MPs seeking to recover their strength before the re-opening of Parliament, or notable personalities of the day such as Harriet Martineau.[27] Despite its pretensions to society, hydropathy came to be linked to a growing, more democratised, domestic tourist trade in the second half of the century. A correspondent for the *Liverpool Mercury* of 1874, discussing health resorts, noted that they had chosen the Llandudno Hydropathic Establishment (which boasted a compressed air bath) for their annual sojourn as 'Ilkley was stale; Ben Rhydding had lost its charms; Buxton was too near, and Torquay too far'.[28] Ben Rhydding was, by that time, one of several such resorts in a competitive market. By the end of the century many hydropathic resorts had passed into the hands of companies who made them more explicitly part of the commercial holiday sector, a move that sounded the death knell for the 'spa side' of many resorts.[29] This would be the fate of Ben Rhydding, taken over by the Wharfedale Company that also had business interests in cabs and coaches.[30] (It would finally be transformed into a golf hotel with the addition of a golf course to the site in the 1880s and, although the building itself was demolished in the mid-twentieth century, the golf course remains today.)

The final quarter of the nineteenth century, then, saw compressed air baths become more widely available to those beyond the elite Ben Rhydding set – and beyond resorts. Brompton Consumption Hospital in London had opened in 1842, admitting consumptive patients who were not generally permitted at other hospitals due to concerns about contagiousness. The Hospital was dependent upon charity donations, fundraising and wealthy benefactors to support its treatment of both in- and out-patients. In 1879–80, the Hospital acquired two compressed air baths, which were used to treat a variety of respiratory conditions and employed as a preventative measure for those with a 'consumptive' tendency or in the early stages of tuberculosis. Brompton extended compressed air treatment to the hospital patient. Their baths, contained in a basement alongside three Turkish baths, offered a rather less picturesque experience than the bath at Ben Rhydding, although the baths at the latter were formative in the Hospital's decision to obtain its own. Staff from Brompton had investigated both the Ben Rhydding and Malvern compressed air baths before committing to installing any themselves.[31] The patient's encounter with the bath at Brompton was a rather medicalised and industrialised one, during which they would find little to connect them with the outside world. Indeed, Brompton's Compressed Air Bath Sub Committee expressed concern that one of their baths had no means of communication with the outside at all and suggested that, 'If nothing better can be arranged the whistle attached to the air escape could be used for the purpose [of attracting attention] and a code of signals arranged accordingly.'[32]

Despite the rather medical, and isolating, encounter with the machine in the Brompton Hospital basement, the quality of the air that was pumped into the bath remained important, with the emphasis on clean and 'natural' air. Brompton's regulations specified that air was to be pumped in 'from the outside of the building and not from the heated passages or engine room' where the bath was situated.[33] In this sense, we might view the architecture of the hospital itself as a component of medical technology, interacting with the immediate environment – and indeed hospitals for respiratory illness are often explicitly located and designed with this environmental interaction in mind.[34] The compressed air bath was imagined by its advocates as a machine that enhanced the natural atmosphere by offering a controlled microcosm of it. Charles F. Taylor, in advertising his compressed air apparatus in New York, described his baths as extensions of the ordinary atmosphere: 'We move about at the bottom of an immense sea of air … In fact, we are in a never-ending bath.'[35] The compressed air bath was inextricably bound to the external environment, as was the sitter within it who was encouraged to consider their interaction with the environment outside as well as inside the bath. Compressed air treatment did not effect change on its own, but was to be used in combination with hydropathic treatments, prescribed diets and gymnastics. At Ben Rhydding, for instance, a short course of exercise on the surrounding moor was recommended as a vital part of the patient's treatment.[36] The bath was not a machine in stark contrast to nature, but something dependent upon it and used in conjunction with it.

Encounters physical and spiritual

In the present day, golfers on Ben Rhydding's neatly manicured course will find themselves in close proximity to the rugged outline of Ilkley's 'Cow and Calf rocks', much the same view that the nineteenth-century visitor to the hydropathic hotel would have encountered. The geographical placement of compressed air baths reflected the conviction that the air in the bath had not only to be compressed, but taken from as wholesome an environment as possible. Baths appeared across Europe at resorts already renowned for their naturally healing properties: near Reichenhall's salt springs or in the historic spa town of Malvern. By the end of the century several city centre baths existed, many of them part of 'aerotherapeutic' institutions such as J.A. Fontaine's Établissement Médico-Pneumatique on the Rue de Châteaudun in Paris (which also provided a detailed quote for supplying and transporting a bath to the Brompton Hospital).[37] These city centre establishments often had a more industrial flavour than their resort counterparts. Maurice Dupont's Paris establishment, for example, drew its air from the street mains that sourced air from the 'heights of Belleville' and distributed it to various mechanical apparatus in the area including pneumatic street clocks.[38] Although city centre aerotherapeutic institutes reduced the opportunity for contemplating the landscape – an illustration of Fontaine's establishment depicts something resembling a factory, with multiple baths in one large room (see Figure 5.2) – they did not dispense entirely with the appeal to the surrounding environment.[39] J.L. Stone's air-cure establishment in New York installed an observatory at the top of the building that allowed visitors to survey the city, combining their medical treatment with a round of sightseeing.[40]

In looking at the illustration of Fontaine's establishment, one imagines that it was an intriguing sight for the visitor. As John Kasson notes in *Civilizing the Machine* (1977), the contemporary view of machinery was not necessarily negative. Spectators could take great

Figure 5.2   J.A. Fontaine's Parisian Établissement Médico-Pneumatique.
Source: BIU Santé, Paris. www.biusante.parisdescartes.fr/histmed/image?09374.

pleasure in watching machines at work as material representations of progress, industry and vigour.[41] Rather than depicting trains, for example, as 'sudden, shocking intruder[s] upon a fantasy of idyllic satisfaction',[42] Aileen Fyfe argues that industry was as much an attraction for the sightseer as was the scenic, pastoral landscape. Travellers descended down mine shafts to view workings underground, or perused the interiors of factories in the 'industrial visit':

> The industrial visit was supposed to be a rational inquiry into modern economic processes and technologies, but it could combine more visceral attractions: the heat, light, and smoke of blast furnaces could be seen as a sublime spectacle, while a descent into a Cornish tin mine provided a thrill of excitement.[43]

Although such 'industrial tourism' would decline over the course of the nineteenth century – discouraged by evolving health and safety legislation and owners reluctant to have their production schedules interrupted by gawping onlookers – industry was still a compelling spectacle for many people. The compressed air bath offered the sitter the novel opportunity of experiencing the machine from the inside, but also the chance to restructure and reframe their engagement with the natural landscape.

The bath was unusual in its potential for viewing nature: the prevailing fashion for such encounters in the nineteenth century was to do so in buildings made of glass. Glasshouses allowed visitors to immerse themselves in a somewhat artificial and scaled-down version of the 'vital landscape'.[44] Rather like the glasshouse, the compressed air bath contained and privatised the external environment. Bath proprietors were keen to emphasise that the air introduced was 'the ordinary atmospheric air':[45] it was essentially unchanged yet at the same time was enhanced by its mode of delivery. 'The *air of heaven, without any change in it, is forced* into the apartment,' wrote one enthusiastic compressed air bath user.[46] Thus, the compressed air bath worked on a similar principle to the glasshouse filled with plants and animals: the visitor was immersed within an artificial landscape to undergo a mediated encounter with nature. This idea of a mediated encounter is often identified in relation to the nineteenth-century train journey. Seated within a marvel of modern engineering, the train passenger was able to enjoy novel views of nature, with the passing landscape presented as numerous tableaux through the window.[47] Like the train window – or, later, the car window as discussed by Tim Cole in

Chapter 7 of this volume – some compressed air baths offered the sitter the opportunity to indulge in a novel form of sightseeing by presenting new views to them through the small bath windows. Milliet, describing his Lyons baths (both housed in 'elegant cases'), boasted that from inside patients could look out over 'the plains of Dauphiny, between the Rhone and the Alps'.[48] Just as the window of the train re-framed the landscape, Milliet's baths offered a way of viewing natural beauty while seated inside a product of modern engineering prowess. 'Like a divining rod, the machine ... unearth[ed] the hidden graces of landscape',[49] as the sitter sat in quiet contemplation, breathing in the concentrated essence of the vista that was laid out before them. Many bath sitters, though, were just as interested in the mechanical encounter, like Fyfe's 'industrial visitors'. Compressed air baths were a novelty as well as a serious medical treatment, incorporated into touring schedules and sightseeing lists. Margaret Dyne Jeune, an Oxford lady visiting Malvern in 1860, went with her friends to experience the compressed air bath at Townshend House, which held between 10 and 12 people. She pronounced it an 'unpleasant' and underwhelming experience, and noted of the operation that 'The scene altogether was ludicrous and its recital caused much amusement in the circle, which the paucity of events caused to be very easily amused.'[50]

Whether experienced as ludicrous or more edifying, new forms of interaction with nature and technology were bound up with new identities. Like the glasshouse, the compressed air bath could provoke reflection not only on nature but also on the self. While it is easy to imagine the bath as an unnerving experience for sitters (and this was the experience of many, as the next section discusses), a good number saw the apparatus in more positive terms. For these individuals, the thrill of being within the belly of a working, *breathing*, machine outweighed their anxiety. The bath was frequently said to arouse a 'feeling of spiritual wellbeing, levity and liberty'.[51] A testimonial apparently from one Emma E. Bailey who had used a compressed air bath in New York said: 'Next to the Gospel, this subject stirs my soul, for, as spiritually I found my new life with the Gospel of Christ, so physically I became all new by means of the Condensed Air Treatment.'[52] Although we should approach such testimonials with caution – often included in pamphlets that were designed to advertise an institution – it would not be surprising if clients spoke of the bath in such reverential terms: the hydropathic resorts where most baths were situated often had strong evangelical leanings (particularly in America), on the part of both proprietors and their clients. The 'conversion narrative' was also a common motif in nineteenth-century alternative medicine, with patients and doctors describing their frustration with 'allopathic' medicine before chancing upon a new and life-changing treatment.[53]

A particularly striking account of the compressed air bath comes from Amanda T. Jones, an American author, inventor and psychic. In her *Psychic Autobiography* (1910), Jones recounted her stays at Buffalo's aerotherapy institute in the 1860s and 70s. In describing her experiences in the air bath, she associated it with an enhancement of perception, both sensory and extra-sensory:

> I used to sleep in [the] air bath, waking with all my senses clarified, – the Psychic with the rest. And so it came about one day in April that I awoke and saw what proved to have for me momentous meaning. There stood, within my reach, a large and very heavy wooden cross – unlovely yet illuminated of itself, as though from inner light.[54]

This led to the vision of a home for abandoned children and a subsequent drive for fundraising, and was not the only time that Jones linked her use of the air bath to personal enlightenment. The second instance was less religious in tone, but relied on the clearing of the head that the air bath was said to facilitate as well as inspiration from the practical workings of the bath itself: 'Waking ... out of my usual air-bath slumber ... I said ... "I see how fruit can be canned without cooking it. The air must be exhausted from the cells and fluid made to take its place."'[55] This was the birth of her patented vacuum canning method, the 'Jones Process'. For Jones, the compressed air bath represented that technology–environment interaction that Jon Agar identifies in the opening chapter of this volume: environment as a source of inspiration for new technologies. The virtues of salubrious environments like the Alps had led doctors to look towards a technological alternative, which they had found in the compressed air bath. Once this atmosphere was contained, however, the bath was transformed into an environment in and of itself, offering renewed inspiration and fuelling further technological development. The microcosm of the healthful atmosphere that was contained within the bath at Buffalo worked on Jones' scientific imagination to inform her conception of vacuum canning. And, although historically the spa visit had been aligned with sociability, the existence of a compressed air bath at health resorts could allow for a very personal and solitary 'dialogue with nature' – and higher realms – as Jones' reminiscences suggest.[56]

Decorating the compressed air bath

Some compressed air-bath users, like Jones, seemed completely at ease with their position inside the machine, and others enjoyed passing the time by admiring the surrounding landscape. But how, as large mechanical objects in that landscape, were baths presented? The compressed air bath was a rather formidable piece of machinery. As the *Dundee Evening Post* put it: 'The outside resembles more than anything else the turret of a modern man-of-war.'[57] Entering this large metal chamber, within which they were to be enclosed for two hours, could be an unnerving experience for sitters, many of whom had to be reassured by bath operators as to the safety of the treatment.[58] Although the *Evening Post* mocked the apparatus – 'One step further and the fashionable cure will be taken in the Greathead compressed air shield at the "working face" of the newest Thames tunnel' – it also discussed the bath in more positive terms as the 'newest' invention and 'latest thing'.[59] The bath was both impressive and frightening, progressive and primitive.

As the *Evening Post* suggested in its reference to industrial endeavours like the Thames tunnel, compressed air had attendant dangers. Placing people within an airtight metal room and subjecting them to changes in air pressure was a risky exercise. (Some seemed to relish this and viewed the bath as a form of endurance test: a medical student of the 1850s boasted of how long he had stayed in a bath at his own request.[60]) Ralph Grindrod, describing his Malvern bath, appealed to the bulk of the apparatus itself as reassuring: 'The strength of the machinery prevents any mischief from mechanical causes.'[61] Doctors made much of the soothing properties of the bath, which could induce blissful slumber due to the 'perfect stillness and absence of noise'.[62] Most baths were placed at a distance from the engine that powered them, removing the noise of the machinery, but in cases where this was not possible it could be a less than relaxing experience. In *A Corner of Spain* (1898), American novelist Miriam Coles Harris described her trip to an aerotherapy establishment in Málaga where she was prescribed regular sittings in the compressed air bath:

> After you have entered, a horrid clanking noise accompanies the screwing up of the door, which is done by two attendants; you feel that you are past help … While in the cage the sound of the machinery by which the air is forced in is most unpleasant; you have a feeling that your head is being blown off.[63]

The physical isolation of the bath (here imagined as a 'cage') was worsened for the author by the knowledge that she was inside an elaborate and potentially dangerous piece of machinery: 'A pencil and paper are given to you by which to communicate with the outer world. If you are very ill and want to be let out, you must write your request on the paper and hold it up against a sort of port-hole.'[64]

It was partly to minimise anxieties like those experienced by Harris that proprietors furnished the interiors of their baths. Adding carpets, chairs and tables and lining interior walls with wood panelling, draped silk and pictures served to avoid the impression of confinement within an unfamiliar metal contraption. The bath at Ben Rhydding included a couch for weak patients and 'a contrivance for passing in and out small articles, such as letters, without disturbing the pressure of the air inside'.[65] Most baths were large enough only for a single person, but larger ones – like Malvern's – could seat up to 12 people at once, who were provided with chess and other games to pass the time together, creating the illusion of 'a pleasant and agreeable conversational gathering in an ordinary room'[66] (nevertheless, one suspects that the illustration provided in Grindrod's *Malvern* (1871) – reproduced in Figure 5.3 – was rather optimistic regarding the bath's spaciousness).[67] As Charles Lee of Buffalo's aerotherapy institute put it: 'If I have decorated with a certain degree of elegance these apartments … it has not been so much for the purpose of concealing the nakedness of the metal, as for the sake of surrounding those who resign themselves to this temporary sequestration with cheerful objects.'[68] Furnishing bath interiors was a kindness for those who had to sit within them, but it was also an essential part of treatment that could contribute to the bath's efficacy. Grindrod suggested that many of the symptoms reported by patients after their first bath – oppressed breathing and a feeling of heat – were largely due to the simple idea of being confined in a small space.[69] In imitating a drawing room, it was thought that patients would be put at ease and less cognisant of their position inside a working machine.

Yet it was important that patients were not entirely tricked by this staging: they still needed to be observed by an attendant and to behave in a way that would maximise the benefits of treatment. The windows that allowed the patient to gaze out at a beautiful landscape served another practical purpose: they allowed a doctor or attendant to monitor the person inside. The porthole-style windows made clear the bath's role as a multiple observational space: the sitter who gazed out at the landscape of Ben Rhydding from one side of the bath might turn around to see an attendant peering in at them from the other. Within the eyeline

Interior of Compressed-Air Chamber.

Figure 5.3   The interior of Malvern's compressed air bath as depicted in Ralph Barnes Grindrod's *Malvern: Its Claims as a Health Resort* (1871). Source: Wellcome Library, London.

of doctors looking through the windows hung thermometers and other equipment that provided vital readings of the atmosphere inside. Sitters thus had to balance their ease within this uncanny domestic space with strict medical requirements that militated against any full absorption into the fantasy. They were to sit upright whenever possible and to take full, deep inspirations; any reading or other activity was to be done carefully, with attention continually paid to the breathing and posture. In stark contrast to the gas experiments of Humphry Davy and his circle in the late eighteenth century, the bath was not a place for levity. At Ben Rhydding a sign apparently reminded sitters that they were 'not to laugh ... in the Air-bath!'[70] At many establishments, the lung capacity of patients was measured before and after baths, their blood pressure tested and pulse taken. When not occupied by patients, baths could function as more explicitly experimental spaces: clinical assistants at Brompton carried out observations on each other, also administering bleedings to observe the quality of the blood in a compressed environment.[71]

The exterior design of compressed air baths was just as crucial as their interior dressing. In placing baths at existing health resorts, proprietors like Macleod and Grindrod did not wish to erase or replace the established virtues of their institutions, but rather to augment them. An engraving of the small building that housed Ben Rhydding's bath depicts an elaborate and rather decorative affair (see Figure 5.4).[72] Situated a short walk from the main building, it was set back from the path and framed by trees and plants (similarly, the bath at Malvern's Townshend House was located just off the 'winter promenade'[73]). Its wooden exterior recalled a Swiss chalet, with the addition of some small turrets and arches. Elsewhere in the grounds of Ben Rhydding was a 'gothic shrine' in honour of Vincent Priessnitz – revered as the 'founder' of hydropathy – that connected the establishment to hydropathic traditions in Europe.[74] The exterior view of Malvern's compressed air bath was rather different (see Figure 5.5).[75] In a rear view of the building, the metal body of the bath was not disguised but remained open to view, looking much like a silo attached to a farm building. Yet there was also some attempt to ornament the apparatus, with a neat roof and decorative eaves added. As Kasson notes, machinery offered possibilities as well as challenges to the nineteenth-century builder or engineer. Although decorative embellishments to apparatus could be seen as an attempt to make 'acceptable' something that was incongruous with the natural environment, they also signalled the assimilation of the machine into contemporary culture.[76] Adding elaborate gothic arches to railway yards, for example, made the industrial aesthetically appealing, often drawing attention to rather than disguising it. In the case of the compressed air bath such embellishment also served to associate the apparatus with longer-standing traditions of travel for health. In housing Ben Rhydding's compressed air bath within a faux Swiss chalet, the establishment was demonstrating a keen awareness of the other resorts to which their wealthy clients might take themselves.

COMPRESSED-AIR BATH.

Figure 5.4 The building that housed Ben Rhydding's compressed air bath, from R. Wodrow Thomson's *Ben Rhydding: The Asclepion of England* (1862).
Source: author's own collection.

Why make an arduous journey abroad when you could immerse yourself in an equally salubrious climate on the moors of Yorkshire? In some cases, however, the compressed air bath was something to be used in conjunction with travel for health. Julian J. Hovent, a Belgian enthusiast of aerotherapy, advised a course of both rarefied and compressed air via mechanical means 'when a doubt exists as to whether a patient should be sent to the mountains or to the seaside'.[77] The bath was not always a simple replacement for a journey abroad, but for those who found themselves too weak or financially stretched to do so it was a promising alternative that drew upon the natural resources of the British environment.

Exterior of Compressed Air Chamber.

F SEARS & C°   16. DUKE S⁺ L'POOL

Figure 5.5   The exterior of Malvern's compressed air bath, from Ralph
Barnes Grindrod's *Malvern: Its Claims as a Health Resort* (1871).
Source: Wellcome Library, London.

Conclusion

This chapter has discussed the compressed air bath's position in health
resorts and hospitals of the nineteenth century: its geographical location, its relationship to traditions of medical climatology and its place
in the contemporary cultural imagination. Within the bath sitters were
able to undergo a mediated encounter with an element of the natural

environment – either breathing an 'enhanced' version of the external atmosphere while viewing its source through a window or undertaking a more recognisably medical treatment in the basement of a hospital. In both cases, sitters would have been acutely aware of their own bodily relationship with the air as the weight of compression altered their breathing, doctors monitored them through windows or – if they were lucky enough to be undertaking the treatment at an expensive resort – they looked out on to the rolling hills of Ben Rhydding or the valleys of the Alps.

The compressed air bath has clear historical descendants (most obviously the decompression chamber) and continued to be used well into the twentieth century: Brompton Hospital's bath was still in use in the 1920s.[78] It is likely that the collection of metal for the war effort in the 1940s put an end to those baths still in operation (this was the fate, for example, of the stunning 'Steel Ball Sanatorium' in Cleveland, Ohio, which accommodated multiple patients undergoing hyperbaric treatment[79]). The large amount of space required to house a bath had long been a barrier to their implementation: Brompton quickly found that in placing theirs adjacent to three Turkish baths, the two types could not be operated at the same time as the hot air of the Turkish affected the compressed.[80] Several attempts to introduce portable apparatus for the inhalation of compressed air were made, but these still took up significant space and were costly pieces of equipment beyond the reach of most patients.[81]

The compressed air bath demonstrates that – despite the deleterious atmosphere of many areas in the nineteenth century, marred by factory smoke or coal smuts – air could still be a healing element that held the potential for relief from a range of medical conditions. Christopher Hamlin suggests that air and its therapeutic use in the nineteenth century 'defies conventions' of the medical past: it sits uneasily with lesion-based medicine and complicates, rather than refines, cause–effect relationships.[82] Indeed, in placing compressed air baths in health resorts *already* renowned for their healing properties, it was difficult to say whether improvements in health were a result of the bath or of the resort more generally. This chapter has not sought to determine the bath's therapeutic efficacy, however, but to use it to think about machine-mediated contact with the environment in the nineteenth century. The bath, by offering such an encounter with the environment, collapsed boundaries between the natural and the cultural. By situating baths in established resorts, prescribing them as one element of a broader hydropathic or aerotherapeutic regime, and explicitly drawing upon airs that were considered more 'healthful' than others, it was difficult to object-ively assess the therapeutic impact of the machine itself. Indeed, the way in which doctors spoke of the treatment – something to be undertaken in combination with exercise on the moors, one part of a broader hydro-pathic regime – suggests that they saw the bath not so much as a panacea, but rather as a modification of the natural environment.

The immediate external environment was an integral part of the compressed air-bath system. As such the bath is an example of theme (1) as elaborated in Agar's opening chapter to this volume: 'Environment as an input into a technological system'. 'Healthy' atmospheric air provided the rationale for the bath's existence and played an integral part in its functioning: the bath was a means of concentrating and enhan-cing the air's 'natural' efficacy. By taking the air of the Yorkshire moors and compressing it, baths like that at Ben Rhydding also illustrate Agar's theme (2): 'Environment as something natural made into, or a compo-nent within, a technological system'. Doctors and resort owners repeat-edly emphasised the naturalness of the air introduced into their baths, while taking pains to note that this was air simultaneously altered by its mode of delivery. Without the air, the bath was merely a metal chamber, and without the compression, one could have achieved the same thera-peutic effect by taking a walk. The air of the surrounding area was a vital component of the machine, channelled through tubes, pumped into chambers and, finally, released once again into the external atmosphere. In this way, we might see the bath as something assimilated into the environment – becoming part of a circular process as it took in, altered and expelled air – just as we may view the environment as something assimilated into the machine.

The compressed air bath, though primarily a piece of medical tech-nology, could also offer non-medical benefits: the spiritual awakening of Amanda Jones, the novel sightseeing experience of Margaret Dyne Jeune and the new views of the landscape that several resorts appealed to in their advertising. Baths offered a slightly different experience according to their size, set-up and location. When placed at resorts like Ben Rhydding that already capitalised on their picturesque surroundings, the compressed air bath could serve to 'unearth the hidden graces of land-scape'.[83] Looking out of the bath's windows on to the Yorkshire moors while breathing in the concentrated air that was being drawn directly from them, sitters forged new contacts with their surrounding envir-onment, appreciating and experiencing it in new ways. In this respect, the compressed air bath highlights Agar's theme (4): 'Environment as something alongside an artificial world'. Seated inside a metal chamber that was – despite doctors' best efforts at 'dressing' these spaces – quite clearly a man-made and rather industrial contrivance, the scene beyond the bath windows was collapsed and framed. Small port-holes formed windows on to a landscape that was both immediately present (in the air pumped into the machine) and out of reach (beyond a thick metal door that could only be opened once depressurised). Separated from the landscape before them, the sitter was compelled to look at that landscape anew. Thus the bath, beyond its medical effects, could transform one's access to, and personal relationship with, the external environment. The compressed air bath was a literal example of a 'machine in the garden' and yet, 'Sometimes, the machine made the garden better.'[84]

Acknowledgements

Research leading to these results received funding from the European Research Council under the European Union's Seventh Framework Programme, ERC Grant Agreement number 340121. I am grateful to participants of UCL's Technology and Environment in Modern Britain workshop, attendees at the Wellcome Unit for the History of Medicine at Oxford seminars, Mat Paskins, and the anonymous reviewers for their insightful comments on earlier versions of this chapter.

Notes

1   L.D. Fleming, *The Air Cure; or Atmospheric Therapeutics* (Rochester: Wm. S. Falls, 1867), 3.
2   Leo Marx, *The Machine in the Garden: Technology and the Pastoral Ideal in America* (New York: Oxford University Press, 1964).
3   Nan Fairbrother, *New Lives, New Landscapes* (London: The Architectural Press, 1970), 4.
4   See, especially, Dolly Jørgenson, Finn Arne Jørgenson, Sara B. Pritchard and Kevin C. Armitage, *New Natures: Joining Environmental History with Science and Technology Studies* (Pittsburgh: University of Pittsburgh Press, 2013).
5   William Cronon, 'The Trouble with Wilderness Or, Getting Back to the Wrong Nature', *Environmental History* 1 (1996): 7–28.
6   Cronon, 'Trouble with Wilderness', 15.
7   Eric Schatzberg, 'Technik Comes to America: Changing Meanings of Technology before 1930', *Technology and Culture* 47 (2006): 486–512.
8   Joel D. Howell, *Technology in the Hospital: Transforming Patient Care in the Early Twentieth Century* (Baltimore: Johns Hopkins University Press, 1995), 8.
9   On this see Tamara Siroone Ketabgian, *The Lives of Machines: The Industrial Imaginary in Victorian Literature and Culture* (Ann Arbor: University of Michigan Press, 2011).
10 Will B. Mackintosh, 'Mechanical Aesthetics: Picturesque Tourism and the Mechanical Revolution in Pennsylvania', *Pennsylvania History: A Journal of Mid-Atlantic Studies* 81 (2014): 88–105.
11 Mackintosh, 'Mechanical Aesthetics', 95, 98. Also, see Wolfgang Schivelbusch, *The Railway Journey: The Industrialization of Time and Space in the 19th Century* (Berkeley: University of California Press, 1986).

12 Janet Browne, 'Spas and Sensibilities: Darwin at Malvern', *Medical History* 10 Supplement (1990): S102–13, S103.

13 James Baird, *Ben Rhydding: Its Amenities, Hygiene, and Therapeutics* (London: A.G. Dennant, 1871), 6.

14 Baird, *Ben Rhydding*, 46.

15 James Bradley, Marguerite Dupree and Alastair Durie, 'Taking the Water-Cure: The Hydropathic Movement in Scotland, 1840–1940', *Business and Economic History* 26 (1997): 426–38.

16 The idea of surrounding a patient with compressed air was not entirely novel: Thomas Henshaw put forward such a proposal in the seventeenth century. See John L. Phillips, *The Bends: Compressed Air in the History of Science, Diving and Engineering* (New Haven: Yale University Press, 1998), 198. The credit for the first compressed air bath is generally accorded to Tabarié, though there was some dispute at the time whether he had been pipped to the post by Charles Pravaz or Victor Junod; in 1852 Tabarié and Pravaz were both awarded 2,000 francs for their work by the Academy of Sciences in Paris. Anon., 'Compressed Air Baths', *The Manufacturer and Builder* 7 (1875): 91.

17 See, for example, Thomas Hayden, 'On the Respiration of Compressed Air', *Proceedings of the Royal Irish Academy: Science* 1 (1870–4): 199–208, 202.

18 Hayden, 'On the Respiration of Compressed Air', 201.

19 Archibald Simpson, *On Compressed Air as a Therapeutic Agent in the Treatment of Consumption, Asthma, Chronic Bronchitis, and Other Diseases* (Edinburgh: Sutherland & Knox, 1857), 11.

20 Anon., 'A Day at Ben Dhrypping', *The Leisure Hour* 9 (2 February 1860): 73.

21 Anon., 'Society Notes: Philadelphia County Medical Society. The Pneumo-Therapeutic Institute of Brussels', *The Times and Register* 22 (27 June 1891): 535.

22 A Graduate of the Edinburgh University [James Baird], 'A Graduate of the Edinburgh University', *Ben Rhydding: The Principles of Hydropathy and the Compressed-Air Bath* (London: Hamilton, Adams & Co., 1858), 85.

23 Ralph Barnes Grindrod, *The Compressed Air Bath: A Therapeutic Agent in Various Affections of the Respiratory Organs, and other Diseases* (London: Simpkin, Marshall & Co., 1860), 15.

24 C. Theodore Williams, 'The Treatment of Bronchial Asthma', *The American Journal of the Medical Sciences* 96 (1888): 129–40.

25 Chirurgus, 'The Compressed Air-Bath', *Medical Times and Gazette* 2 (27 October 1860): 422.

26 Alastair J. Durie, 'The Business of Hydropathy in the North of England, c.1850–1930', *Northern History* 39 (2002): 46–7.

27 Baird, *Ben Rhydding*, 12; Anna M. Stoddart, *Elizabeth Pease Nichol* (London: J.M. Dent & Co., 1899), 172.

28 Anon., 'The Llandudno Hydropathic Establishment', *Liverpool Mercury* (26 February 1874).

29 Bradley et al., 'Taking the Water-Cure'.

30 Durie, 'Business of Hydropathy', 43.

31 Royal London Hospital Archives, RLHBH/A/12/32 Report of Sub Committee on Compressed Air Baths (19 August 1879).

32 Royal London Hospital Archives, RLHBH/A/12/40 Report to the Pneumatic Cabinet Sub-Committee by C. Theodore Williams and Charles Bliss (1889). Report of the Sub-Committee on Compressed Air and Turkish baths, memo by C. Theodore Williams, Douglas Powell and John Tatham (undated).

33 Royal London Hospital Archives, RLHBH/A/12/32 Report of Sub Committee on Compressed Air Baths (19 August 1879).

34 Annmarie Adams, Kevin Schwartzman and David Theodore, 'Collapse and Expand: Architecture and Tuberculosis Therapy in Montreal, 1909, 1933, 1954', *Technology and Culture* 49 (2008): 908–42, 912.

35 Charles F. Taylor, 'Dr Taylor's "Compressed Air-Bath"', *The Water-Cure Journal and Herald of Reforms* 25 (1857): 51.

36 John Tomlinson, *Tomlinson's Handy Guide to Ben Rhydding, Bolton Abbey, and the Neighbourhood* (London: Robert Hardwicke, 1864), 42.

37 Royal London Hospital Archives. RLHBH/A/12/32 Report of Sub Committee on Compressed Air Baths (19 August 1879), letter from Dr Fontaine (5 August 1879).

38 Professor Dujardin-Beaumetz, 'Hygienic Therapeutics: A Lecture on Aerotherapy (trans. E.P. Hurd)', *Therapeutic Gazette* 12 (15 May 1888): 292.

39 Paul Bert, 'L'anesthésie par le protoxyde d'azote: travaux récents', *Revues scientifiques publiées par le journal 'La République Française'* 1 (1880): 318.

40 J.L. Stone, *Morbi Aere Vitali Curantur, or, Philosophy and Application of Condensed Air as a Curative Agent* (Rochester: Rochester Union and Advertising Company's Printing, 1881), 35.

41 John F. Kasson, *Civilizing the Machine: Technology and Republican Values in America, 1776–1900* (Harmondsworth: Penguin, 1977), 139, 141.

42 Marx, *Machine in the Garden*, 29.

43 Aileen Fyfe, 'Natural History and the Victorian Tourist: From Landscapes to Rock-Pools', in *Geographies of Nineteenth-Century Science*, ed. David N. Livingstone and Charles W.J. Withers (Chicago: University of Chicago Press, 2011): 371–98, 384.

44 William M. Taylor, *The Vital Landscape: Nature and the Built Environment in Nineteenth-Century Britain* (Burlington: Ashgate, 2004), 1.

45 Thomas Poyser and Dr Milliet, 'On the Treatment of Chronic and Other Diseases by Baths of Compressed Air', *Association Medical Journal* 1 (9 September 1853): 798.

46 Grindrod, *Compressed Air Bath*, 9. Emphasis in original.

47 Nicholas Green, *The Spectacle of Nature: Landscape and Bourgeois Culture in Nineteenth-Century France* (Manchester: Manchester University Press, 1990), 2; see also note 11.

48 Poyser and Milliet, 'On the Treatment of Chronic and Other Diseases': 798. This article contained a translation of a short tract by Milliet published in the same year, *The Medico-Pneumatic Establishment of Air-Baths at Lyons*.

49 Marx, *Machine in the Garden*, 234.

50 Margaret Jeune Gifford (ed.), *Pages from the Diary of an Oxford Lady 1843–1862* (Oxford: The Shakespeare Head Press/B. Blackwell, 1932), 109.

51 J.H. Etheridge, 'Compressed Air: Synopsis of a Lecture Delivered before the Rush Medical College Students, Dec. 1872', *Chicago Medical Journal* 30 (1873): 166–71.

52 Stone, *Morbi Aere Vitali Curantur*, 25.

53 Marijke Gijswijt-Hofstra, 'Conversions to Homoeopathy in the Nineteenth Century: The Rationality of Medical Deviance', in *Illness and Healing Alternatives in Western Europe*, ed. Marijke Gijswijt-Hofstra, Hilary Marland and Hans De Waardt (London: Routledge, 1997), 161–82.

54 Amanda T. Jones, *A Psychic Autobiography* (New York: Greaves, 1910), 285.

55 Jones, *Psychic Autobiography*, 339.

56 Green, *Spectacle of Nature*, 131.

57 Anon., 'Compressed Air Baths', *Dundee Evening Post* (18 October 1900).

58 For example, Grindrod, *Compressed Air Bath*, 30.

59 Anon., 'Compressed Air Baths', *Dundee Evening Post*.

60 Edward Haughton, 'On the Use of the Compressed Air Bath', *Dublin Hospital Gazette* 5 (15 February 1858).

61 R.B. Grindrod, *Malvern: Past and Present: Its History, Legends, Topography, Climate, etc.* (Malvern: H.W. Lamb, 1865), 141.

62 Anon., 'Etude clinique de l'emploi et des effets du bain d'air comprimé dans la traitement de diverses maladies selon les procédés de M. Emile Tabarié; par M.E. BERTIN. Paris, 1855', *The British Journal of Homoeopathy* 14 (1856): 127.

63 Miriam Coles Harris, *A Corner of Spain* (Boston: Houghton, Mifflin & Company, 1898), 69–71.

64 Harris, *Corner of Spain*, 69–70.

65 A Graduate of the Edinburgh University, *Ben Rhydding*, 81.

66 Grindrod, *Compressed Air Bath*, 23.

67 R.B. Grindrod, *Malvern: Its Claims as a Health Resort with Notes on Climate, in Its Relation to Health and Disease, Also an Exposition of the Physiological and Therapeutic Influence of Compressed Air* (London: Robert Hardwicke, 1871), illustration between 130 and 131.

68 Charles A. Lee, *The Physiological and Therapeutical Effects of Compressed Air Baths* (Buffalo: Jones S. Leavitt, 1868), 29–30.

69 Grindrod, *Compressed Air Bath*, 37.

70 A Graduate of the Edinburgh University, *Ben Rhydding*, 16.

71 C. Theodore Williams, 'Lectures on the Compressed Air Bath and its Uses in the Treatment of Disease', *British Medical Journal* 1 (25 April 1885): 824–8.

72 R. Wodrow Thomson, *Ben Rhydding: The Asclepion of England: Its Beauties, Its Ways, and Its Water-Cure*, second edition (Ilkley: John Shuttleworth, 1862), illustration between 40 and 41.

73 Grindrod, *Malvern: Past and Present*, 142.

74 Thomson, *Ben Rhydding*, 7.

75 Grindrod, *Malvern: Its Claims as a Health Resort*, illustration between 128 and 129.

76 Kasson, *Civilizing the Machine*, 160.

77 Dr Hovent, 'Pneumo-Therapeutics', *The Medical News* 59 (18 July 1891): 64.

78 Anon., 'The Outlook', *British Journal of Tuberculosis* 18 (1924): 89.

79 Morgan Choffin, 'The Cunningham Sanatorium', *Cleveland Historical*. Last accessed 1 April 2017. https://clevelandhistorical.org/items/show/378

80 Royal London Hospital Archives, RLHBH/A/12/40 Report to the Pneumatic Cabinet Sub-Committee by C. Theodore Williams and Charles Bliss (1889), Report of the Sub-Committee on Compressed Air and Turkish Baths, Memo by C. Theodore Williams, Douglas Powell and John Tatham (undated).

81 For instance, L. Waldenburg, 'On a Portable Pneumatic Apparatus for the Mechanical Treatment of Diseases of the Lungs and Heart', *British Medical Journal* 1 (11 April 1874): 477–8.

82 Christopher Hamlin, 'Surgeon Reginald Orton and the Pathology of Deadly Air: The Contest for Context in Environmental Health', in *Toxic Airs: Body, Place, Planet in Historical Perspective*, ed. James Rodger Fleming and Ann Johnson (Pittsburgh: University of Pittsburgh Press, 2014), 23–49.

83 Marx, *Machine in the Garden*, 234.

84 Mackintosh, 'Mechanical Aesthetics', 103.

6

The Agriculture Gallery: displaying modern farming in the Science Museum

David Matless

Hand sowing to helicopter

Until January 2017, visitors to the Science Museum's Agriculture Gallery could look up to view agricultural progress wrought in iron. In 1952, the year after the gallery opened, curator William O'Dea described new exhibits in the *Museums Journal*:

> Above the cases on the long wall of the gallery there is a novel decorative feature, 100 ft long … Scenes from Egyptian, medieval and modern agriculture were made in wrought iron to drawings by Ralph Lavers, ARIBA, and are displayed against a curved fibrous plaster background illuminated by fluorescent lamps. The technique is akin to that of the cyclorama and the effect is quite lively. The wrought-iron work, executed by J. Starkie Gardiner, Ltd., Merton Road, SW18, is a remarkable piece of craftsmanship.[1]

Designer and architect Lavers, who had strong interests in classical archaeology and Egyptology, had in 1947 designed the aluminium and steel Olympic torch used at the 1948 London Olympic Games. For the Science Museum metal was turned to another ancient-modern spectacle, the cyclorama moving from a right-hand end of silhouettes of human and oxen-drawn ploughs, and seed broadcasting in ancient Egyptian agriculture (the scenes based on Egyptian tomb drawings), through the flailing, hand sowing, scything, harrowing and bird-scaring of medieval English husbandry, humans and horse in harness (the scenes based on illustrations from the British Museum's fourteenth-century Luttrell Psalter), to a modern left-hand end of tractor ploughing, willow pollarding and helicopter crop spraying. Eyes moving right to left, technology would proceed, from ancient to modern, hand sowing to helicopter. A clean-lined, vividly silhouetted, strikingly modern deployment of wrought iron marked a new display of farming.

The Agriculture Gallery of the Science Museum in London opened in 1951 to display the history and present condition of farming, predominantly English farming. New technologies were then transforming agriculture in what would be termed at the time a 'second agricultural revolution'. While subject to some discussion by Jane Insley, notably in relation to its use of dioramas, and by Andrew Nahum and David Rooney in terms of the history of the Science Museum, the Agriculture Gallery deserves fuller scrutiny than it has received. Indeed Rooney noted in 2010 that the gallery had ushered in 'a new paradigm of museum display and lighting technique that is still fresh today'.[2] Until its removal in 2017, the gallery was a surviving relic of a powerful conjunction of science, landscape and modernity, and as with other Science Museum galleries provides insight into the exhibition of the modern in the post-war decades.[3] This chapter seeks to convey the institutional and cultural context of the Agriculture Gallery's development, and the nature of its displays, which offered museum visitors a progressive story of the past and a vivid display of present and future.

The modern agricultural narratives presented in the gallery carried an environmental patriotism. If later critiques of modern farming could themselves mobilise patriotism, as when in 1980 Marion Shoard's influential *The Theft of the Countryside* identified the farmer as the 'executioner' of 'a vital part of our national identity',[4] visions of modern agricultural landscape were also often explicitly national and patriotic, post-war developments presented as extending national wartime achievement, tapping into an English tradition of improvement, and linked to a national capacity for science and technology. Agricultural landscape imagery, far from being a symptom of nostalgia and national decline,[5] could articulate a vision of a dynamic and technological country; 'country' in the sense of both countryside and nation.[6] The recent UK decision to leave the European Union, and the consequent debate over agricultural policy, makes it especially pertinent to examine narratives of English farming in the period before EEC accession in 1973, and the Agriculture Gallery gives one route into the place of agriculture in the post-war English imagination. Given the prominence of questions of national identity in recent political debate, the resurgence of concern over and for Englishness, and the likely debates to come over agricultural policy, the post-war story becomes newly resonant.

Establishing a new Agriculture Gallery

The Agriculture Gallery was a product of the post-war settlement, in terms of museum funding and intellectual outlook. The gallery was established in 1951 in the Museum's new Centre Block, and was until 2017 the last display curated in the post-war period to remain in the Museum, a unique survival of, at the time, modern and innovative curatorial practices. Windowless, and 'incorporating the latest techniques in artificial lighting',[7] the gallery offered a confident statement of a mid-twentieth-century vision of the modern, following the wartime transformation of farming, and survived as a valuable historical artefact of the post-war period, a time when agricultural modernity was celebrated, in terms of both food production and landscape enhancement. The gallery stood for 65 years as a record, and indeed a relic, of the public communication of such technological optimism.

The initial Agriculture Gallery had a predominantly arable focus, but this was extended in 1965 with a full-size dairying display, described in Assistant Keeper of Agricultural Machinery and Implements Lesley West's 1967 account of 'An Agricultural History Museum', in the US Agricultural History Society's journal *Agricultural History*. The dairy exhibits, including 'a full-size reproduction of an early nineteenth-century dairy complete with dairymaid, and in direct contrast … a full-size working demonstration of a modern milking parlour and dairy', complete with milking cow ('it is mechanized, giving movement to the head and tail'),[8] were later removed as part of wider museum alterations and reorganisation,[9] and the discussion of exhibits in this chapter primarily concerns those arable displays that survived into the twenty-first century.

O'Dea's 1952 account of 'The Science Museum's Agricultural Gallery' explained the gallery's beginnings after wartime storage, and its initially restricted coverage:

> The collection of agricultural implements and machinery at the Science Museum, South Kensington, had been crated away in store for ten years when it was decided, early in 1950, that it should have 5,000 sq. ft. in the rejuvenated and extended galleries of the museum that were to be available in 1951. Restrictions

have again postponed the completion of the museum extensions so that the decision then taken only to show arable farming until new space became available is one that might not be so easy to justify now.[10]

In 1967 West also accounted for the arable focus, and explained the initial gallery organisation (1965 had brought a 'complete facelift', with new displays, models and labelling, though with the 'basic case layout' the same):

> It was felt that within the space available only arable farming could be treated properly, and in view of the importance of agriculture to the economic position of this country and the potential export market for agricultural machinery, that the excellent historical material available should serve as an introduction to a contrasting section illustrating modern developments on the farm. On this basis the gallery was divided into three bays: the first dealing with the development of methods of tillage, the second with sowing, reaping, threshing, binding, winnowing, and milling, and the third depicting work on the modern farm.[11]

Gallery displays included models, wrought-iron friezes, technical implements and machinery, with 'a number of fibrous plaster figures of full and quarter scale' made by 'Norman Cornish, Battersea High Street, SW11', and other improvised features: 'The bristles from broom heads provided the raw material for cornfields.'[12] Dioramas showed contemporary and historical agricultural practices, varying according to seasons and agricultural sectors; these were the first major deployment of this display technique in the Science Museum. O'Dea described the 'scenic backgrounds, prepared for us by contractors (A.E.L. Mash and Associates, St James's Place, SW1)', who also 'made most of the models'.[13]

Displays drew in part on pre-existing Museum agricultural collections of objects and models, accumulated since the late nineteenth century, for example, showing model carts, and plough models acquired during the 1920s; A.J. Spencer and J.A. Passmore's 1930 guide to the Science Museum's *Agricultural Implements and Machinery* holdings had traced developments from the 'primitive tool' through 'intermediate types' to 'modern machinery' in arable and dairy farming, and milling.[14] The establishment of the gallery also allowed the Museum to solicit donations of new machines and models from agricultural engineering companies, indicating a close relationship between the Museum as a state cultural institution, and a modernising agricultural industry. O'Dea commented:

> The agricultural collections had been due for attention in 1939, but immense strides were made in mechanization during and just after the war and it was clear that it would have been unwise to reopen the collections without a major degree of modernization. One-third of the space available was therefore reserved for modern exhibits – and that before a single item had been promised.[15]

In agriculture, as in other sectors, the Museum could serve as a point of conjunction for state, scientific, artistic, commercial and engineering interests. The Museum worked with the Agricultural Engineering Association, O'Dea describing recruiting agricultural firms to provide models on a uniform scale of 1:12, circulating a persuasive brochure:

> The brochure was made an awkward size and the two pages were dry-mounted on boards so stiff that they could not easily be torn up or even got rid of. We circulated a dozen or more of these intimidating documents to selected firms and the result was quite amazingly good.

The Museum put firms in touch with model makers, and 'In the end we obtained about 100 models, all to the same scale, from nearly a score of firms.'[16]

Models often displayed their maker's name, dioramas foregrounding engineering firms as names of scientific progress, effectively advertising their product. Thus a diorama of threshing was fronted by labels noting the donated models: Taskers Trailer, Ransomes Straw Baler, Avery Sack Scales, Ransomes Threshing Machines (see Figure 6.1). Lists dated 1962 in the Museum archive show 17 firms that had already donated models, including major companies such as Ford, Ransomes, David Brown, Massey Ferguson and International Harvester. New exhibits are also specified that 'may be required afresh from AEA members':

> *Pre-harvesting*: potato planter, transplanter, knapsack sprayer, drainage and ditching machinery, water and organic irrigation equipment, helical digger.
> *Harvesting*: baler, combine harvester, potato harvester, hay conditioner, hay mower and crimper.

Figure 6.1 Detail of 'Threshing' diorama.
Source: photograph by the author, May 2010.

Crop handling: grain dryer, bulk grain hopper.
Digging machinery: post hole borer, post driver.
Shearing machinery: sheep shearing machine.
Dairying: milking parlour – full scale, the farm dairy – full scale.[17]

West noted in 1967 that the renewal of displays after an expansion of gallery space in 1961 included replacement of many modern models, with 'enthusiastic cooperation on the part of the agricultural machinery manufacturers'.[18] In April 1963 the Museum followed up requests for the new with a letter from West to *Farmers Weekly* asking if readers might have old dairy equipment for the new dairy display; historic milking units, churns, pails and cheese moulds: 'Should any of your readers be able to assist the Museum regarding the whereabouts of any of the above equipment, I would be most grateful if they would write to me.'[19]

Time for modern farming

The Agriculture Gallery makes sense within, and gives an insight into, the cultural framing of farming in the post-war period. The wartime modernisation of agriculture through mechanisation, scientific application, the use of chemicals for fertilisers and pest control, and state support and regulation, was sustained in peace time.[20] The post-war decades saw the farmer cast as a modern technological custodian of the country, guided by the state to ensure food supply, with the relationship between government and farming set by the 1947 Agriculture Act, guaranteeing prices, enhancing the protection of farm tenancies, giving subsidy and promoting efficient production.[21] Agricultural modernity was celebrated for its food output, scientific method and landscape enhancement.

This 'second agricultural revolution' has received little cultural historical scrutiny. Agricultural histories tend, with few exceptions, to stop at the Second World War, those studies addressing the post-war decades focusing on assessments of productivity and the mechanisms of farm management.[22] The work of Abigail Woods, however, sets post-war agricultural change in animal husbandry within broader debate over the nature of modernity, and the modern outlook on nature; Matthew Holmes' chapter in this volume indicates parallel themes around plant biotechnology.[23] Woods discusses indoor and outdoor 'progressive' pig production, and the role of scientific expertise, arguing for 'a more historically situated understanding of agricultural modernity', including attention to its own 'romantic' ideals.[24] Philip Conford's *The Development of the Organic Network: Linking People and Themes, 1945–95* also contains insightful cultural analysis of the vision of 'agricultural efficiency and industrial food', which the organic movement set itself against: 'the visions of the age to come at times verged on the realms of science fiction (though these visions have since been reduced to the prosaic by reality)'.[25]

Mid-twentieth-century visions of agricultural technological progress have also been overlooked in popular accounts produced since the late 1960s, where the emphasis has been on modern agriculture as a source of environmental degradation, as ecologically destructive and essentially utilitarian.[26] Such accounts, however, downplay the cultural and indeed aesthetic appeal of agricultural modernisation, and it is important to recover narratives of agricultural modernisation in order to appreciate their cultural power, and thereby help understand how an agricultural revolution was able to proceed with, at first, relatively little public contest. Change could be presented as in harmony with longstanding traditions of agrarian improvement, indeed as a successor to the 'first' agricultural revolution of the eighteenth century, yet the English countryside could also become the site for a modern environmental version of what David Nye, in the US context, terms a 'technological sublime'.[27] Whatever retrospective view is taken on the productive or destructive effects of agricultural modernity, it is important to understand the power of its transformative visions.

The romantic ideals of agricultural modernity are manifest in the Agriculture Gallery, with its combines and tractors, model and full scale. Indeed, in its content and style of presentation the gallery echoes the narratives and imagery found in a wide range of media in the period, whether in popular literature, industry publications or broadcasting. Academic voices could also generate imaginative narrative, as when, in his 1945 book *Problems of the Countryside*, C.S. Orwin concluded by imagining a Rip van Winkle figure waking up 'a generation later', i.e. in the late 1960s, to find a country transformed by agricultural progress. Orwin's figure encountered 'a spaciousness and order ... which was new', shaped by new crops and mechanisation: 'Everywhere there was the suggestion of technical changes, all of which seemed to promote a greater activity on the land'.[28] Orwin, Director of the Agricultural Economics Research Institute at the University of Oxford, celebrated the potential of agricultural modernisation, envisaging wartime improvements in agricultural production being extended in peacetime through planning. A new spacious order would characterise many of the Agriculture Gallery displays.

Agriculture in mid-twentieth-century Britain moved to the modern. *The Future of Agriculture*, as a 1943 collection introduced by Minister of Agriculture R.S. Hudson outlined, was one of mechanisation and scientific application, of tractors in harness and machine milking. Advertisements within *The Future of Agriculture* anticipated the Science Museum displays in presenting Ransomes ploughs 'behind the tractor', straight furrows progressing, and all-electric model dairies, the farmworker a new technician.[29] There is a parallel here with Ralph Harrington's discussion in Chapter 3 of this volume of the bulldozer as a technology of environmental modernity. The agricultural future was also set out as advancing from the past, whether in industry publications or popular literature, including that aimed at children. Thus on the covers of Margaret and Alexander Potter's 1944 Puffin children's book *A History of the Countryside*, back cover tractors succeed front cover old manual labour, and, inside, a pre-war landscape of 'tumble down farms' is succeeded by wartime mechanical revitalisation and reclamation: 'From gorse bracken thistles to potatoes oats and rape.'[30] Weeds are subdued, productive order comes.

The Agriculture Gallery carries specific connection to a wider children's culture of agricultural landscape. The extensive use of dioramas in the gallery followed on from the museum's use of this display technique in its Children's Gallery, established in 1931, which, due to its popularity with adult visitors, was also referred to as the 'Introductory Collections', featuring dioramas of transport, domestic lighting and power alongside mechanical models. Mining dioramas were added after the war.[31] It is notable that many of the Agriculture Gallery diorama cases were low to the ground, and would have been visible to a young child unaided. West noted further appeal to children in animal models: 'The realistically modelled plastic horse incorporated into the display is a great favourite with the many young visitors to the Museum. So much so that his nose has to be painted at regular intervals, as with constant patting the surface coat wears thin.'[32] If the Museum tapped into the expertise and resources of engineering companies, there was also an echo of the developing British production of toy model farm vehicles, reaching its peak in the 1950s and 1960s to dominate the world market, and comprehensively documented in the rich volumes on *Farming in Miniature* produced by Robert Newson, Peter Wade-Martins and Adrian Little.[33] A child might have looked into the agricultural dioramas and been reminded of their toys at home. The resonances between the visual culture of the museum, and that of child's play, are strong, suggestive of landscapes of novelty, wonder and control, miniature spaces fostering an ordered imaginative geography.

A celebratory popular children's visual culture of farming was also evident in magazines and broadcasting. Thus the children's educational magazine *Look and Learn*'s special March 1964 'Focus on the Farmer's Year' showed cover and inside imagery of arable and livestock farming that would not have been out of place in the gallery. On the cover a boy and girl walk with their dog across fresh stubble to see a combine harvesting wheat; the dog spies a foreground rabbit fleeing the machine. The centre spread shows a main image of 'a typical farm of the eighteenth century', its colour and content echoing the gallery's historic dioramas, surrounded by vignettes of the new technologies of today: tractor ploughing, seed drills, mechanical milking, muck moving, beet harvesting, hedge trimming, harrowing, pea vining, baling, combine harvesting. The densely populated eighteenth-century field contrasts with contemporary solo operatives working 'the indispensable machinery, all colours, shapes and sizes'.[34] Labour is saved, production smoothed. Children's television could also bring the modern farm into the urban, rural or suburban home, with programmes such as the BBC's *Camberwick Green*, broadcast from 1966, enrolling new farming into an English landscape ideal. *Camberwick Green* presented the 'modern mechanical farm' of 'go-ahead farmer' Jonathan Bell as at one with a pastoral vision of the country, narrator Brian Cant singing as Bell moved his machinery:

> A go ahead farmer is Jonathan Bell
> Who works his farm and works it well
> He doesn't hold much with the good old days
> In modern times use modern ways
> Electric mechanical all that is new
> Which does the work that men used to do
> He swears by it all and he proves it too
> On his modern mechanical farm.

The modernity in such representations of the country is often overlooked in nostalgic retrospect; when *Camberwick Green* series creator Gordon Murray died in June 2016, an obituary, referring to Murray's series of 'Trumptonshire' programmes (*Camberwick Green*, *Trumpton*, *Chigley*), noted that 'It was not immediately clear when these dramas were set', and on the basis of a doctor driving a vintage car suggested 'it was probably before the first world war'.[35] *Camberwick Green*'s traditionalist Windy Miller, himself devoted to topical 1960s concerns of free-range chickens and home-made cider (and thereby subject to jokes from the modern farmer), achieves retrospective prominence ahead of progressive Bell. The communication of agricultural modernity (and its alternatives) to children could, however, help make new farming an accepted part of the scenery for the wider public. Children (and parents) at the Agriculture Gallery might view the dioramas, and recall their favourite shows.

History for the modern

The Agriculture Gallery combined displays of modern farming with presentations of agricultural history, in dioramas and historic machinery. The Agriculture Gallery put historic narratives into public display, offering the metropolitan adult or child museum visitor a modern country, a landscape of new science and bright order, yet one where history underpinned the present in a story of technological progress. Rather than display the past as the out-of-date, the Science Museum showed the historic modern, anticipating the now.

High on the gallery walls, above the diorama cases and at the same level as the cyclorama, murals by A.R. Thomson, RA, pictured the historic progress of farm machinery. The 1948 Olympics again intrudes into gallery formation; if Ralph Lavers had designed the Olympic torch, Alfred Thomson (1894–1979), known primarily as a portrait artist, muralist and war artist for the RAF, had won a gold medal for painting at the Games, the last time such medals were awarded. In the mid-1950s Thomson pictured 'Jethro Tull 1674–1740 Inventor of a Seed Drill & Pioneer of Rowcrop Farming'.[36] The mural shows Tull with his new machine, watched by people of varying social classes curious as to modern novelty, as in the background distance seed is hand broadcast in a manner destined to become obsolete. In 1964 Thomson added a second mural, showing the late nineteenth-century advance of a reaper-thresher as a precursor to the modern combine harvester (a full-size example of which stood nearby); 24 horses pull the machine as a side arm reaps ripe corn. Thomson also registered gallery staff, a later label noting: 'The lady in red in the left-hand corner … is based on Mrs Lesley West, then Curator of Agriculture.'

Scale models of carts and wagons and threshing machines populated display cases, the models themselves sometimes marking histories of progress. A case of nineteenth-century threshing machines included: 'Garratt's Portable Horse-Driven Threshing Machine. This model was shown at the Great Exhibition in 1851 and embodies patents taken out in 1843, '44 and '50.' The model, 'Lent by Messrs. Garrett & Sons', with an acquisition date of 1894, offered a direct link between the new Agriculture Gallery and the display of technological progress one hundred years earlier at the Great Exhibition. An extensive display of model 'Primitive Hoe Ploughs', representing examples from around the world, lent to the Museum by Major A.S.B. Steinmetz in 1926, also brought the ancient to the modern.[37] The plough models were given a distinctly modern display setting, in brightly lit display cases with plain, light backgrounds, put in harness to silhouette animals. O'Dea described a display economical in both design style and cost, ploughs pulled by 'cheaply made, bent, soldered and black enamelled wire outline figures of draught animals', including horse, ox and elephant, made in the Museum workshop.[38]

Figure 6.2 Detail of 'Manuring and Potato Planting, 1850' diorama. Source: photograph by the author, November 2015.

History was also set in progressive harness in dioramas from medieval to Victorian; medieval oxen ploughing, horse ploughing, steam ploughing, progress in technology vividly rendered. A diorama of 'Manuring and Potato Planting, 1850' (see Figure 6.2) showed women, backs bent, facing away from the viewer, hand-planting in a just-ploughed and manured field, a farmhouse, barn and church beyond on the painted backdrop. The full diorama label gives a precise narrative of socio-technological history, and enrols the scene into the broader narrative of gallery displays:

> This exhibit is followed by a series of dioramas devoted to mechanised methods in agriculture. By contrast this scene shows the amount of field labour required to plant a potato field in 1850.
>   One man is ploughing while two others cut and load manure. A woman leads the manure cart from which a man forks a heap at intervals into the middle furrow of three. A woman follows, dividing each heap among three furrows. Three other women then spread each small heap along the furrow length. Three more women carry baskets of seed potatoes and lay sets along the manured rows.

Another three women are refilling their baskets from a load of seed potatoes, and will replace the first three when their baskets are empty. Finally a ploughman divides the ridges to cover in the seeded furrows.

Contrast this with the manure loader/distributor in the next scene and the mechanical potato planters, both full size and in model form, shown elsewhere in the gallery. The two drivers and three loaders can cover as much area as the 14 workers of 1850, and in a much shorter working day.

Some 14 figures populate the 1850 field. The diorama of modern potato harvesting, shown in the 'autumn' section of a display on 'Mechanised Arable Farming', featured a reduced, entirely male workforce, the only female figure in the scene painted sitting at leisure on a background fence, the label noting that 'much of the harder manual labour has been taken out of farming'. The 'next scene' referred to in the 1850 label showed a 'Massey-Ferguson Tractor With Front End Loader', a single male operator shifting manure into a vividly varnished heap. The hard labour of the past, the aching backs of female Victorian planters or medieval peasant ox ploughers, eases into the modern world.

Exhibits of more recent history showed the entry of new machines to the British agricultural field. A full-size Fordson tractor, of the type supplied by the US to boost food production late in the First World War, was displayed in the gallery, 'Lent by the Ford Motor Co.', the label stating:

> The world's first mass-produced tractor rolled off the assembly line in Dearborn, Michigan, USA, on 8th October, 1917. This particular example is numbered 1857 and was probably one of the first batch to be delivered to this country. Within six months of initiating production, the entire British order of 7,000 tractors had been delivered.

A diorama of 'Tractor Ploughing, 1917' (see Figure 6.3) featuring a model Fordson prompts a further social narrative, a female tractor driver watched over a gate by a male soldier, perhaps returned from conflict, roles reversed for the duration: 'The introduction of these machines, most of which were driven by women, gave a tremendous impetus to the progress of farm mechanization in this country.' A painted backdrop showed telegraphic connection, a church spire and oast houses behind. The field carved by machines, the mud carefully modelled, the diorama could also call to mind other less optimistic images of wartime mud,

Figure 6.3 'Tractor Ploughing, 1917' diorama.
Source: photograph by the author, March 2013.

such as Paul Nash's 1918 Western Front painting 'We Are Making a New World', held in the Imperial War Museum. Nash's mordant title could lend an un-ironic label to 'Tractor Ploughing, 1917'.

In its presentation of past progress the Agriculture Gallery echoed wider initiatives in the field of farming history. The gallery was indeed established in the same period as the academic discipline of agricultural history, with the British Agricultural History Society (BAHS) and its journal the *Agricultural History Review* established in 1953. The BAHS held its preliminary meeting, attended by 420 people, at the Science Museum in 1952; a visit to the Agriculture Gallery would have been a likely part of the meeting.[39] The gallery also sits alongside other agricultural displays inaugurated in 1951. The 1951 Festival of Britain on the South Bank in London featured agricultural displays in the 'Land and People' exhibition, including modern machinery, and the Science Museum gallery echoes the ethos of the Festival in presenting a modern country building on past achievement; the Museum would itself host an Exhibition of Science as part of the Festival.[40] However, 1951 also saw the establishment of the Museum of English Rural Life (MERL) in Reading, opened to the public in April 1955, a predominantly historical collection of agricultural artefacts offering a different presentation of farming, focusing on the past rather than the present, unlike the Science Museum's emphasis on progress from past through present to future.

Keeper John Higgs, also a key organising figure in the BAHS, presented MERL as in part an exercise in salvage: 'the rapid technical advances of the past few years have made it more than ever necessary to save examples of the equipment of the past before it is too late'.[41] MERL remains an important institution, for both its displays and its archival and library resources. Agriculture also featured in folklife museums such as the Welsh Folk Museum at St Fagans, opened in 1948, where the emphasis, as in parallel early twentieth-century European museums, notably in Scandinavia, was on tradition and folklore, rather than modernisation; Higgs cited such museums as an inspiration to MERL.[42] Unlike other agricultural and rural life museums, then, the Science Museum's Agriculture Gallery was distinctive in presenting a story of ongoing progress rather than a lost past, and in telling a farming story to museum visitors in London.

Vividly new

In the Agriculture Gallery modern farming became a metropolitan public spectacle. The gallery displayed new farming in various forms, including full-size machinery such as the Fordson tractor noted above, and a red Massey Ferguson combine harvester, elements of whose machinery could be set in motion. A similar red combine featured in model form in an adjacent 'Summer' corn harvest diorama, a McCormick International rather than Massey Ferguson, accompanied by red tractors, trailers and balers (see Figure 6.4). The farm labourer becomes machine operative, harvest taken in with ease. Visitors moved from full scale to model in a few paces, viewing harvest operations, with labels explaining various tractor specifications for visitors so inclined.

Other dioramas took in farmyard and barn, or showed operations varying by season. In one scene of 'Early Summer' haymaking, four men worked to store and dry baled hay, aided by the Lister Multi-Level Elevator (carrying bales to a higher level for stacking in a corrugated iron barn), the Lister Moisture Extraction Unit ('A mobile crop drier consisting of an air-cooled 40 h.p. diesel engine driving a large axial flow fan'), and the John Deere Baler ('Will bale and load up to 7 tons of straw or 9 tons of hay per hour'). Several dioramas showed a Kent landscape, with signature oasthouses, modern machines working the garden of England, landscape thereby enhanced rather than diminished. A diorama of 'Tillage' allowed push-button interaction, one press of a central button making four tractors circle a central island, each performing a different operation

Figure 6.4 Detail of 'Summer' diorama.
Source: photograph by the author, November 2015.

(ploughing, cultivating, harrowing, rolling), grooves and dust made and overridden in movement, into a tunnel and out again:

> This demonstration is intended to give some idea of what happens to the earth under some of the various processes to which it is subjected. As it has been necessary to find a material that could be made to return quickly to its original state each time the tractors revolve, and as small-scale operations are difficult to manage, the demonstration is only intended as a general guide.

Colour and lighting made the dioramas of contemporary agriculture present farming as vividly new, a bright order of modern practice, the typical adult visitor's eye level making the scene prospective in both commanding overview and projected future. Varnish gave a shine even to the dung shovelled by new tractors. If the tractor was by the early 1950s a not unfamiliar sight for many visitors, some operations on display were distinctly novel, most notably in the diorama showing crop spraying. If the cyclorama helicopter gave one wrought-iron evocation of this element of the agricultural future, a grounded, vividly detailed version appeared in

Figure 6.5 Detail of crop spraying diorama showing the 'Allman High/Low Volume Sprayer'.
Source: photograph by the author, December 2016.

a diorama showing the tractor-pulled 'Allman High/Low Volume Sprayer' (see Figure 6.5). The styling of modern chemical farming in the English landscape in this 1951 display is striking, with no contradiction appearing between the most modern farming techniques and an idyllic English scene, and little sense of any risk to labour. John Sheail notes how the deaths of seven agricultural workers from Dinitro-ortho-cresol (DNOC) poisoning between 1946 and 1950 helped prompt the 1952 Agriculture (Poisonous Substances) Act, with regulations requiring operatives applying dangerous chemicals to wear protective clothing.[43] The chemical being applied in the gallery diorama is unspecified, but the implication is that chemical farming need do no harm to either operative or environment. A man is seated, entirely unprotected, on an open Massey Ferguson tractor, trees in blossom nearby, cottages beyond, and two figures watching from an arched stone bridge. Technology becomes novel spectacle in traditional landscape: 'Allman High/Low Volume Sprayer. This tractor mounted sprayer is for the application of selective weed killers, insecticides or fugicides [sic]. The drift guard on the boom prevents damage to surrounding orchards, etc. Capacity 120 gallons; operating pressure 0–600 lb. per sq. in.' The tractor model is noted as donated by Massey Ferguson (United Kingdom) Ltd, the sprayer model by E. Allman & Co., Ltd.

In the 1960s the gallery could become a focus not only for the display and celebration of the new, but for critiques of agricultural modernity. The use of pesticides and herbicides became a focus of public concern following the 1963 publication of Rachel Carson's *Silent Spring*,[44] while Carson provided a foreword to Ruth Harrison's 1964 *Animal Machines*, a key British text criticising 'the new factory farming industry', highlighting the conditions of battery and broiler chickens, and intensively reared beef cattle, pigs and veal calves.[45] Further research is required here to ascertain the extent and nature of any public criticism of the gallery displays, and the dairy industry featured from this period was indeed not a focus for criticism in *Animal Machines*, but criticism could certainly occur. A letter dated 18 January 1970 from P.H. Reeve, secretary of the London-based Union of Animal Societies, devoted to 'farm animal welfare', addressed to Keeper Lesley West, reported that four Union representatives had visited the Museum's animal agricultural displays and found them 'no longer up to date' and 'seriously misleading'. Reeve asked for an impartial display (and thereby by implication an exposure) of factory farming:

> Over 90% of laying poultry are nowadays kept in intensive indoor conditions; virtually 100% of broiler chickens produced in this country are kept in battery cages. This industry is very large. Yet, you have no display of poultry units or of battery cages. The majority of pigs are kept in conditions very much more intensive than those shown in your display. You show no veal calf units at all.
>
> We make it clear that we think you should impartially represent modern farming techniques. At the moment, we consider your displays more like a public relations exercise on behalf of ideal farmers. You show the conditions of a horse in the last century. Do the chickens justice by showing the conditions of them in the latter part of this century. I should be pleased to come and discuss the matter with you.[46]

Reeve suggested an expanded coverage to show the agricultural truth, implying the Museum might be cautious over showing agricultural modernity in its more contested guises. Reeve's comment on public relations indeed finds an echo in West's 1967 account of the gallery:

> Soon after the gallery was opened a representative of the Farmers Union, which at the time had just spent the equivalent of $100,000 in an attempt at educating the British public to the fact that farming was no longer a business that technologists might hesitate to enter, on seeing the new gallery, flatteringly expressed the view that the Science Museum had succeeded better on a smaller budget.[47]

1951–2017

The Agriculture Gallery represented a particular conjunction of technology and environment: displaying the capacities of new technology to transform environments, using new techniques to create a new display environment, prompting public debate around technology and environment. The gallery was a documentation and celebration of technological and scientific capability, reflecting connections between a national museum and a vital national sector. Wartime experience and post-war planning shaped British farming and its representation in the gallery. New farming was presented in model form, 'model' here denoting both the miniature and the ideal.[48]

After the Agriculture Gallery's opening in 1951, regular additions were made in the first two decades, but after 1970 the gallery received very few additional exhibits, with the dairying display removed for the development of other galleries. Insley notes a minor revamp of the display in 2003,[49] but the arable parts of the gallery, shaped in the 1950s and 1960s, survived into the twenty-first century, an old modernity hanging on, a fascinating snapshot from just before that key shift in the public image of agriculture under environmental critique.

In its later years the Agriculture Gallery offered modern landscape in suspension, and this suspended quality could make the gallery a peculiarly compelling space, a modern that was not modern any more, which had clung on un-updated, yet which marked a moment when curation and farming and fine-detailed modelling of mud and manure, figures and machines, met, and made a show of the new. O'Dea noted of the gallery that 'The reactions of the public, including the farmers who visit the museum, have been most gratifying.'[50] Here, for the 1950s museum curator, for the casual passer-by or the visiting agriculturalist, was a space for today. With the gallery's passing, we lose memory of a significant past landscape of modernity.

This chapter is one attempt at a record of the gallery, but before its closure the Science Museum made a short film on the gallery's history, and its plans for a future display on twenty-first-century farming. The film was presented by broadcaster Tom Heap, known for his reports on contemporary farming and countryside issues on the BBC's popular *Countryfile* programme. I acted as an 'expert' commentator in the film, along with former Science Museum curator John Liffen, who gave memories of the gallery, and Mary Cavanagh of the Museum's exhibitions team, responsible for developing content for a new gallery on modern agriculture. The resulting short film, made by Stuart Reeves, is available on the Museum's website, and on YouTube.[51] The process of film-making, and conveying the past visitor experience to present and future online viewers, gave new insights into the gallery space: the low level of the dioramas making them visible to children, the effect of the combine in operation after the relevant button was pressed. Recording the displays, especially the dioramas, for posterity, underlined the ways in which they had become effective time capsules, miniatures of an older modern.

The dismantling of the gallery closes the exhibits for direct experience, though the object displays will survive in store, with some potentially re-exhibited, and a few dioramas will be preserved, including the 1850 potato field and the 1917 tractor ploughing. Otherwise, aside from the film, a significant mid-twentieth-century display of modern and historical agricultural technologies, which captured notable dimensions of the relationship between technology and environment in modern Britain, will be gone.

Acknowledgements

Thanks to Charles Watkins of the University of Nottingham and Tim Boon of the Science Museum for discussions on the material covered in this chapter, to Matthew Kelly, Jon Agar and Jacob Ward for response to an earlier draft, and to audiences at the Science Museum, Institute of Historical Research and UCL for their comments.

Notes

1   William O'Dea, 'The Science Museum's Agricultural Gallery', *The Museums Journal* 51 (1952): 300.

2   David Rooney, '"A Worthy and Suitable House": The Science Museum Buildings and the Temporality of Space', in *Science for the Nation*, ed. Peter Morris (Basingstoke: Palgrave Macmillan, 2010): 157–75, 174. Jane Insley, 'Little Landscapes: Agriculture, Dioramas and the Science Museum', *International Commission for the History of Technology* 12 (2006): 5–14. Jane Insley, 'Little Landscapes: Dioramas in Museum Displays', *Endeavour* 32 (2008): 27–31. Andrew Nahum, 'Exhibiting Science: Changing Conceptions of Science Museum Display', in Morris, *Science for the Nation*, 176–209.

3   Jennifer Rich, 'Sound, Mobility and Landscapes of Exhibition: Radio-Guided Tours at the Science Museum, London, 1960–1964', *Journal of Historical Geography* 52 (2016): 61–73.

4   Marion Shoard, *The Theft of the Countryside* (London: Temple Smith, 1980), 9.

5   Martin Wiener, *English Culture and the Decline of the Industrial Spirit 1850–1980* (Cambridge: Cambridge University Press, 1981).

6   David Edgerton, *England and the Aeroplane: An Essay on a Militant and Technological Nation* (London: Penguin, 2013). David Matless, *Landscape and Englishness* (London: Reaktion Books, 2016).

7   Lesley West, 'An Agricultural History Museum', *Agricultural History* 41 (1967): 269.

8   West, 'Agricultural History Museum', 272.

9   Insley, 'Little Landscapes', 12.

10 O'Dea, 'Science Museum', 299.

11 West, 'Agricultural History Museum', 269.

12 O'Dea, 'Science Museum', 301.

13 O'Dea, 'Science Museum', 301.

14 A.J. Spencer and J.B. Passmore, *Agricultural Implements and Machinery* (London: HMSO, 1930), 3.

15 O'Dea, 'Science Museum', 300.

16 O'Dea, 'Science Museum', 300.

17 Science Museum Archive, File Sc. M 100/2.

18 West, 'Agricultural History Museum', 271.

19 Letter, 22 April 1963, Science Museum Archive, File Sc. M 100/2/34/2.

20 Brian Short, Charles Watkins and John Martin (eds) *The Front Line of Freedom: British Farming in the Second World War* (London: British Agricultural History Society, 2007). Brian Short, *The Battle of the Fields: Rural Community and Authority in Britain during the Second World War* (Woodbridge: Boydell & Brewer, 2014).

21 Brian Holderness, *British Agriculture since 1945* (Manchester: Manchester University Press, 1985). Peter Self and Herbert Storing, *The State and the Farmer* (London: George Allen & Unwin, 1962).

22 Paul Brassley, 'Output and Technical Change in Twentieth-Century British Agriculture', *Agricultural History Review* 48 (2000): 60–84. Paul Brassley, David Harvey, Matt Lobley and Michael Winter, 'Accounting for Agriculture: The Origin of the Farm Management Survey', *Agricultural History Review* 61 (2013): 135–53.

23 Matthew Holmes, 'Crops in a Machine', Chapter 8 this volume.

24 Abigail Woods, 'Rethinking the History of Modern Agriculture: British Pig Production, c.1910–65', *Twentieth Century British History* 23 (2012): 165–91. Abigail Woods, 'A Historical Synopsis of Farm Animal Disease and Public Policy in Twentieth-Century Britain', *Philosophical Transactions of the Royal Society B* 366 (2011): 1943–54. Abigail Woods, 'Science, Disease and Dairy Production in Britain, c.1927 to 1980', *Agricultural History Review* 62 (2014): 294–314.

25 Philip Conford, *The Development of the Organic Network 1945–95* (Edinburgh: Floris Books, 2011), 41.

26 Shoard, *Theft*.

27 David Nye, *American Technological Sublime* (Cambridge, MA: MIT Press, 1994).

28 Charles Orwin, *Problems of the Countryside* (Cambridge: Cambridge University Press, 1945), 105.

29 Laurence Easterbrook (ed.), *The Future of Agriculture* (London: Todd, 1943).

30 Margaret Potter and Alexander Potter, *A History of the Countryside* (Harmondsworth: Puffin, 1944).

31 Kristian Nielsen, '"What Things Mean in Our Daily Lives": A History of Museum Curating and Visiting in the Science Museum's Children's Gallery from c.1929 to 1969', *British Journal for the History of Science* 47 (2014): 505–38.

32 West, 'Agricultural History Museum', 273.

33 Robert Newson, Peter Wade-Martins and Adrian Little, *Farming in Miniature: A Review of British-Made Toy Farm Vehicles up to 1980: Volume 1 Airfix to Denzil Skinner* (Sheffield: Old Pond, 2012). Robert Newson, Peter Wade-Martins and Adrian Little, *Farming in Miniature: A Review of British-Made Toy Farm Vehicles up to 1980: Volume 2 Dinky to Wend-al* (Sheffield: Old Pond, 2014).

34 'Focus on the Farmer's Year', *Look and Learn* 113 (14 March 1964): 4. (supplement).

35 Stuart Jeffries. 'Gordon Murray [obituary]', *Guardian* (1 July 2016): 33. The 'Jonathan Bell' song is aired in several episodes of *Camberwick Green*, whenever the character is involved in the plotline. Episode five, where Farmer Bell is the main character, was first broadcast on 31 January 1966. Series information is available on the enthusiast 'Trumptonshire Web' website. Last accessed 18 December 2017. www.t-web.co.uk/trumpgo.htm.

36 This and other quotations otherwise unreferenced are from the labels on display in the Agriculture Gallery until its closure in 2017.

37 The Steinmetz and Garrett models are described in Spencer and Passmore, *Agricultural Implements*, 49–50, 69.

38 O'Dea, 'Science Museum', 301.

39 Joan Thirsk, 'The British Agricultural History Society and *The Agrarian History of England and Wales*: New Projects in the 1950s', *Agricultural History Review* 50 (2002): 155–63.

40 Harriet Atkinson, *The Festival of Britain* (London: IB Tauris, 2012). Peter Morris (ed.), *Science for the Nation* (Basingstoke: Palgrave Macmillan, 2010).

41 John Higgs, 'The Museum of English Rural Life', *Nature* 175 no. 4468 (18 June 1955): 1061.

42 Higgs, 'Museum', 1061. Gaynor Kavanagh, 'Mangles, Muck and Myths: Rural History Museums in Britain', *Rural History* 2 (1991): 187–203.

43 John Sheail, 'Pesticides and the British Environment: An Agricultural Perspective', *Environment and History* 19 (2013): 87–108.

44 John Clark, 'Pesticides, Pollution and the UK's Silent Spring, 1963–64: Poison in the Garden of England', *Notes and Records*, 71 (2017): 297–327.

45 Rachel Carson, *Silent Spring* (London: Hamish Hamilton, 1963). Ruth Harrison, *Animal Machines* (London: Vincent Stuart, 1964).

46 Letter, 18 January 1970, Science Museum Archive, File Sc. M 100/2/39/1.

47 West, 'Agricultural History Museum', 271.

48 Soraya De Chadarevian and Nick Hopwood (eds), *Models: The Third Dimension of Science* (Stanford: Stanford University Press, 2004).

49 Insley, 'Little Landscapes', 13.

50 O'Dea, 'Science Museum', 301.

51 From January 2017 the film was available online at www.sciencemuseum.org.uk/visitmuseum/plan_your_visit/exhibitions/agriculture, and at the Museum's YouTube page: www.youtube.com/user/sciencemuseum. Last accessed 1 August 2017.

7

About Britain: driving the landscape of Britain (at speed?)

Tim Cole

In 1951, the Festival of Britain office published a series of 13 guidebooks that purposefully took motorists away from the main exhibition on London's South Bank, *About Britain*.[1] The fifth volume, covering the *Chilterns to Black Country*, began – like all the other guidebooks – with a lengthy 'verbal portrait' written, in this case, by the historian W.G. Hoskins. Introducing the area, Hoskins explained to both the domestic and foreign visitor that it was a relatively small patch of England stretching from north Staffordshire to south Berkshire, which was 'a distance of 130 miles: only three to four hours in a car driven by a man determined to see nothing but the hard road in front of him'.[2] In order to see more than simply the road ahead, however, Hoskins urged his readers to leave their car in a region where

> The miles are few, the hills are low, the horizons often restricted, to those accustomed to the greater scale of other countries. ... walking is the best way to enjoy such landscapes: but if this is too arduous, and a car must be used, it should be used with great restraint. England is so small, and its detail, especially in the Midlands, so intricately woven, that the traveller in a hurry will see nothing.[3]

This was typical Hoskins. In his earlier volume on *Midland England* in the Batsford *Face of Britain Series*, he was quick to claim that the 'quality of the Midland counties cannot be apprehended from a speeding train or a car', hence the need to 'walk, or cycle, or use a car with great restraint if at all, to enjoy what Midland England has to offer', although he made it perfectly clear that his own preference was to walk.[4] As David Matless notes, Hoskins advocated slow travel, 'preferably on foot or bicycle, and if a car is used … it should be to potter around rather than to drive through the landscape'.[5]

Hoskins' admonition to drive sparingly and slowly echoed sentiments more widely shared in mid-twentieth-century Britain.[6] To give just one example, John Prioleau – the long-serving motoring correspondent for the *Spectator* – informed those following his 'weekend signposts to the open road' that the road along the river Taw from Eggesford to Barnstaple afforded the chance to see 'one of the most beautiful valleys in the south'. 'It is all peaceful country with that warm look which gladdens the heart of the Devonshire lover,' Prioleau gushed, before alerting his readers that, 'any temptation to drive fast must be most sternly repressed'. He thought it 'not likely that the occasion will arise', but advised them to 'be on your guard against it for you will miss countless treasures unless you keep to a positive crawl'.[7] Both Prioleau and Hoskins can be seen as part of a broader move by writers in inter-war Britain to construct what Catherine Brace has dubbed, 'a moral geography of speed'. This set up an opposition between the 'morally repugnant' act of rushing through Britain at speed and the superior experience of a 'slow, considered, lingering encounter with the countryside'.[8] This binary was one among several that sought to frame not simply better and worse ways of encountering natural landscapes, but ultimately good and bad citizenship.[9]

The 'moral geography of speed' created in the pages of travel literature was an act of informal regulation (and self-regulation) in a context of both state liberalisation of motoring speeds and the development of ever more powerful engines. In 1903, the Motor Cars Act increased the maximum speed limit on British roads to 20 miles per hour. Although widely flouted, this speed limit persisted throughout the 1910s and 1920s until it was abolished in the 1930 Road Traffic Act. A brief period of formal deregulation ended with the 1934 Road Traffic Act that introduced a maximum speed limit of 30 miles per hour in 'built-up areas'. However, once out of the towns and cities, drivers could put their foot to the floor. It was only 'moral geographies' that were stopping them.

There is plenty of evidence that 'moral geographies' of speed were constructed in inter-war and early post-war British topographical literature and I do not set out here to dismiss this useful categorisation that points to the importance of considering class as central to constructions of the environment, which is a theme I briefly return to at the end. However, I do want to suggest the need to nuance our thinking. Rather than assuming that Britain was seen as a homogenous whole, I point to some of the ways that contemporaries imagined Britain as a series of micro-landscapes with varied topographies and at different scales. This should come as no surprise. Someone like Hoskins represented a much broader movement in mid-twentieth-century British topographical writing that emphasised – as Catherine Brace herself has pointed out – the import-ance of regionalism.[10] In this chapter I bring this recognition that place mattered to nuance overly monolithic readings of 'moral geographies' of speed. Rather than all landscapes being seen as best approached slowly and on foot, contemporaries developed more complex micro-geog-raphies that imagined landscapes differing in both scale and nature and so, therefore also, varying speeds of encounter. However, as I show in closing when I turn from driving Britain to the Quantocks, there was not always agreement over the scale of a particular landscape and therefore nor was there consensus over the means and speed of encounter.

Focusing attention on the links between scale, topography and speed brings this chapter into conversation with a number of typologies that frame the volume as whole.[11] Thinking about cars and roads is per-haps immediately suggestive of the most obvious of typologies that the volume introduces – that of the relationship between technology and the environment being primarily one marked by environmental change and damage by outputs of technological process. There were certainly critics of both driving at speed and new road-building, not simply in inter-war and post-war travel literature, but also – as I show in the Quantocks – on the ground. However, there were counter-voices that saw cars – and road-building – as offering new means of encounter through the tech-nology of the windscreen. And here the chapter engages primarily with two of the other typologies introduced and examined across this volume.

First, and perhaps more significantly, I seek to add to understandings of the environment as something represented through technology. In part my chapter draws on texts like the *About Britain* guides that deployed media technologies of printed text, photographs and maps, to navigate the motorist. But at a more fundamental level I am interested in considering the car windscreen as a mediating tech-nology that afforded a particular way of seeing and experiencing the environment. Cars opened up not only new areas, but also new ways of seeing: what Lynda Nead has dubbed 'motorised vision'.[12] The car was a travel technology that afforded freedom not only to go off timetable and off track, but also to choose whether – and where and when – to speed, slow down, or stop. This led, Wolfgang Sachs has suggested, to a breaking of 'the rigid perspective as seen through the train window because, freed from the tracks, it could change direction *and speed* at the driver's will'.[13] Pointing to the recollections of one of the early writers on motoring, James Hissey, that he would 'dally where the scenery, places and people pleased me, and, by grace of the speedy car, I would hasten over comparatively uninteresting stretches of country', Nead suggests that, 'it was the capacity for both styles of motion that was the special charm of the motor car in these years'.[14] In what is very much initial thinking, I seek to pick up Gillian Rose's call for scholars to explore 'the potentialities of specific technologies for representing landscapes differently', and to ask where (and why) the windscreen – especially the windscreen at speed – offered a kind of cinematic screen on and through which landscapes were constructed as visual spectacle.[15]

However, it is not simply the case that technologies represented landscapes differently, but as I argue here, different landscapes were seen as requiring different (speeds of) technologies. Here the chapter meshes with another of the typologies that this volume addresses. As the editors argue, the idea of the environment as something untouched by artifice makes little sense in the British context. However, imagining a landscape as 'wild' lay, in part at least, behind the decisions over where to speed up or slow down. Rather than a monolithic preference for slow travel, a number of different landscapes were seen as ones that could and should be sped through. These included those that were imagined at scale and seen to be 'wild'. Ideas of what has been memorably called in the American context 'windscreen wilderness' can be seen emerging (as well as being contested) in inter-war and post-war Britain.[16]

Speed, scale and topography: driving *About Britain*

As he advised his readers that central England was best approached on foot, Hoskins worked from the starting point that this landscape was one that operated at an intimate scale. No doubt with at least half an eye on American tourists in Britain during its Festival year, Hoskins informed those visiting 'from more spacious lands' that they needed to 'acquire a new scale of measurement in England' and 'look for depth, rather than grandeur of height or breadth of scene', because 'in the main the beauty and interest of the English scene – town or country – lies in its quality rather than in its size'.[17] This sense that Britain was an island that played out at a different scale from the vast North American continent was one articulated by others. J.B. Priestley suggested that one of Britain's 'charms' is that it is immensely varied within a small compass. We have here no vast mountain ranges, no illimitable plains, no leagues of forest, and are deprived of the grandeur that may accompany these things. But we have superb variety. A great deal of everything is packed into little space. I suspect that we are always faintly conscious of the fact that this is a smallish island, with the sea always round the corner. We know that everything has to be neatly packed into a small space. Nature, we feel, has carefully adjusted things – mountains, plains, rivers, lakes – to the scale of the island itself.[18]

In this world in miniature, Priestley argued that North American topographical features – a 12,000 foot high mountain, 400 mile long plain, or 'a river as broad as the Mississippi' – would be monstrously out of place. But it was not simply that landscapes at such a scale would not fit within the British context. They did not fit anywhere. Priestley was dismissive of America, where 'the whole scale is too big, except for aviators'. America was a country, he claimed, where

There is always too much of everything. There you find yourself in a region that is all mountains, then in another region that is merely part of one colossal plain. You can spend a long, hard day in the Rockies simply travelling up or down one valley. You can wander across prairie country that has the desolating immensity of the ocean. Everything is too big; there is too much of it.[19]

In contrast to the continental scale of the United States, which demanded exploration from the air by plane, Britain was a country that played out at a much smaller scale and therefore a day's travel by car or on foot could – and did – offer up considerable variety. As an example of experiencing this 'country of happy surprises' on the ground, Priestley described a journey 'down into the West Country, among rounded hills and soft pastures', through 'the queer bit of Fen country you have found in the neighbourhood of Glastonbury' before 'you suddenly arrive at the bleak tablelands of Dartmoor and Exmoor, genuine high moors, as if the North had left a piece of itself down there'.[20] Hoskins offered an experience of variety on an even smaller scale. 'Except on the loneliest moors – and even there quite often,' he wrote in the introduction to the *About Britain* guide to the *Chilterns to Black Country*, 'there are hardly ten square miles in England where one could not spend a whole day in leisurely exploration, provided only that one knows where to look and goes prepared beforehand.'[21] To ground this claim, he took his readers to the 'ruined Gothic tower' on Mow Cop, which was the birthplace of Primitive Methodism, and then on to the canal tunnel at Kidsgrove before finishing up in the town of Biddulph with its fantastical gardens at Biddulph Grange.[22] 'At this point we are still within four miles of Brindley's canal-tunnel at Kidsgrove, where we began this modest and rather uninviting tour,' Hoskins noted.

> We have moved within a tiny radius indeed, at no point ... as much as four miles from the summit of Mow Cop, which would have dominated the skyline the whole day long; and it would take the interested traveller a whole day to see this piece of country properly.[23]

Although it is clear that Hoskins tended to prefer slow encounter on foot at an intimate scale, he was not completely opposed to making use of a car. Indeed, when he directed his reader to explore 'ten square miles' he advocated their selective use of a car for more than simply pragmatic reasons. The sights around Mow Cop were set within an industrial and urban landscape that was – he asserted – best navigated at speed. 'Before we reach Mow Cop,' Hoskins wrote, 'we enter the dreary little colliery town of Kidsgrove, destined, if ever a town was, to be passed through rapidly.'[24] It was only 'once one is away from the debris of industry', that Hoskins pronounced walking to be 'the best way to enjoy' rural England.[25]

While urban Britain was best driven, Hoskins suggested that this was done for different reasons in different places. In the case of a 'dreary little colliery town' like Kidsgrove, this was because 'small industrial towns are depressing spectacles: they have all the aridity and ugliness of the large cities without their titanic vitality and scale to redeem them'.[26] But while Kidsgrove, and its counterparts, were 'dreary' and 'depressing' enough to be sped through, large cities like Birmingham and Stoke-on-Trent were best experienced from the car because of their scale, which rendered them visual spectacles to be witnessed from and through the car windscreen. In direct contrast to his rendering of 'ten square miles' of rural middle England, 'the Birmingham-Black Country mass' was 'about 270 square miles in area'. Rather than encouraging his readers to either avoid this 'urban mass' altogether, or leave their cars and explore it on foot, Hoskins directed motorists to take in the 'superb general view of this industrial concentration from the main roads slightly to the west of it'. From their elevated viewpoint – which came close to creating their own aerial photograph of industrial urbanism – they were offered a visual spectacle through their car windscreen of factory-chimneys and cooling-towers, gasometers and pylons, naked roads with trolley-bus wires everywhere, canals and railway-tracks, greyhound racecourses and gigantic cinemas; wide stretches of cindery waste-land, or a thin grass where the hawthorns bloom in May and June – the only touch of the natural world in the whole vast scene; plumes of steam rising all over the landscape, the pulsing sounds of industrial power coming across the dark waste; and the gaunt Victorian church-spires rising above the general level, or completely blackened towers receding into the smoky distance. This is the *Black Country*, well and truly named.[27]

Drivers were also encouraged to visit Stoke-on-Trent and its 'seven miles of concentrated ugliness and dirt', whose 'ugliness is so demonic that it is fascinating to look down upon … from the marginal hills'. From here was a view of

hundreds of bottle-shaped kilns, black with their own dirt of generations, massed in groups mostly on or near the hidden canal, with square miles of blackened streets of little brick houses, and chapels, churches, spires and towers, tall chimneys of iron and steel works, steam from innumerable railway lines that thread their way through the incredible tangle of junctions: as a spectacle it should never be missed. Whatever the time of day or night, winter or summer, it is worth seeing. The Potteries at night are a show-piece. But any time or day will do: each season has its own special value in this spectacle.[28]

'What impresses one about the whole spectacle is its satanic ugliness, but no less its terrific vitality,' Hoskins concluded.[29] His rendering of these large industrial cities as terrible spectacles of what Matless terms 'the industrial sublime' made them into visual experiences best appreciated from behind (the safety of) the windscreen.[30]

But this offering up of certain landscapes at the scale of visual spectacle was not solely limited to the urban. There was a hint at least that some parts of the rural landscape – what Hoskins dubbed 'some of the loneliest moors' – were landscape at a scale that demanded a different speed and means of motion. This comes through in another volume in the *About Britain* series – the eighth volume, to the *East Midlands and the Peak* – where Hoskins also wrote the introductory verbal portrait. Pointing to the radically different geology found within this region, Hoskins took his readers on a virtual tour of the bedrock found across this region. As he did, he at least hinted that different areas should be experienced at different speeds of motion. Thus the 'wide clay vale' to the east of Lincoln was 'good solid farmland, satisfying to contemplate, better by leaning over a gate than from a moving car'.[31] Once into the chalk Wolds, however, it seemed that he – and his readers – picked up the pace in a landscape that he admitted was 'an acquired taste in scenery'. 'Those who like the dramatic and colourful in landscapes, and nothing else, will find it dull,' Hoskins warned, 'but it is capable of showing some pleasant changes of detail to the observant eye,' he reassured his readers, specifically this time from the moving motor car. Here Hoskins' readers would discover a landscape:

> Much of it, where the chalk is exposed, resembles the Chiltern country and the southern chalklands generally: those long deli-cate curves along the skyline, the huge fields with their pale pastel colours, the flattened 'tumps' of beech-trees at intervals on the highest ground, otherwise and an almost treeless landscape and one that looks quite empty of human life. One can travel by car for miles and see nobody in the fields, and hardly a single house in the biggest view.[32]

Reaching the southern edge of the Wolds, however, his readers needed to leave their cars once again as 'more trees appear, the views close in and become more intimate, the ground rather more tumbled and the streams more frequent'.[33]

Rather than Hoskins being resolutely opposed to viewing land-scape in motion, he differentiated between landscapes that were to be driven through, and those that were to be walked in. Although there was a broad rural–urban split to his thinking, it is also clear that he saw the possibility of exploring different rural landscapes at different speeds of motion, working with different understandings of scale and topography. Granted Hoskins was, as David Matless points out, more at home in a railway carriage on a branch-line than he was in a speeding motor car, but a striking passage of Hoskins' view of the landscape through the carriage window shares much with other writers who saw the possibil-ities offered by the technology of the car windscreen.[34] Writing of seeing Rutland from the train, Hoskins noted how

> [t]he railway has been absorbed into the landscape, and one can enjoy the consequent pleasure of trundling through Rutland in a stopping-train on a fine summer morning: the barley fields shaking

in the wind, the slow sedgy streams with their willows shading meditative cattle … the warm brown roofs of the villages half buried in the trees, and the summer light flashing everywhere. True that the railway did not invent much of this beauty, but it gave us new vistas of it.[35]

His sense of the picturesque possibilities of travel technologies is something that was more widely shared, and also extended to the motor car. For J.B. Priestly, there was the potential for a kind of transformation wrought through motoring at speed that could awaken hidden beauty. Writing in his introduction to *The Beauty of Britain*, Priestley defended driving through the British landscape, and specifically the potential of driving at speed, against his imagined critics. 'I shall be told that the newer generations care nothing for the beauty of the countryside, that all they want is to go rushing about on motor-cycles or in fast cars,' Priestley noted, before reassuring his readers that, 'Speed is not one of my gods; rather one of my devils.' However, this particular 'devil' – speed – was one that he argued must be given 'its due'. As he went on to explain,

> I believe that a swift motion across a countryside does not necessarily take away all appreciation of its charm. It depends on the nature of the country. With some types of landscape there is a definite gain simply because you are moving so swiftly across the face of the country. There is a certain kind of pleasant but dullish, rolling country, not very attractive to the walker or slow traveller, that becomes alive if you go quickly across it, for it is turned into a kind of sculptured landscape. As your car rushes along the rolling roads, it is as if you were passing a hand over a relief map. Here, obviously, there has been a gain, not a loss, and this is worth remembering. The newer generations, with their passion for speed, are probably far more sensitive than they are thought to be. Probably they are all enjoying aesthetic experiences that so far they have been unable to communicate to the rest of us. We must not be too pessimistic about young people if they prefer driving and gulping to walking and tasting.[36]

In contrast to Nead's claims that 'speed did not suit the picturesque; the tempo of the modern pastoral was more leisurely and required seeing the landscape as a sequence of moving pictures … indistinct impressions that rendered the picturesque redundant', it would seem that for at least some, speed created a new picturesque – what Brace suggests 'might even constitute an entirely new aesthetic experience'.[37] Priestly saw speed offering the opportunity to either transform landscapes or enable them to be seen anew through the windscreen.

What that experience of driving at speed was like was something that James Hissey – writing in the early decades of the twentieth century – sought to convey. Although dismissive of the mere 'hurrygraphs' glimpsed by motorists who 'rush at full speed from town to town, from hotel to hotel', Hissey confessed that he could not resist the temptation to speed on one memorable occasion.[38] In *Untravelled England*, he wrote of throwing caution to the wind and speeding through the Cotswolds. 'One horizon succeeded the other in rapid, bewildering succession,' Hissey recalled. 'Our eyes were on the distance – only that could we discern clearly – the wonderful distance that ceaselessly came rushing to us. For a time a strange illusion took place; it was as though the car were standing still, and the country it was that went hurtling past.' This was presumably the experience of the 'hurrygraph' that Hissey was so critical of, and yet he retained fond memories of this journey, informing his readers that, 'A rush at full speed in a motor car over a lonely road, and through a deserted country, wide and open, is an experience to be ever afterwards remembered.'[39]

Although Nead describes this journey as 'more akin to the phantom ride' and therefore in a sense removed from its topographical context, I would place it as an experience very much rooted in a more widely shared understanding of landscape.[40] Hissey's memorable journey 'through a deserted country, wide and open' – or what he described elsewhere as 'the wild, sweeping wolds' – points to the way that he drew upon, and replicated, more widely shared distinctions between the upland Cotswolds and the valleys.[41] The former were landscapes that could be navigated at speed.[42] As Hissey's language of 'wild' Wolds and Hoskins' rendering of 'loneliest moors' suggests, upland landscapes were imagined at a scale and character that invited different speeds and means of motion. When Christopher Trent differentiated between 'most of England' that was 'like a garden laid out on a vast scale' and the 'relatively small parts of the north and west' that 'remain as nature designed them', he suggested that both should be accessed by different kinds of roads driven at different kinds of speed. In the case of 'garden' England in the south and east,

> the influence of man on nature over many hundreds of years and the consequent density of the rural population by comparison with many other countries, has resulted in many thousands of byways being available for motorists with adequate surfaces and modest

gradients. Byways link farmhouse with farmhouse and village with village (and there are nearly 20,000 villages in England). They are the perfect medium for exploring the English countryside.

While generally critical that 'the vast majority of motorists know only the high roads', Trent did admit that some of these roads, 'especially in the west country, are beautiful and throw a vivid light on the nature of the countryside'.[43]

Rather than seeing a monolithic application of 'moral geographies' of speed to British landscapes in the mid-twentieth century, there is a need to nuance our thinking and recognise that contemporaries worked with – and constructed – micro-geographies of scale and topography that informed their decisions about how, and at what speed, to access rural and urban space. Even someone like Hoskins, who clearly expressed his preference for slowly walking the landscape, saw some urban and rural landscapes as best experienced from behind the car windscreen. In particular notions of scale were critical here. Those areas in the Lincolnshire Wolds that were to be driven through – at some speed – were areas, as Hoskins was himself well aware and so was quick to point out to his readers, which had been transformed at the time of the eighteenth-century enclosures into large fields, some of which were now almost on an American scale, reaching 'an immense size, sometimes sweeping out of sight over a distant horizon, big enough to impress even an American farmer'.[44]

But it was not simply that Hoskins acknowledged scale. The guidebooks he wrote a 'verbal portrait' for were influential in creating a new sense of scale. Alongside Hoskins' verbal portrait, these guides mapped out day journeys 'by car or coach' of anything between 84 (Stafford–Stoke–Buxton–Uttoxeter–Stafford) and 119 miles (Oxford–Aylesbury–Bedford–Warwick).[45] Hoskins himself had undertaken three of the routes offered up in the *About Britain* guides in July 1950 as he checked those proposed on the ground and sketched out the captions to accompany the strip maps that enabled motorists to experience Britain at scale.[46] But more significantly, these guides also included aerial photographs that sought to stretch the 'distant horizon' for the reader and so offer up visualisations – and imaginings – of the landscape at a different scale.

The insertion of aerial photographs into these texts was something that the CEO of the Festival Tours, Colonel Penrose Angwin, had suggested early on in the planning of these 'new' kind of guides. 'Some oblique air-photographs are likely to be particularly desirable,' he argued, in helping to convey the 'characteristics and functions of the Area' at the regional scale.[47] The use of aerial photographs in the guides meshed with their wider use across the Festival of Britain in creating what Harriet Atkinson describes as a 'double vision' – 'both vertical and horizontal' – that 'offered readers, or visitors, the opportunity simultaneously to experience Britain from a single point on the ground and also from an authoritative, elevated point above it, as powerful and all seeing'.[48] The inclusion of aerial photographs was a dramatic intervention into topographical literature. Comparing Hoskins' 1939 Batsford guide to *Midland England* with his 1951 *About Britain* guides to the *Chilterns to Black Country* and *East Midlands and the Peak*, the visual shift is striking. Rather than seeing the landscape at the scale of the individual building or field as the earlier guide did, aerial photography opened up the landscape not simply from a radically different viewpoint, but also at a markedly different scale – a scale to be flown over, and driven through.[49]

The rubber hits the road: contesting the speed, scale and topography of the Quantocks

Within the variegated geography of England that I have suggested was developed across the twentieth century, the Quantocks was something of a liminal – and therefore contested – space. Although upland moorland in Somerset in the south-west of Britain, it was far smaller than nearby areas like Exmoor or Dartmoor. Whether this was a landscape of wild spectacle to be seen from a car or a more intimate and human landscape to be explored on foot was an ongoing point of debate. This can be seen in the ways that the area was variously imagined and constructed across travel literature in the twentieth century. In part this reflects the shifting technologies of travel, from the train and bicycle to the car. But these imaginings were also constitutive of what was deemed the appropriate technology by which to access these landscapes.

That the Quantocks meant different things to different travel writers comes through clearly in three inter-war publications. In his *Quantock Life and Rambles* published on the eve the Second World War, Edward Smith described the Quantocks as upland moorland on the grandest of scales. This ridge of hills was nothing less than 'a bit of Scotland transplanted into Somersetshire'.[50] Imagined in such terms, it is no surprise that this was landscape to be driven through. Readers were encouraged to take 'the road from Kingston to Enmore' that 'has more to offer, in every season of the year' than any other road Smith could recall.

Although this was driving country, motorists were instructed to pause at the gateways to enjoy far-reaching views of the coast, the mountains of South Wales and Glastonbury Tor.[51]

A radically different imagining of this landscape can be seen in *Somerset Ways* published by the Great Western Railway a decade earlier and the fifth edition of Beatrice Cresswell's *Homeland Handbook* to *The Quantock Hills* published a decade and a half before Smith's guide. Rather than this being a slice of Scotland in Somerset, Cresswell saw this ridge of hills as the 'gentle west country' rather than 'the rugged north'.[52] The author of *Somerset Ways* was in agreement that this was a landscape on a small scale. 'There are those who think that when the earth was made this little corner was reserved for all that was small and perfect,' they wrote, adding that,

> Colour too vivid for the great wastes of Exmoor was set here; shapes softer even than the soft South Downs; woodland dells, the only ones left to us in England where a common man may still see fairies; long tree-clad combes; rivers and streamlets a joy alike to eye and ear.[53]

In a remarkable passage, Cresswell offered a similar rendering of the nature of this place,

> in truth the wildness of the region is but the playfulness of a charming child making pretence to be something extremely terrible, and betraying the jest by its laughter. In the Quantocks there is nothing of that vast wildness which gives almost a touch of terror to some of our English moors. Here Dame Nature is all tenderness. The wind fluttering through the trees seems to fill the leaves with laughter: the wide views from the healthy summits extend over scenes of culture and prosperity. Only in winter, when the trees are bare, do the hills become grey and saddened, but even then whenever possible they become astir with horse and hounds, and the voices of the streamlets are never silent.[54]

Given such a construction of place it comes as no surprise that this was not primarily a landscape to be encountered by motor car, or even bicycle, but on foot.[55] For Cresswell, the Quantocks was a place where 'Nature reveals herself slowly, she will not draw aside her veil too soon, we must know her intimately before we can declare that we have seen her, face to face.'[56] Such a face-to-face encounter was unimaginable through the windscreen of the motor car, especially the glimpse or glance unfolding at speed. It necessitated 'rambles among the combes' if anyone was to 'know something about the hills … have … a glimpse of their beauties, a taste of their delights'.[57]

These different constructions of the scale of the Quantocks – and hence their mode of access – that were developed across the first half of the twentieth century, persisted into late twentieth-century debates over whether to make up the ridge track running across the hills into a spine road. Elsewhere I have written of the ways that these debates in the 1960s and 1970s were tinged with questions of class – particularly whether road-building would bring more of the wrong kind of people into the Quantocks.[58] This was certainly an aspect of these debates – and one that links with the ideas of moral geographies of speed that I have discussed at the outset. They can also be seen as a more parochial battle between two rival local organisations – the Friends of the Quantocks who were firmly opposed to such plans, and the Quantock Right Association that threw itself behind the idea of a spine road up over the hills. For the former, the designation of the area as an Area of Outstanding Natural Beauty in 1957 was reason to restrict vehicle access to the hills.[59] For the latter, there was little point in such designation if people – and they picked out 'invalids and convalescents' – could not get 'on the hills by car to see this outstanding beauty'.[60] During the 1960s and 1970s these two rival organisations clashed, with the Friends of the Quantocks balancing the twin concerns of preservation and access, while the Quantock Right Association was solely interested in safeguarding and extending access.[61]

But as plans rumbled on and were debated by a wider public in the press and at parish meetings in 1972 and 1973, different ways of imaging the scale of this landscape can be seen. Just as Cresswell and Smith constructed two different 'Quantocks', the same can be seen in public debates over plans to build a spine road. For advocates of the spine road, the Quantocks was a landscape of visual spectacle at the same kind of scale as the neighbouring upland area of Exmoor. Such landscapes could – and should – be viewed from within – or just outside – the car. Thus Alderman Archie Clarke called for the creation of 'circular routes for the hills with provision for children and the elderly who would arrive by car', allowing drivers to 'picnic on the sunny side of the hill, taking their old people, children and those who cannot walk any distance with them, and have a viewing point'.[62] For Clarke, 'the days of walkers in tweeds being the main users of the hills were over, and … the motor would have to be accepted and provided for in future policy' and 'the car has come here to stay and you have got to cater for it. They're catering for cars in the Exmoor National Park but we don't seem to have any idea on the Quantocks. The whole place is bunged up and congested.'[63]

Opponents also picked up on Exmoor, but as a warning. Writing to the *Observer*, Auberon Waugh expressed his fears that a spine road along the Quantocks would turn 'one of the few places in Southern England where people can get away from the motor car in conditions of outstanding natural beauty … into the sort of Driverama one already sees in Exmoor, where cars are bumper to bumper in summer', or as another phrased it, 'a promenade for carborne sightseeers'.[64] But it was not just that opponents of the scheme did not want the Quantocks to become like Exmoor. They also imagined it as a place at a different scale. As one put it, the 'relatively small Quantock territory' was a landscape to be walked in, and not driven through.[65] A landscape imagined at this scale was a place 'where the horse or a pair of legs and a walking stick' – rather than a car – should 'remain supreme'.[66]

Those who opposed the scheme did not simply construct the Quantocks as a landscape in miniature. They also imagined it as a cultural landscape of the romantic poets, supremely Coleridge. As one writer to the press noted, 'I often walk on the Quantocks for the exercise, the views, and that special "something" that Coleridge found and by which he was so inspired.' Writing at the height of the conflict, he concluded his polemical charge against off-roaders, 'Coleridge composed some of his best poetry whilst walking on the Quantocks. If he were alive now, I wonder how "The Rapacious Motorist" would read? I have a feeling that it would be banned for obscenity.'[67] In contrast to advocates of the car who saw it as a means of expanding the 'touring radius', and bringing 'every village and hamlet, every farmhouse and cottage, within easy reach … of everyone who owns a motor-car' and not simply those 'hardy', Victorian 'pedestrians' who 'thought nothing of walking thirty miles or more in a day', there were those who emphasised the importance of the pedestrian encounter with the Quantocks, which was a landscape imagined not simply at a particular scale, but also through a distinctive historical-cultural lens.[68]

In his *Portrait of the Quantocks* published in 1964, Vincent Waite reasserted the primacy of rambling and took a side swipe at those 'whose senses are … blunted by the modern craze for motoring speed'.[69] Drawing an analogy with an earlier moment of mass access to the countryside during another transport revolution, Waite reflected that

> Some seventy-five years ago the author of Thomas Poole and His
> Friends wrote of the Quantocks: 'Far be it from the present writer

not to rejoice, as in a great and signal benefit, that railways have thrown open so much of the beauty of hill and moor and sea to the foot of the cheap excursionist, thus enriching the lives of thousands with new possibilities of enjoyment. Nevertheless there is, and ever must be, a special charm in untroddenness, which we cannot but lose with some regret.' This 'present writer' wonders what the 'present writer' of that generation would think of the car-cluttered roads of today which have thrown open so much to the wheel of the motorist. If more metalled roads are made over the hills there is no doubt that a great deal of their charm will be lost.[70]

Waite's desire for 'untroddenness' came close to an elitism that sought to keep the Quantocks for the few rather than the many through his privileging of middle-class notions of solitary experiences of nature.[71]

But while class conflict is part of the story of the battle over the Quantocks as well as broader notions of 'moral geographies' of speed, I have argued in this chapter that we do well to nuance these renderings and to take seriously the ways that writers and publics imagined Britain as a variegated landscape. As I have suggested, behind thinking about speed and technologies of accessing landscape lay ways of thinking about the nature and scale of those landscapes. While many places in Britain were seen as best accessed on foot, there were other landscapes that could – and should – be driven through, including at speed. Rather than simply being rejected outright, the car windscreen was a mediating technology that constructed a new aesthetics of some landscapes in the twentieth century. Driving, particularly at speed, was challenged both in print and on the ground. However, there were also advocates for 'motorised vision'. Writing in the German context, Sachs argues that, 'Around the automobile ... grew new standards as to what is beautiful and important and worthy of effort in life – a construction of reality, so to speak, that casts nature as well as space in a new light and allows experiences and pleasures scarcely known before.'[72] His conclusion that motoring portrayed 'nature ... in a new light' is one that historians of technology and environment do well to consider playing out in culturally specific and nuanced ways in the British case also. As I have suggested in this chapter, talking about the motor car does not simply mean thinking about the obvious relationship between technologies and environments being primarily stories of environmental damage wreaked by new technologies. Rather, there are also stories of how shifting understandings and visions of the environment have been represented through new technologies.

Notes

1   For the early background to this, see National Archives, Kew (hereafter TNA), WORK 25/44/ A5/A4, The Council of the Festival of Britain 1951 Papers, Confidential FB.C. (48) 7, 'Outline of Theme and Programme' (26 August 1948). On the broader aims of encouraging tourist encounter with Britain as a whole, see Becky E. Conekin, *The Autobiography of a Nation: The Festival of Britain* (Manchester: Manchester University Press, 2003), 26–7.

2   William G. Hoskins, 'Chilterns to Black Country: A Portrait', in *About Britain No. 5. Chilterns to Black Country: A New Guide Book with a Portrait by W.G. Hoskins*, ed. Geoffrey Grigson (London: Collins, 1951), 7.

3   Hoskins, 'Chilterns to Black Country: A Portrait', 8–9.

4   William G. Hoskins, *The Face of Britain. Midland England: A Survey of the Country between the Chilterns and the Trent* (London: B.T. Batsford, 1949), 5.

5   David Matless, 'One Man's England: W.G. Hoskins and the English Culture of Landscape', *Rural History* 4, no. 2 (1993): 187–207, 202.

6   See, for example, Pyrs Gruffudd, David T. Herbert and Angela Piccinni, 'In Search of Wales: Travel Writing and Narratives of Difference, 1918–50', *Journal of Historical Geography* 26, no. 4 (2000): 589–604, 594.

7   John Prioleau, *Car and Country: Week-End Signposts to the Open Road* (London: J.M. Dent & Son, 1929), 109.

8   Catherine Brace, 'A Pleasure Ground for Noisy Herds? Incompatible Encounters with the Cotswolds and England, 1900–1950', *Rural History* 11, no. 1 (2000): 75–94, 87.

9   Brace, 'A Pleasure Ground for Noisy Herds?', 75, 90; David Matless, 'Moral Geographies of English Landscape', *Landscape Research* 22, no. 2 (1997): 141–55.

10   Catherine Brace, 'Finding England Everywhere: Regional Identity and the Construction of National Identity, 1890–1940', *Ecumene* 6, no. 1 (1999): 90–109.

11   In particular, see Agar's interconnection (3) in Chapter 1.

12   Lynda Nead, *The Haunted Gallery. Painting, Photography, Film c.1900* (New Haven: Yale University Press, 2007), 137. See also Jonas Larsen, 'Tourism Mobilities and the Travel Glance: Experiences of Being on the Move', *Scandinavian Journal of Hospitality and Tourism* 1, no. 2 (2011): 80–9; David Louter, *Windshield Wilderness: Cars, Roads, and Nature in Washington's National Parks* (Seattle: University of Washington Press, 2006); Christof Mauch and Thomas Zeller (eds), *The World Beyond the Windshield: Roads and Landscapes in the United States and Europe* (Athens, OH: Ohio University Press, 2008); Peter Merriman, 'A New Look at the English Landscape: Landscape Architecture, Movement and the Aesthetics of Motorways in Early Postwar Britain', *Cultural Geographies* 13 (2006): 78–105.

13   Wolfgang Sachs, *For the Love of the Automobile: Looking Back into the History of our Desires* (Berkeley: University of California Press, 1992), 155 (emphasis added).

14   Nead, *Haunted Gallery*, 154–5, citing James Hissey, *Untravelled England* (London: Macmillan, 1906), 6.

15   Peter Merriman, George Revill, Tim Cresswell, Hayden Lorimer, David Matless, Gillian Rose and John Wylie, 'Landscape, Mobility, Practice', *Social and Cultural Geography* 9, no. 2 (2008): 191–212, 200.

16   Louter, *Windscreen Wilderness*.

17   Hoskins, 'Chilterns to Black Country: A Portrait', 9. On the efforts to attraction American visitors, see Harriet Atkinson, *The Festival of Britain: A Land and its People* (London: I.B. Tauris, 2012), 21.

18   Hoskins, 'Chilterns to Black Country', 3.

19   John B. Priestley, 'The Beauty of Britain', in The Pilgrims' Library, *The Beauty of Britain: A Pictorial Survey* (London: B.T. Batsford, 1937), 3–4.

20   Priestley, 'The Beauty of Britain', 5–6.

21   Hoskins, 'Chilterns to Black Country: A Portrait', 9.

22   Hoskins, 'Chilterns to Black Country: A Portrait', 10–13.

23   Hoskins, 'Chilterns to Black Country: A Portrait', 18.

24   Hoskins, 'Chilterns to Black Country: A Portrait', 12–13.

25   Hoskins, 'Chilterns to Black Country: A Portrait', 8–9.

26   Hoskins, 'Chilterns to Black Country: A Portrait', 41.

27   Hoskins, 'Chilterns to Black Country: A Portrait', 25–6.

28   Hoskins, 'Chilterns to Black Country: A Portrait', 27.

29 Hoskins, 'Chilterns to Black Country: A Portrait', 28.

30 Matless, 'One Man's England': 189.

31 William G. Hoskins, 'East Midlands and the Peak: A Portrait by W.G. Hoskins', in *About Britain No. 8. East Midland and the Peak: A New Guide Book with a Portrait by W.G. Hoskins*, ed. Geoffrey Grigson (London: Collins, 1951), 22.

32 Hoskins, 'East Midlands and the Peak', 23.

33 Hoskins, 'East Midlands and the Peak', 25.

34 On the view from the train, see Wolfgang Schivelbusch, *The Railway Journey: The Industrialization of Time and Space in the Nineteenth Century* (Berkeley: University of California Press, 1986).

35 William G. Hoskins, *The Making of the English Landscape* (London: Hodder & Stoughton, 1955), 206, cited in Matless, 'One Man's England', 195; David Matless, *Landscape and Englishness* (London: Reaktion Books, 2016), 373.

36 Priestley, 'The Beauty of Britain', 2–3.

37 Nead, *Haunted Gallery*, 155. See also Lynda Nead, 'The Age of the "Hurrygraph": Motion, Space and the Visual Image c.1900', in *The Edwardian Sense: Art, Design, and Performance in Britain 1901–10*, ed. Morna O'Neill and Michael Hatt (New Haven and London: The Paul Mellon Centre for Studies in British Art and Yale University Press, 2010), 99–113; Brace, 'A Pleasure Ground for the Noisy Herds?', 88.

38 James J. Hissey, *The Road and the Inn* (London: Macmillan, 1917), vii; James J. Hissey, *An English Holiday with Car and Camera* (London: Macmillan, 1908), 162.

39 James J. Hissey, *Untravelled England* (London: Macmillan, 1906), 427.

40 Nead, *Haunted Gallery*, 162.

41 James J. Hissey, *A Leisurely Tour in England* (London: Macmillan, 1913), 338; Brace, 'Finding England Everywhere', 94–100.

42 Brace, 'Finding England Everywhere', 94–100.

43 Christopher Trent, *Motoring on English Byways: A Practical Guide for Wayfarers* (London: G.T. Foulis, 1962), 1.

44 Hoskins, 'East Midlands and the Peak', 50.

45 Geoffrey Grigson, *About Britain No. 5. Chilterns to Black Country: A New Guide Book with a Portrait by W.G. Hoskins* (London: Collins, 1951), 67–80.

46 TNA, WORK 25/256/G1/C2/466, Letter from H.O. Aldhous to Dr W.G. Hoskins (22 July 1950).

47 TNA, WORK 25/57/A5/Q1 'Notes by Colonel Angwin on Paragraph 5 of the Agenda for the First Meeting of the Guide Book Editorial Committee 11 November 1949 – Content of the Books'.

48 Atkinson, *The Festival of Britain*, 144.

49 See, for example, the images in Geoffrey Grigson, *About Britain No. 8. East Midland and the Peak: A New Guide Book with a Portrait by W.G. Hoskins* (London: Collins, 1951), 22–3; and Grigson, *About Britain No. 5*, 10. On aerial photography and the viewpoint of Britain, see, especially, Kitty Hauser, *Shadow Sites: Photography, Archaeology, and the British Landscape 1927–1955* (Oxford: Oxford University Press, 2007), 151–99.

50 Edward H. Smith, *Quantock Life and Rambles* (Taunton: The Wessex Press, 1939), 3.

51 Smith, *Quantock Life and Rambles*, 62–3.

52 Beatrice F. Cresswell, *The Quantock Hills, their Combes and Villages: The Homeland Handbooks Vol. 35. Fifth Edition* (London: The Homeland Association, 1922), 117.

53 Anon., *Somerset Ways, Second Edition* (London: The Great Western Railway Company, 1928), 57.

54 Cresswell, *Quantock Hills*, 10.

55 Anon., *Somerset Ways*, 58.

56 Cresswell, *Quantock Hills*, 57.

57 Cresswell, *Quantock Hills*, 57.

58 Tim Cole, '"Beauty and the Motorway – The Problem for All": Motoring Through the Quantocks Area of Natural Beauty', in *Local Places, Global Processes. Histories of Environmental Change in Britain and Beyond*, ed. Peter Coates, David Moon and Paul Warde (Oxford: Windgather Press, 2016), 171–83.

59 Somerset History Centre, Taunton [hereafter SHC], C/GP/HF/852, Concerning Rights of Way and Signage on the Quantocks 8/2/58-16/1/62, Letter from E.S. Rickards, Clerk of Somerset County Council to Clerk of Over Stowey Parish Council (15 September 1959).

60 SHC, C/GP/HF/852, Letter from W. Richards, Clerk to Over Stowey Parish Council to F.A. Goodcliffe, Somerset Association of Parish Councils (14 January 1960).

61 SHC, D/PC/5/1/1, 'Rule and Regulations of the Quantock Right Association' (undated).

62 SHC, A/CVZ/1/1, 'Beauty and the Motorway – The Problem for All' (2 February 1973); Letter from Alderman A.W. Clarke (16 June 1972).

63 SHC, A/CVZ/1/1, 'Beauty and the Motorway – The Problem for All' (2 February 1973); 'Case for Quantock Car Road. Readers' Letters', Letter from Alderman A.W. Clarke (16 June 1972).

64 SHC, A/CVZ/1/1, 'Spoiling the Quantocks', Letter from Auberon Waugh, *The Observer* (2 July 1972); 'Quantocks Road', Letter from Martyn Skinner, *Bridgwater Mercury* (27 June 1972).

65 SHC, A/CVZ/1/1, 'Curbing Hill-Top Invasion, Editorial', *Bridgwater Mercury* (27 June 1972).

66 SHC, A/CVZ/1/1, 'Guarding Quantock Peace'. Readers' Letters' Letter from J. Skeggs, *Somerset County Gazette* (undated).

67 SHC, A/CVZ/1/1, 'Quantock Paths', Letter from Andrew Puckett (22 March 1974).

68 *Dunlop Guide to Great Britain* (London, n.d. [1920s?]), 804; Trent, *Motoring on English Byways*, 3.

69 Vincent Waite, *Portrait of the Quantocks* (London: Robert Hale, 1964), 9.

70 Waite, *Portrait of the Quantocks*, 20.

71 On this elitism elsewhere, see Elizabeth Baigent, '"God's Earth Will Be Sacred": Religion, Theology, and the Open Space Movement in Victorian England', *Rural History* 22, no. 1 (2011): 31–58.

72 Sachs, *For Love of the Automobile*, 160.

8

Crops in a machine: industrialising barley breeding in twentieth-century Britain

Matthew Holmes

From the 1950s to the 1980s, agricultural production in Britain boomed, heralding what has been termed the 'modern' or 'silent' revolution in agriculture.[1] The rise of industrialised agriculture in the decades after the Second World War owed much to the mechanical and chemical tools traditionally associated with industrialisation: farm machinery, pesticides and improvements in food processing.[2] Genetics and breeding were also of vital importance in increasing the yields of fundamental crop plants.[3] These improvements were partially the result of traditional methods of plant selection and crossing. Yet the range of tools available to plant breeders at state-funded agricultural institutes and private seed companies expanded to include such techniques as industrial-scale hybridisation, artificial chromosome doubling, induced mutation and plant tissue culture.

By the mid-twentieth century, barley had long been an integral part of farming, subject to domestication and artificial selection. The plant was, in itself, a vital piece of agricultural technology.[4] This chapter focuses upon two leading barley varieties, Proctor barley and Golden Promise, which were produced using two seemingly new tools: industrial hybridisation and radiation-induced mutation. Barley is an example of what has been termed an 'industrial plant', an organism tied to 'mechanical processes, modern science and the capitalist goals of industrial society'.[5] As such, the development, and even reception, of new plant breeding technologies for barley in Britain was intimately bound to existing technological systems. Barley was engineered both to suit the needs of the brewing industry and to exploit the by-products of atomic energy.

Harnessing previously unexamined archival material held at the National Institute of Agricultural Botany and the John Innes Centre, this chapter offers a new take on the history of plant breeding and agriculture in modern Britain.[6] The history of barley since the 1950s leads us to two conclusions: first, that plant breeding technologies are intrinsically connected with existing industries and technological systems, and second, that the modern revolution in British agriculture is something of a misnomer.[7] Although significant new crop varieties were bred, these results were more often than not due to the improvement or industrialisation of existing plant breeding technologies.

Industrial hybridisation: Proctor barley

Changing attitudes to hybrid plants

Hybridisation was by no means a new plant breeding technique by the mid-twentieth century, having provided both a source of intellectual contention for the educated and income for florists back in the nineteenth century. Hybrid crop plants, especially cereals, were generally unpopular with farmers: their tendency to produce sterile or inferior seed did not endear them to those who wished to save the seed from each year's harvest.[8] Scepticism towards hybrid crops began to change at the dawn of the twentieth century, as between 1910 and 1935, traditionally bred – either by inbreeding or open pollination – maize varieties in the United States were gradually replaced by hybrids. Hybrid corn faced little opposition during the era of the Great Depression. The principal concern of the American public was to obtain 'ample and affordable supplies of food, clothing and shelter', while farmers used hybrid corn to benefit from the policies of the New Deal and the Agricultural Adjustment Administration.[9] Between 1930 and 1965, the volume of corn production in the United States rose by some 2.3 billion bushels.[10] By this time, hybridisation was also an established technique in British plant breeding. In fact, the renowned barley breeder for Guinness Brewers, Edwin Sloper Beaven, did not see a renewed emphasis placed upon hybridisation by geneticists as 'at all novel'.[11]

Yet if you were a barley grower in 1950s Britain, you might well be forgiven for thinking that you were on the receiving end of a radical development in plant breeding. Under the Directorship of George Douglas Hutton Bell (1905–93), the Cambridge-based Plant Breeding Institute had just released a game-changing hybrid plant: Proctor barley. The new hybrid had outcompeted all comparable varieties in crop trials at the National Institute of Agricultural Botany, which highly recommended the variety for farmers in 1952, and in 1953 awarded Proctor its coveted Cereal Award.[12] Combining high yield and malting quality, Proctor barley occupied approximately 70 per cent of barley acreage in the United Kingdom by 1960. Executive Secretary of the Royal Society D.C. Martin wrote that production of barley in the UK had doubled over a six-year period, on the basis of which the Royal Society named Bell the first recipient of the prestigious Mullard Award in 1967.[13] Official accolades heaped upon Bell by the National Institute of Agricultural Botany and the Royal Society were joined by unofficial accolades from admiring farmers. Some wrote to Bell to personally express their gratitude for his 'wonderful work'.[14] Such acclaim left Bell visibly uncomfortable. In a letter to one such admirer, Bell morosely remarked, 'All I can hope is that the variety [Proctor] lives up to the reputation which it has so quickly acquired.'[15] In a 1954 letter to a member of a Somerset brewing firm, Bell stated that he thought Proctor had been 'taken up' too quickly. Moreover, he recalled doing '[his] best to damp things with Proctor before it had been put on the market'.[16] It is truly extraordinary to find a crop plant so popular that it made its own creator uneasy.

Engineering the industrial hybrid

The high yields that made Proctor popular with growers were the end result of something new: Britain's first industrialised hybridisation programme. In 1933 an attempt was made to cross tough Scandinavian barley varieties with established British barley varieties. This programme involved 'several departures from the then accepted practice of devising and handling hybridisation programmes'. These departures from the norm included a larger number of hybrid crosses than were usually performed, longer crop trials and a harsh selection procedure based on the conformity of hybrids to a preconceived and idealised morphological model. Five years of crop trials resulted in five hybrid varieties, one of which was Proctor.[17] In other words, the traditional hybridisation process had been extended, intensified and launched with a specific goal in mind.

Producing hybrids like Proctor now involved a whole new level of technical difficulty, expense and labour. In 1957, Dr J.H. Oliver of the Briant and Harman Brewing Company described the complexities of the new 'hybridisation process', which was in many ways testing the endurance of the barley breeder. Hybridisation of barley involved the removal of the plant's anther (to prevent self-fertilisation), followed by the delicate task of artificial introduction of pollen from the desired cross. With this complete, the de-anthered and pollinated plant had to be further protected from pollination by insects. All in all, Oliver proclaimed, 'what might be termed the process of fertilisation is simple, but the trouble started is considerable'.[18] With this process being conducted on a far larger scale than before, it is little wonder that farmers lacked the time and resources to carry out the hybridisation. The task was instead left in the hands of specialist research centres like the Plant Breeding Institute or, to a lesser extent, private firms.

Industrialisation moved hybridisation several degrees further away from the traditional tools of selection and crossing. Traditional breeding could potentially be carried out by farmers, or by small private breeders, whose individual experience and skill were vital to grow distinctive crop varieties in diverse growing conditions.[19] Hybrids were completely different: they degenerated over time, different varieties were practically indistinguishable from each other and their outward appearance gave no clues as to how they would grow. Expert input was already important to successfully breed and grow hybrids.[20] Investment and labour on a vast scale was now equally vital to grow hybrid barley on an industrial scale. It was such changes that could lead President of the National Farmers' Union Henry Plumb to declare that 'We have experienced since the War a silent revolution in British agriculture from a craft-based industry to one based on science and high productivity.'[21] Similarly, in 1970 seed merchant T. Martin Clucas described how the modern plant breeder, while 'still an artist, like his predecessor', was now 'aided by science and technology'.[22]

Crosstalk with the brewing industry

Part of the rationale behind developing a new hybridisation programme during the 1930s was to blend the malting characteristics of British barley – which made it ideal for the brewing industry – with the hardiness of Scandinavian varieties.[23] Throughout the twentieth century, British agriculture and food processing had become more and more integrated. Thus, when structural changes to the brewing industry occurred, which favoured 'cheaper and larger supplies of sub-optimal feeding barley' for malting, barley breeders responded.[24] The idealised model barley developed by hybridisers in the 1930s was one ideally suited to the needs of the brewing industry. As one of the outcomes of this programme, Proctor combined high yields with the ability to be used as a malting barley.

Acclaim from the brewing industry for Bell and Proctor barley even outshone that from farmers. In February of 1955, the Director of the Norfolk Agricultural Station wrote directly to Bell, delightedly informing him that the Malting Barley Competition at the Stalham Farmers' Club had been decisively won by Proctor.[25] At a meeting of the Yorkshire Section of the Brewers' Guild in 1957, J.H. Oliver declared 'that Proctor was the most remarkable hybrid barley for brewing purposes that had ever been bred'.[26] Proctor slotted seamlessly into the British brewing industry, as a biological component of that industry overtly engineered to suit its needs. Other, complicating factors were also at play in the widespread uptake of Proctor, including the growing profitability of livestock fattened on barley and a fall in oat (a rival animal feed) acreage.[27]

Proctor barley, and the industrial hybridisation system that produced it, in part enjoyed its success due to the targeted attempt by Bell and the Plant Breeding Institute to meet the utilitarian needs of the brewing industry. The planned and hands-on production of Proctor also appealed to some members of the brewing community on another level. Some in the industry, like J.H. Oliver, did not put much stock in laboratory scientists, particularly geneticists, as suitable experts on hybrid varieties. After all, respected barley breeder Edwin S. Beaven had declared that 'the geneticist will generally offer an explanation of the plant breeder's results after they have been ascertained'.[28] By 1957 it was not Beaven's scepticism of hybridisation that had endured, but his mistrust of scientific experts.[29] Following in this tradition, Oliver therefore put the decision of whether Proctor was a useful development or not in the hands of British brewers and maltsters. 'It would be unwise, indeed unfair,' he wrote, 'to think that it [Proctor] can be left to Lyttel Hall [The Brewing Research Foundation laboratory in Nutfield].'[30]

A favourable perception of hybrids within the brewing industry was important in ensuring Proctor barley's commercial success. It was an advantage for Proctor to be a large-scale, field-tested, industrial technology: precisely because such an approach, and the organisms produced by it, fitted with the existing beliefs of those in the brewing industry. Laboratory science, and the belated explanation of the heredity phenomenon by geneticists, did not pass muster with either British barley breeders or brewers. By contrast, the hands-on and planned production of hybrid barley to meet certain specifications was far more appealing, on both utilitarian and intellectual grounds.

Hybrid crops and population growth

Enthusiasm for hybrid crop varieties of all types became increasingly evident as Proctor barley dominated the British barley market for much of the 1950s and 1960s. When the National Institute of Agricultural Botany held its annual Crop Conference in 1970, hybrids came top of the agenda. Their primary advocate was one of the Institute's Field Officers, K.E. Haine, who led proceedings by stating that the 'outstanding development resulting from basic research [in plant breeding] has been the use of F1 (first-generation) hybrids'. In front of an audience of farmers, breeders and representatives of the food industry, Haine was delighted to report that hybridisation had been applied to the vegetable kingdom, with a large number of brassica F1 hybrids undergoing trials at the Institute's test sites, including 50 varieties of that holiday favourite, Brussel sprouts.[31]

Yet in a following discussion, representatives of the wider agricultural community voiced concerns about hybrids. A representative of a food company, Mr How of Ross Foods, argued that seasonal fluctuations in the performance of hybrid varieties were a matter of concern for growers. A member of the National Agricultural Advisory Service, Mr Brown, announced that he preferred the consistency of older crop varieties. This remark was presumably intended as a rebuke against the tendency of hybrid crops to degenerate into parental types over successive generations. Faced with this backlash, Haine admitted that hybrid varieties had their faults. However, he was certain that these would be ironed out in the future.[32] Attempting to roll out hybrid crops across British agriculture was clearly not a straightforward matter, especially when there was no pressing demand from an established industry for a crop plant with specific characteristics tailored to their needs.

Calls for hybridisation to revolutionise agriculture continued throughout the 1970s. In an era marked by neo-Malthusian fears of an unsustainable growth in world population, high-yielding hybrids were viewed as a means to feed a hungry world.[33] It is perhaps no coincidence that a key bastion of support for hybridisation, the National Institute of Agricultural Botany, had been on the receiving end of Malthusian prophecies from the late 1960s. In what was not an uncommon event, Vice President of the National Farmers' Union, David H. Darbishire, addressed members of the Institute in 1972, announcing the onset of a global population crisis that threatened the very survival of humankind.[34] Similar voices were heard elsewhere. At the 1977 meeting of the British Association for the Advancement of Science, Professor Bleasdale of the National Vegetable Research Station joined a chorus of voices that urged industrialised countries to use their own progress in agricultural production to supply the Third World. Hybrid crops were a key part of Bleasdale's plan to kick-start what he termed Britain's own 'Green Revolution'.[35] Contemporary fears and technological ambition had combined to support the continued production of hybrid crops.

Ecological consequences

The pursuit of industry-led goals by the breeders of Proctor barley had consequences beyond the malting floor or conference room. Even the most carefully engineered crop plants were still subject to environmental and evolutionary forces. From the mid-1950s, overwhelmingly popular crop varieties like Proctor altered the makeup of Britain's arable farms. Smaller seed merchants complained that they could no longer keep up with the demand for the small number of varieties that dominated Britain's cereal acreage.[36] The transition to industrialised monocultures, many of which were made up of hybrids, had begun to generate serious concern within the British agricultural community by the 1970s.

At the National Institute of Agricultural Botany's 1970 Crop Conference, K.E Haine recognised that 'If a disease … severely affects an F1 hybrid, the result may be disastrous, with every single plant showing symptoms of the disease.'[37] A lack of genetic variation between individual plants made fields of hybrid crops extremely vulnerable to disease: a fact painfully brought home to hybrid corn growers in the United States, who lost 15 per cent of their crop to an outbreak of southern leaf corn blight in 1970.[38] D.R. Marshall, a member of the Commonwealth Scientific and Industrial Research Organisation, described how the 'major technological revolution in farming' had left important crop groups genetically uniform and 'markedly vulnerable to disease and pest epidemics'.[39] By 1981, previously unknown plant diseases were reported by the National Institute of Agricultural Botany's Plant Pathology Branch, including 'net blotch of barley'.[40]

Two solutions were pursued to counter the growing deluge of new diseases in industrialised monocultures: one low-tech, one high-tech. The low-tech solution recommended by the National Institute of Agricultural Botany involved a return to old farming practices, growing a wide variety of cereals on arable farms and avoiding vulnerable monocultures.[41] This approach was preferable, some claimed, to attempting to breed new crop plants and entering into a genetic arms race with pathogens.[42] This attitude was not shared by a private seed firm, Milns Seeds of Chester, which favoured a high-tech solution to counter the rise of new diseases. In its 1969 Seed Catalogue, the firm promised its customers that pathogens would be defeated by the application of the latest plant breeding technologies: not just hybridisation, but artificial mutations induced by gamma radiation.[43]

Barley goes nuclear: Golden Promise

Power over nature

A distinctive, albeit overlapping, pathway to the manipulation of barley was the use of mutation breeding. As with hybridisation, creating new crop plants by artificially inducing mutation was by no means a new idea. The Dutch botanist and founder of mutation theory, Hugo de Vries, articulated the ambition of manipulating mutation for human benefit in a lecture series delivered at the University of California in 1904: 'Indeed, if it once should become possible to bring plants to mutate at our will and perhaps even in arbitrarily chosen directions, there is no limit to the power we may finally hope to gain over nature.'[44] New tools soon became available to fulfil this ambition, including X-rays and colchicine.[45] Yet practical results were few and far between. When the General Electric Research Laboratory in New York attempted to use X-rays to induce beneficial mutations in plants during the 1920s and 1930s, the only marketable product to emerge from the laboratory was a single variety of ornamental lily.[46]

A long history of poor practical returns did not deter British biologists from embarking upon their own investigations into the possibilities of mutation breeding. The John Innes Horticultural Institute, a leading centre of genetics and plant breeding based in Norwich, played host to a 1952 symposium on the uses of 'Chromosome Breakage'. Addressing attendees, Director of the Institute Cyril Dean Darlington announced that 'experimental gene mutation and chromosome breakage' were 'comparable with … the discoveries of Mendelian experiment … they have created a new branch of technology and are in the process of creating a new branch of science'.[47] The symposium appeared to represent just that, with speakers on agriculture, horticulture, animal genetics and medicine. Yet in 1953 Darlington left the JIHI for the Sherardian Chair of Botany at Oxford University. A significant factor in his move was Watson and Crick's 1953 discovery of the structure of DNA. Darlington could sense a 'molecular revolution sweeping across the biological world', which would not favour researchers based in horticultural institutions.[48]

At the 1955 meeting of the British Association for the Advancement of Science, a session on 'Genetics and plant breeding' was addressed by Dan Lewis, Head of the Genetics Department at the John Innes Horticultural Institute. Although genetics could not 'produce plants to order', Lewis saw great potential in the use of X-rays to release hitherto unrealised variability in crop plants.[49] At the 1957 British Association meeting, radiation and its agricultural applications were also the focus of a member of the Plant Breeding Institute, R.N.K. Whitehouse. Although 'few mutation-bred strains [of crop plants] have yet reached the market', Whitehouse declared, this was simply 'because insufficient time has elapsed'.[50] If anything, British plant breeders had fallen behind on the international stage. Mutation-bred barley had been released on to the Swedish market; barley with improved straw-strength and mildew resistance had been developed in Germany and Austria; in the United States, mutation breeding had been applied to wheat, oats, peas and peanuts.[51]

Gamma ray breeding

Despite Whitehouse's optimism, one of the main tools of chromosome manipulations, X-rays, had proved to be largely ineffective at producing large numbers of useful mutations in plants. Moreover, with the departure of Darlington from the JIHI, ambitions for a new science of chromosome breakage seemed to be dead in the water. Salvation seemed to beckon for mutation breeders with the development of atomic technology. Radioactive isotopes, produced as a by-product of nuclear fission, formed the raw material of atomic age mutation breeding programmes. On 8 December 1953 at the United Nations General Assembly in New York, US President Dwight D. Eisenhower gave his famous 'Atoms for Peace' speech. Speaking 'against the dark background of the atomic bomb', Eisenhower claimed that 'experts would be mobilized to apply atomic energy to the needs of agriculture, medicine and other peaceful activities'.[52]

In the wake of the atomic bomb had come the realisation that radiation must adopt a publicly acceptable face: agriculture and medicine were two means of making atomic research palatable to the general public.[53] For its post-war advocates, nuclear power possessed the potential to 'return mankind to an idyllic, prelapsarian bliss'.[54] In Britain, none endorsed this potential with more enthusiasm than Muriel Howorth (1886–1971). An acolyte of Nobel laureate Frederick Soddy, Howorth sought to promote the atom at every opportunity, founding the Atomic Gardening Society in 1959.[55] On a more practical level, isotopes were made available for plant breeders through the Atomic Energy Research Establishment at Harwell.

By 1949 the Establishment had gained a reputation as the world's 'leading isotope exporter'.[56] The powerful gamma rays emitted by these isotopes seemed to promise a far more effective means of altering chromosomes and inducing mutation in plants than X-rays.

Faith in the power of gamma rays was clearly held at Milns Seeds, which as early as 1957 began a breeding programme using gamma radiation at its Plant Breeding Station in Chester. The respected and long-established firm did not seek to hide the means used to accomplish this marvel, openly stating in its Seed Catalogue that grains from an existing barley variety, Maythorpe, had been 'exposed to gamma-rays and the resulting material carefully screened'.[57] The most promising mutant plants were then multiplied and placed in field trials, where variety 759/4 – later to be known as Golden Promise – proved to be the most successful.[58] Two years after the release of Golden Promise on to the commercial market in 1965, Milns could boast that its performance 'in all parts of the country has led to increasing demand and widespread popularity'.[59] The reign of Proctor barley suddenly appeared to be in jeopardy when, in 1967, Milns Seeds reported that Golden Promise gave growers an even higher yield per acre.[60] Golden Promise was especially popular with brewers in Scotland, comprising some 10 per cent of barley seed sales from 1973 to 1984.[61]

The positive reception of Golden Promise by the brewing industry encouraged Milns Seeds to pursue 'exciting possibilities' using a combination of irradiation and hybridisation. Milns did admit that its varietal improvement programmes 'may result in the use of cereals that are somewhat unorthodox in appearance'.[62] The unorthodox cereal, another barley variety named Midas, was released in 1971. According to the company's Seed Catalogue of that year, Midas was the result of 'a cross involving a very short strawed mutation, produced by irradiation, together with Proctor and North American selections with Mildew resistance'. The barley was described as a 'variety with very unusual and distinct characteristics and all the essential factors'.[63] On the face of it, mutation breeding at Milns had proved a remarkable success. In little over a decade, a private breeding firm had successfully applied gamma radiation to barley breeding and produced two highly popular varieties.

Competition from hybrids

Not all were impressed by the newfound interest in mutation breeding. The driving force behind Proctor barley, George Douglas Hutton Bell, had been a longstanding sceptic. During the late 1940s, Bell noted that contemporary efforts to use chemicals, X-rays and other techniques to produce mutations lacked practical value: 'the general experience has so far been that the new forms induced by the various treatments have little or no economic value as far as the improvement of crop plants is concerned'.[64] Some 20 years later, in the midst of the Atomic Age, Bell's attitudes towards mutation breeding remained dismissive. Although he was not opposed to the concept of manipulating chromosomes to breed better plants, his vision of how this should be achieved was very different to that of Milns Seeds.

Bell used an address delivered to the Royal Society in March of 1968 to explain his vision of plant breeding through 'artificially controlled hybridisation'. Chromosome manipulation could be used to overcome barriers to hybridisation, although Bell preferred established techniques over the use of gamma radiation; for instance, the chemical colchicine. Complex hybridisation programmes and extensive backcrosses could also be used as a form of 'chromosome substitution'.[65] There was little need for gamma radiation, as variation and chromosome manipulating could both be accomplished using existing tools. In its 1969 seed catalogue, Milns Seeds announced its intention to develop 'new hybrids', which would incorporate the 'dwarf material [referring to the short straws of Golden Promise] obtained by induced mutation'.[66] New barley varieties like Midas were therefore produced by a combination of mutation breeding and hybridisation.

Despite the efforts of Milns Seeds, barley varieties produced by traditional breeding and hybridisation continued to dominate British agriculture, even when new plants altered through radiation were readily available. Proctor barley continued to appear alongside both Golden Promise and Midas in Milns Seeds catalogues throughout the 1960s. The firm's 1969 catalogue described the then-ageing Proctor as 'a well tried variety'.[67] As late as 1971, the sister firm of Milns, Marsters Seeds, informed its customers that Proctor barley was sold out: Golden Promise, however, remained in stock.[68] The continuing popularity of the hybrid may simply have been a vestige of its overwhelming uptake during the 1950s. Barley growers may have felt more comfortable and confident managing a field of Proctor.

It seems that Golden Promise did not achieve widespread success with growers and brewers in virtue of superior genetics alone. The barley was more of an effective 'all-rounder', with reasonable malting quality, good yields and tolerance of adverse weather.[69] During a 1978 Conference at the National Institute of Agricultural Botany, an afternoon discussion turned to the subject of 'why Golden Promise had done so well commercially, but had not done so in NIAB trials'. One of the Institute's field officers recalled that the characteristics of Golden Promise had been so well documented that 'farmers were able to treat the variety accordingly, and thus grow it successfully'.[70] The commercial success of Golden Promise may have been down to farmers being provided with useful information on how to grow it, rather than the application of a novel form of Atomic Age biotechnology.

Measuring fallout on the farm

Mutation-bred crops like Golden Promise were intimately tied to atomic technology, which provided the radioactive isotopes necessary for their existence. The crossover between atomic technology and biology was readily apparent in government-funded institutions like the Agricultural Research Council Radiobiological Laboratory in Wantage, Berkshire and the Medical Research Council Radiobiology Unit at Harwell, Berkshire. We have already seen how the latter provided isotopes to plant breeders on an international scale.[71] At Wantage, researchers carried out their own mutation breeding programmes on ornamental plants during the early 1960s. Their efforts even resulted in new variety of chrysanthemum, named Cream Sweetheart.[72]

Other experiments at Wantage focused on crop plants, including the irradiation of rye samples contributed by the Birmingham University Genetics Department in 1963.[73] Yet by the closing years of the 1960s, research priorities had changed. In 1968 the first in a new series of research papers produced by Wantage Laboratory appeared in the journal *Radiation Botany*. These papers reported upon large-scale experiments conducted by the Laboratory, which were designed to assess the impact of radioactive fallout on British agriculture. Crop plants, including barley, wheat, oats, potatoes and legumes, were subjected to high levels of gamma radiation to 'assess possible effects of environmental contamination, for example near-in fallout from atomic weapons'.[74] The main objective of these investigations was 'to provide information which would assist in civil defence assessments of the possible consequences of catastrophic discharges of radioactivity onto agricultural land'.[75]

To accurately gauge the potential impact of nuclear fallout across Britain, Wantage conducted its irradiation experiments on a nationwide scale. Strontium-90 was sprayed on to pastures on test sites at Rothamsted Experimental Station and the University of Reading.[76] In 1974, cereals and potatoes exposed to radiation at Wantage were grown at the National Institute of Agricultural Botany's regional crop trials centres, Headley Hall in Yorkshire and the Norfolk Agricultural Station.[77] These tests were representative of a wider concern surrounding the long-term consequences of radioactive fallout for the environment. Only a year before tests began, high levels of Strontium-90 had been discovered in milk following the Windscale fire.[78] Such concerns were, however, international. The Brookhaven National Laboratory in the United States published its own analysis of how maize and common fruit and vegetables would react to nuclear fallout in 1970.[79]

Gamma radiation had moved from being viewed as a powerful plant breeding tool to a serious risk to agriculture, whether delivered by nuclear war or nuclear accident. This is not to say that crop breeding using gamma radiation simply halted during the late 1960s. Quantifying the extent to which mutation breeding has continued into recent times is fraught with difficulty. Fearful of entanglement with the divisive debate over genetically modified foods, many mutation breeders are now disinclined to even articulate the techniques by which their varieties are produced. There is no longer a clear picture of how many agricultural and horticultural plant varieties are bred using radiation.[80] It is likely that since the 1960s a 'burgeoning interest in other means of genetic manipulation' has led biologists and breeders alike to abandon radiation-induced mutation as a plant breeding technology.[81]

Conclusion

Barley, like many other crops, clearly displayed all the characteristics of an 'industrial plant' during the mid-twentieth century. Changes to the biology of barley went hand-in-hand with changes to industry, or the emergence of new technological systems. The restructuring of the brewing industry to favour high-yielding hybrid barley was seized upon by breeders, who endeavoured to meet the demand of brewers through the transformation of hybridisation into an industrial process. The fruit of this labour emerged in the form of Proctor barley, whose creators made use of a long-existing technique, industrialising hybridisation to closely integrate barley with the needs of brewers. Mutation-bred varieties like Golden Promise and Midas were also tailored to the needs of the brewing industry, but were also reliant upon the development and availability of atomic technology.

Controversy and the promotional activities of barley breeders in twentieth-century Britain also demonstrate the importance of attitudes and ideology. It is hard to imagine a hybrid barley like Proctor being received with such enthusiasm if moral outrage at the very concept of hybridisation had existed, or if brewers had condemned mediocre malting barleys in favour of higher quality varieties. Larger, politicised forces were also at work in the development and use of new plant breeding technologies. The hybrid benefited from a long association with Mendelian principles, a history that took on new meanings during the Cold War. Hybrids became symbols of Western science and genetics, standing against the pseudo-science of Lysenko and the Soviet Union. Similarly, mutation breeding possessed clear political dimensions, its advocates seeking to ally the field with advances in medicine and the success of the Green Revolution to ensure a positive reception.[82]

Yet ideology will only get a new plant breeding technology so far, as demonstrated by the case of hybrid varieties during the 1970s. Lacking a purpose beyond that of general crop improvement, hybrids were contested by both agricultural scientists and those in the food industry. Moreover, others remained unconvinced of their superiority to traditional varieties.[83] Ideological or political support for a plant breeding technology could easily dissipate, or be quickly transferred to a more promising technique. Hence the utopian rhetoric of mutation breeding and the desire to 'grind genes in a mortar and cook them in a beaker' was usurped by the rise of new genetic biotechnologies, including cell fusion and recombinant DNA technology.[84] Ideology or political convictions did not necessarily determine the actions of plant breeders.[85]

Just as it would be an incomplete picture to explain the uptake of crop plants or plant breeding technologies in terms of soulless industrial processes or automated systems, so it would be equally incomplete to ascribe their success or failure purely to political or ideological leanings. Proctor barley was able to achieve such a remarkable level of success partly due to the increasingly favourable reception of hybrid plants, and partly thanks to its careful incorporation into the existing industrial process of brewing. Similarly, Golden Promise emerged in an age more or less favourable to the use of radiation in agriculture, while seeking to fulfil the same economic criteria that had ensured the success of Proctor. In both cases, plant breeding innovations chimed with existing attitudes and technological systems.

To what extent Proctor barley, Golden Promise and the technologies behind their creation were truly innovative is open to question. Both were bred using existing techniques, namely hybridisation and mutation breeding. It seems that the modern revolution in British agriculture is a misnomer, as the revolution was not exactly revolutionary.

Instead, it consisted of reforming and refining existing technologies. What changed is that plant breeding altered to either meet the needs of, or benefit from, existing technological systems. Hybridisation was conducted on a huge scale and with an existing end-point in mind, to better meet the needs of a restructured British brewing industry. Meanwhile, mutation breeding received a much-needed boost from the availability of radioactive isotopes. These were not necessarily radical transformations, but did help bring both plant breeding technologies into alignment with Britain's technological systems. These new biological innovations were well received as they were closely integrated with existing industry and technology: in terms of both material needs and attitudes.

The history of barley breeding in twentieth-century Britain further illustrates the difficulties of translating biological innovation into practical products for agriculture. In light of the case studies presented in this chapter, it becomes clear that forcing a major paradigm shift in how plant breeding is conducted is even more problematic than we might already imagine. This chapter has addressed the development of new varieties of crop plants and their uptake in agriculture. The interaction between technology and environment covered in this chapter roughly falls under the second of the combinations identified by Jon Agar in Chapter 1, as environment – in this case, barley – as something natural made into, or a component within, a technological system. Yet as a cultivated crop plant, barley had long been manipulated by humankind. Barley had started upon the road to an industrial component, a trend accelerated by the arrival of new plant breeding technologies. As plant breeding and the crop varieties it produces are part of an interconnected network of both ecological and technological systems, significant material and ideological barriers exist to attempts to change how we produce our food. It can be tentatively suggested that such barriers, twinned with the challenge of integration into existing systems, have played a role in the turbulent reception of genetic biotechnology in Britain.

Notes

1 Quentin Seddon, *The Silent Revolution: Farming and the Countryside into the 21st Century* (London: BBC Books, 1989); Kenneth Blaxter and Noel Robertson, *From Dearth to Plenty: The Modern Revolution in Food Production* (Cambridge: Cambridge University Press, 1995), 36. The history of a museum-based presentation of agricultural modernity and transformation is discussed by David Matless, 'The Agriculture Gallery: Displaying Modern Farming in the Science Museum', Chapter 6 this volume.

2   Paul Brassley, 'Output and Technical Change in Twentieth-Century British Agriculture', *The Agricultural History Review* 48 (2000): 60–84.

3   Valerie Silvey, 'The Contribution of New Varieties to Cereal Yields in England and Wales between 1947 and 1983', *Journal of the National Institute of Agricultural Botany* 17 (1986): 155–68.

4   By the twentieth century, barley could already be considered as much a part of 'technology' as 'nature'. For more on plants as technology, see Dominic J. Berry, 'Plants are Technologies', Chapter 9 this volume.

5   Mark J. Smith, 'Creating an Industrial Plant: The Biotechnology of Sugar Production in Cuba', in *Industrializing Organisms: Introducing Evolutionary History*, ed. Susan Schrepfer and Philip Scranton (London: Routledge, 2004), 86. This examination of barley as an industrialised organism also draws upon Ann N. Greene, 'War Horses: Equine Technology in the Civil War', in Schrepfer and Scranton, *Industrializing Organisms*, 143–65. See also Ann N. Greene, *Horses at Work: Harnessing Power in Industrial America* (Cambridge, MA: Harvard University Press, 2008).

6   The archives of the NIAB have formed the basis of a number of PhD theses at the University of Leeds, including Berris Charnley, 'Agricultural Science, Plant Breeding and the Emergence of a Mendelian System in Britain, 1880–1930' (PhD diss., University of Leeds, 2011), and Dominic J. Berry, 'Genetics, Statistics, and Regulation at the National Institute of Agricultural Botany 1919–1969' (PhD diss., University of Leeds, 2014).

7   The integration of plant breeding technology and existing technological systems in this chapter is inspired by Helen Anne Curry, 'Atoms in Agriculture: A Study of Scientific Innovation between Technological Systems', *Historical Studies in the Natural Sciences* 46 (2016): 119–53. Both Curry and Greene draw upon Thomas P. Hughes, *Networks of Power* (Baltimore: Johns Hopkins University Press, 1983).

8   Robert C. Olby, 'Horticulture: The Font for the Baptism of Genetics', *Nature Reviews Genetics* 1 (2000): 65–70. During the nineteenth century, breeders did have to occasionally defend their hybrid productions against accusations of impiety, transgressing of the laws of nature and the objections of botanists. See Noel Kingsbury, *Hybrid: The History and Science of Plant Breeding* (London and Chicago: University of Chicago Press, 2009), 94–6.

9   Donald N. Duvick, 'Biotechnology in the 1930s: The Development of Hybrid Maize', *Nature Reviews Genetics* 2 (2001): 73; Deborah Fitzgerald, *The Business of Breeding: Hybrid Corn in Illinois, 1890–1940* (Ithaca: Cornell University Press, 1989). Hybrid corn received another boost to its reputation in 1958, when Soviet premier Nikita Khrushchev toured the farm of Roswell Garst on a visit to the United States. In front of a crowd of journalists, Khrushchev held aloft an ear of hybrid corn, an event later interpreted as marking the triumph of Western genetics over Lysenkoism. See Loren R. Graham, *What Have We Learned about Science and Technology from the Russian Experience?* (Stanford: Stanford University Press, 1998), 19.

10 Jack Ralph Kloppenburg, *First the Seed: The Political Economy of Plant Biotechnology, 1492–2000* (Cambridge: Cambridge University Press 1988), 91.

11 Paolo Palladino, 'Science, Technology and the Economy: Plant Breeding in Great Britain, 1920–1970', *Economic History Review* New Series 49 (1996): 119.

12 G.P. Morris, 'Descriptions of Wheat and Barley Varieties', *Journal of the National Institute of Agricultural Botany* 7 (1953): 460. Director Frank Horne subsequently considered the Institute's 1952 endorsement of Proctor to be one of the most important moments of his 25-year career. See Frank R. Horne, 'Some Aspects of Crop and Seed Improvement since 1945', *Journal of the National Institute of Agricultural Botany* 12 (1971): 400.

13 D.C. Martin, Royal Society Press Notice, 18 July 1967, File 1, Box 26, Plant Breeding Institute [hereafter PBI] G.D.H. Bell Collection, John Innes Centre Library and Archives [hereafter JIC]. The Mullard Award recognises innovations of economic benefit to Britain.

14 Walter K. Sternfeld to G.D.H. Bell, 16 August 1953, File 3, Box 26, PBI Proctor Correspondence, JIC.

15 G.D.H. Bell to Walter K. Sternfeld, 19 August 1953, File 3, Box 26, PBI Proctor Correspondence, JIC.

16 G.D.H. Bell to H.L. Thompson, 29 June 1954, File 3, Box 26, PBI Proctor Correspondence, JIC.

17 George Douglas Hutton Bell, 'Plant Breeding for Crop Improvement in Britain: Methods, Achievements and Objectives', *Proceedings of the Royal Society of London. Series B, Biological Sciences* 171 (1968): 148.

18 Dr J.H. Oliver, 'Proctor', reprinted from the *Brewers' Guild Journal*, April 1957, File 1, Box 26, PBI G.D.H. Bell Correspondence, JIC.

19 Deborah Fitzgerald, 'Farmers Deskilled: Hybrid Corn and Farmer's Work', *Technology and Culture* 34 (1993): 328–9.

20 Fitzgerald, 'Farmers Deskilled', 342.

21 Henry Plumb, 'Address', *Journal of the National Institute of Agricultural Botany* 14 (1977): 363.

22 T.M. Clucas, 'The Contribution of Plant Breeders to Vegetable Production in the 70s', *Journal of the National Institute of Agricultural Botany*, 12 (Supplement, 1970): 48.

23 Bell, 'Plant Breeding for Crop Improvement', 147–8.

24 Palladino, 'Science, Technology and the Economy', 120.

25 F. Rayns to G.D.H. Bell, 17 February 1955, File 3, Box 26, PBI Proctor Correspondence, JIC.

26 Dr J.H. Oliver, 'Proctor', reprinted from the *Brewers' Guild Journal* (April 1957), File 1, Box 26, PBI G.D.H. Bell Correspondence, JIC.

27 Palladino, 'Science, Technology and the Economy', 120; Blaxter and Robinson, *From Dearth to Plenty*, 130–1.

28 Dr J.H. Oliver, 'Proctor', reprinted from the *Brewers' Guild Journal* (April 1957), File 1, Box 26, PBI G.D.H. Bell Correspondence, JIC.

29 Both Palladino and Kingsbury portray Beaven as part of an early twentieth-century backlash against Mendelian genetics. See Paolo Palladino, *Plants, Patients and the Historian: (Re)membering in the Age of Genetic Engineering* (Manchester: Manchester University Press, 2002), 79–81; Kingsbury, *Hybrid*, 173–4.

30 Dr J.H. Oliver, 'Proctor', reprinted from the *Brewers' Guild Journal* (April 1957), File 1, Box 26, PBI G.D.H. Bell Correspondence, JIC.

31 K.E. Haine, 'Improved Varieties of Brassica Crops in NIAB Trials', *Journal of the National Institute of Agricultural Botany* 12 (Supplement, 1970): 1–4.

32 J.D. Reynolds, 'Improved Varieties of Peas and Carrots in NIAB Trials', *Journal of the National Institute of Agricultural Botany* 12 (Supplement, 1970): 16.

33 For an overview of the neo-Malthusian movement, see Mauricio Schoijet, 'Limits to Growth and the Rise of Catastrophism', *Environmental History* 4 (1999): 515–30.

34 David H. Darbishire, 'Address', *Journal of the National Institute of Agricultural Botany* 12 (1972): 519–23.

35 J.K.A. Bleasdale, 'Britain's Green Revolution', in *Advances in Agriculture: Proceedings of Section M Agriculture, 139th Annual Meeting of the British Association for the Advancement of Science*, ed. W.A. Hayes (Birmingham: University of Aston, 1978), 1–2.

36 R. Dudley, 'Report of the Second Crop Conference', *Journal of the National Institute of Agricultural Botany* 7 (1954): 198.

37 Haine, 'Improved Varieties of Brassica Crops', 2.

38 Kloppenburg, *First the Seed*, 122.

39 D.R. Marshall, 'The Advantages and Hazards of Genetic Homogeneity', *Annals of the New York Academy of Sciences* 287 (1977): 17–18.

40 Quarterly Report to Council, September–October 1981, Box C-3, Document No. 766, Archives of the National Institute of Agricultural Botany [hereafter NIAB].

41 'Varietal Diversification in Cereals', Sixty-Second Report and Accounts 1981, NIAB.

42 Darbishire, 'Address', 521.

43 Milns Seeds Seed Catalogue Spring 1969, GRSRM: 2002.165.285, Museum Library, Museum of Norfolk Life, Gressenhall Farm and Workhouse [hereafter Gressenhall Library].

44 Hugo de Vries, *Species and Varieties: Their Origin by Mutation* (Chicago: The Open Court, 1905), 688.

45 On the long history of mutation breeding in the United States, see Helen Anne Curry, *Evolution Made to Order: Plant Breeding and Technological Innovation in Twentieth-Century America* (Chicago and London: University of Chicago Press, 2016).

46 Helen Anne Curry, 'Industrial Evolution: Mechanical and Biological Innovation at the General Electric Research Laboratory', *Technology and Culture* 54 (2013): 746–81.

47 Cyril Dean Darlington, 'The Problem of Chromosome Breakage', in *Symposium on Chromosome Breakage*, John Innes Horticultural Institution (London and Edinburgh: Oliver & Boyd, 1953), v.

48 Oren Solomon Harman, *The Man Who Invented the Chromosome: A Life of Cyril Darlington* (Cambridge, MA: Harvard University Press, 2004), 206–7. Nearly three decades later Darlington produced a brief history of plant breeding, which made no mention of mutation

breeding. See Cyril Dean Darlington, 'Genetics and Plant Breeding, 1910–80', *Philosophical Transactions of the Royal Society of London. Series B, Biological Sciences* 292 (1981): 401–5.

49   Watkin Williams, 'Genetics and Plant Breeding', *Nature* 4485 (1955): 719.

50   Anon., 'Radiation and Biology', *Nature* 180, no. 4587 (1957): 629.

51   Anon., 'Radiation and Biology', 629. The session also featured papers on the clinical applications of radiation and its impact on human health.

52   A full transcript of the speech, plus audio and visual recordings, is available on the IAEA website: www.iaea.org/about/history/atoms-for-peace-speech. Last accessed 18 December 2017.

53   Nicolas Rasmussen, *Gene Jockeys: Life Sciences and the Rise of Biotech Enterprise* (Baltimore: Johns Hopkins University Press, 2014), 22–3.

54   Paige Johnson, 'Safeguarding the Atom: The Nuclear Enthusiasm of Muriel Howorth', *British Journal for the History of Science* 45 (2012): 553.

55   Johnson, 'Safeguarding the Atom', 553.

56   Karin Zachmann, 'Peaceful Atoms in Agriculture and Food: How the Politics of the Cold War Shaped Agricultural Research Using Isotopes and Radiation in Post War Divided Germany', *Dynamis* 35 (2015): 312. In the United States, the Oak Ridge nuclear reactor supplied isotopes for laboratories, private companies and medical clinics. See Angela Creager, *Life Atomic: A History of Radioisotopes in Science and Medicine* (Chicago: University of Chicago Press, 2013), 181.

57   Milns Seeds Seed Catalogue, Spring 1967, GRSRM: 2002.165.274.1, Gressenhall Library.

58   It has been suggested that private breeding firms could ill afford to be inventive research centres in the post-war era, as 'the mutagenic effects of colchicine, mustard gas and then radiation seemed too far removed from the business of breeding'. Milns was clearly a rare exception to this rule, with its inventiveness rewarded by financial gain. Palladino, *Plants, Patients and the Historian*, 62.

59   Milns Seeds Seed Catalogue, Spring 1967, GRSRM: 2002.165.274.1, Gressenhall Library.

60   Milns Seeds Seed Catalogue, Spring 1967, GRSRM: 2002.165.274.1, Gressenhall Library.

61   Silvey, 'The Contribution of New Varieties', 162.

62   Milns Seeds Seed Catalogue, Spring 1969, GRSRM: 2002.165.285, Gressenhall Library.

63   Milns Seeds Seed Catalogue, Spring 1971, GRSRM: 2002.165.290, Gressenhall Library.

64   George Douglas Hutton Bell, *Cultivated Plants of the Farm* (Cambridge: Cambridge University Press, 1948), 190–1.

65   Bell, 'Plant Breeding for Crop Improvement', 170. For more on Bell's scepticism regarding mutation breeding and his ongoing preference for hybridisation, see Jacob Darwin Hamblin, 'Quickening Nature's Pulse: Atomic Agriculture at the International Atomic Energy Agency', *Dynamis* 35 (2015): 396–7.

66   Milns Seeds Seed Catalogue, Spring 1969, GRSRM: 2002.165.285, Gressenhall Library.

67   Milns Seeds Seed Catalogue, Spring 1969, GRSRM: 2002.165.285, Gressenhall Library.

68   Marsters Seeds Seed Catalogue, Spring 1971, GRSRM 2002.165.289, Gressenhall Library.

69   Blaxter and Robertson, *From Dearth to Plenty*, 130.

70   Anon., *Cereals in the West and South: Papers Given at Fellows Conferences at Harper Adams Agricultural College and the Hampshire Agricultural College, Sparsholt, January 1978* (Cambridge: National Institute of Agricultural Botany, 1978).

71   Harwell also conducted research on the effects of radiation on plants, albeit with less frequency than Wantage.

72   H.J.M. Bowen et al., 'The Induction of Sports in Chrysanthemums by Gamma Radiation', *Radiation Botany* 2 (1962): 303.

73   C.W. Lawrence, 'Genetic Control of Radiation-Induced Chromosome Exchange in Rye', *Radiation Botany* 3 (1963): 89–94.

74   C.R. Davies, 'Effects of Gamma Irradiation on Growth and Yield of Agricultural Crops – I. Spring Sown Wheat', *Radiation Botany* 8 (1968): 29.

75   C.R. Davies 'Effects of Gamma Irradiation on Growth and Yield of Agricultural Crops – III. Root Crops, Legumes and Grasses', *Radiation Botany* 13 (1973): 134.

76   F.B. Ellis, 'The Contamination of Grassland with Radioactive Strontium – II. Effect of Lime and Cultivation on the Levels of Strontium-90 in Herbage', *Radiation Botany* 8 (1968): 269–84.

77   C.R. Davies and D.B. Mackay, 'Effects of Gamma Irradiation on Growth and Yield of Agricultural Crops – VI. Effects on Yields of the Second Generation in Cereals and Potato', *Radiation Botany* 13 (1973): 137–44.

78 Lorna Arnold, *Windscale 1957: Anatomy of a Nuclear Accident*, second edition (Basingstoke: Macmillan, 1995), 36–7.
79 Hamblin, 'Quickening Nature's Pulse', 394.
80 A.M. Van Harten, *Mutation Breeding: Theory and Practical Applications* (Cambridge: Cambridge University Press, 1998), 17.
81 Curry, *Evolution Made to Order*, 153.
82 Zachmann, 'Peaceful Atoms', 308; Hamblin, 'Quickening Nature's Pulse', 407–8.
83 Norman W. Simmonds, *Principles of Crop Improvement* (New York: Longman, 1979).
84 Hermann Joseph Muller, 'Variation Due to Change in the Individual Gene', *The American Naturalist* 56 (1922): 15.
85 Olga Elina, Susanne Heim and Nils Roll-Hansen, 'Plant Breeding on the Front: Imperialism, War, and Exploitation', *Osiris* 20 (2005): 178–9.

9

Plants are technologies

Dominic J. Berry

Introduction

As the opening chapter of this volume makes clear, historians of technology and the environment have already shifted towards seeing nature
and technology as complexly integrated. My chapter concerns the
extent of that integration. I tackle one issue in particular, that of how
to understand the organism as technology. This question is a source
of lingering uneasiness. For example, the recent and provocative *The
Illusory Boundary* is dedicated to integrating technological and environmental history. However, the authors of its final survey chapter, Hugh
S. Gorman and Betsy Mendelsohn, while emphasising the above shift
also highlight an attendant ambiguity, that in this new scholarship 'it is
not always clear where the machine ends and nature starts'.[1] Meanwhile
co-editor of that volume, Martin Reuss, does not address organisms directly, but concludes that as a result of this work the 'Imagined boundaries between technology and environment shift, splinter, and dissolve
into meaninglessness'.[2] Given his misgivings about the organism as technology, as articulated in a 2001 email list discussion that many historians
have considered important for building the 'envirotech' space, does
his conclusion indeed hold for biological things?[3] I argue that when it
comes to organisms, historians have not reached meaninglessness, that
rendering such a distinction 'meaningless' is not really the aim, but that
we can and should analyse organisms as technologies. Doing so expands
the scope of historical enquiry by revising unhelpful assumptions while
also making historical discussions relevant to a wider (non-historical)
readership.

Writing of organisms and environments as though comparable to, or analogically related to, or essentially the same as, technological things and systems, is not inherently reductive or impoverished in comparison with other forms of analysis and writing. It absolutely can be, of course, but not out of necessity. Prejudices towards this analytical perspective are born of assumptions about the consequences of defining something as a technology. Defining something as a technology does not erase its other identities, or make it fundamentally easier to understand, other than to open up the kinds of question that might be asked and the parallel cases that might be explored. Again, that people do often define organic things as technologies in order to achieve precisely such a simplification is also true. In sum, at the same time as I am addressing concerns about understanding organisms as technologies, I am also challenging the idea that technology cannot or should not inspire the same kinds of writing, analysis or reflexivity that is more commonly directed to environmental things.[4] I would not expect any of the latter to come as a surprise to historians of technology, but for the person approaching this volume with a primary interest in the environment, I hope my unpacking is of use. Conversely, for the reader more heavily invested in the history of science and technology, I want to directly connect this volume to the social, political and scientific context in which it is published.

That plants are technologies is today a widely held position by economists, scientists, lawyers, companies and biological engineers.[5] It is also very widely held by historians, though what these different people are trying to achieve when making the case that plants are technologies rarely aligns. Even within historical scholarship there is considerable diversity in how such an argument can or should be made, and little effort to compare or synthesise accounts.[6] In part this problem is caused by the multiple ways in which technology has been defined by actors in the past and can be defined by historians and philosophers in the present.[7] When it comes to organisms, for different historians it is sometimes the case that only certain plants become technologies (by being technologised), other times plants are integrated into a technological system (leaving the plant's status unaccounted for), other times plants or animals are nature's technologies[8] (wheat as a solar-powered explosive, cows as turbo-charged milk makers[9]). Matthew Holmes in the present volume (Chapter 8) develops yet another sense of plants as technologies, looking at breeding methods and emphasising how some scientists and breeders responded to the expectations of broader industrial systems in ways that ensured their plants incorporated that same industrial ideal. Most commonly plants are made technologies simply and straightforwardly by the fact of human intervention. The latter approach is suggested in the opening chapter of this volume, channelling Edmund Russell: technologies are 'modified' environments just as 'nature has, to varying extents, been engineered … Likewise, organisms have "become tools when human beings use them to serve human ends"'.[10]

Without further clarification of what this statement is intended to mean we risk being read as supporting a narrower range of social, economic and political positions than research at the intersection of technology and environment might actually inspire. Historians cannot ignore the fact that 'plants are technologies' has not only emerged from the pursuit of their own scholarly agendas, but is also deeply bound up with political, economic and scientific developments in our times. 'Plants are technologies', or a similar sentiment, is repeated in almost ritualistic fashion within an ever expanding range of public and policy debates, be they on genetically modified organisms, agricultural industrialisation, biodiversity, biotechnology, intellectual property or the environment. Indeed for at least one section of contemporary bioscience the concession that organisms are technologies has been a founding principle, provoking reactions from ethicists, social scientists, innovation policy makers, philosophers, governments and everyone in between.[11] My chapter directly addresses the position that plants are technologies out of frustration with its ubiquity.

I will demonstrate what 'plants are technologies' can do for the intersection of technology and environment by adopting it, while also making my meaning explicit. I mean that, as with any technology, plants: (1) constitute suitable subject matter for broad debate as to their adequacy in meeting a range of social, economic and political goals; (2) are used to accrue different forms of expertise and a concomitant social status by different experts; and (3) are social and cultural artefacts. Many more specific meanings of technology could be included, these are merely the three tackled here. Do these three meanings belong only to technology, or do they apply to almost anything?[12] Perhaps they do, but then, as I am already committed to techno-environmental history-making, the breadth follows quite naturally. If one wishes to set themselves the challenge of finding 'what and only what' they can learn from study of 'the environment' or 'the technology' they are free to do so, and no doubt much of interest awaits to be discovered. But these makings of environment and technology independent of one another will still remain co-optable in a range of different social and political debates. The role of history and historical scholarship in these contemporary discussions is of immediate importance. My primary aim is to raise the game regarding 'biology as technology' and the chapter can be used as such, to be compared and contrasted with other accounts, so that vacuous gestures towards 'biology as technology' as somehow inherently meaningful are more easily spotted, so that nobody can say 'wheat is a 10,000-year-old technology' again without also having to consider all that statement's implications, and so that different political, economic and social loadings of 'biology as technology' become all the easier to identify.

The history I have selected concerns the potato in Britain at the turn of the twentieth century in the hands of farmers, breeders, state-funded investigators and a Mendelian. This study informs how different organisms found in modern Britain relate to one another, and their significance as part of the environment as totems of progress, or degeneracy, or something otherwise (on competing visions for agricultural modernity, see David Matless, Chapter 6 this volume). There are three sections dedicated to three technological themes. First, the potato's significance as a site of governance, just as other technologies are. This section is dedicated to what kinds of organism are considered suitable for inclusion in British fields and how such decisions are made. One payoff in our own time is that organic things are found to invite social, legal and regulatory intervention regardless of any designation as high technology (or 'biotechnology'), while the case also demonstrates how specific plant technology negotiations come to matter broadly by intruding on more fundamental social and political arrangements. Second, the potato's significance as a tool for making expertise, just as other technologies are. This section concerns how the environment is known and by whom. Here our historical plant technology case helps emphasise the need to open up the governance and investigation of social, technical and environmental issues.[13] Third and finally, that their significance as technical artefacts did not and does not alter their status as social and cultural artefacts, just as being a technology does not alter that status for other technologies. Here I focus on the techno-cultural significances of potatoes in terms of their commercial breeding. By the end I hope to have diminished uneasiness, or at the very least, sufficiently sabotaged the position that 'plants are technologies' for those with more narrow social, economic and political goals.

We are about to transition into the case study. Numerous reviewers warned me that it comes as something of a shift in tone following the introduction. I have not found a satisfactory solution. Perhaps then, another way to prepare you for what follows would be to say: 'So you think biology is technology? OK. Good. Let's see what that means.'

Potato governance

Plants have been recognised and dealt with as technologies suitable for the attention of the modern state for well over a century. Here I lean on Esa Ruuskanen (Chapter 2) and also build by recognising the need to look simultaneously at numerous progressive agendas, be they for the improvement of the land, or methods and tools or organisms. If we look at the UK government's primary agricultural publication, the *Journal of the Board of Agriculture*, at the turn of the twentieth century, we find that the potato, its nature and capacities were being tested by agricultural investigators working in a wide range of sites in the UK and abroad. Reports include the experimental efforts of the Agricultural Department of the Irish Land Commission spraying a mixture of copper sulphate and lime (most widely referred to as Bordeaux mixture) on potato plants as a preventive to disease, the French authority M. Girard looking into the potato's 'meat-producing value' as feed for livestock, and local reports of tests of varieties and planting methods.[14] 'The objects in view', explained a report on work undertaken by Cheshire County Council in 1898, 'were to test the productiveness, character, and yield of a number of varieties; to test the advantages of planting whole sets, cut sets, and sets of different sizes; and to try the effect of artificial manures when applied with farmyard manures.'[15] These articles are indicative of the culture of agricultural research in Britain at the turn of the twentieth century, with little in the way of nationally funded research, and in which multiple small and locally oriented sites pursued independent investigations. The Board of Agriculture (BoA), formed in 1889, played a role in disseminating the results of such initiatives, and was soon able to offer small amounts of sponsorship, awarding newly available funds to County Councils to pursue agricultural investigations and improve education.[16] Further growth in support of a system for British agricultural development followed in the coming decades, and was seized upon by those looking to institutionalise agricultural science.[17]

Around 1900 the annual potato crop in England covered somewhere between 380,000 and 400,000 acres, 27,000 to 30,000 in Wales, which when combined with Scotland amounted to a total of around 500,000 acres throughout Britain.[18] Producers could be found all across these countries, though high levels of potato farming were concentrated in England in the eastern counties, East Anglia, Yorkshire and Lancashire, while Scotland was also an essential producer of seed potatoes. As we shall see, investigations into the potato pursued by cooperative organisations, colleges and local authorities commonly addressed diseases and their outbreaks, eventually resulting in legislation dealing with the zoning and containment of organisms within infected regions, and restrictions on their sale.[19] In the case of potato wart disease, the state came to build on and eventually take over activities begun in private. Wart, sometimes referred to as Black Scab, drastically reduced potato yield, sometimes to nothing. Some of the best-recognised commercial varieties were susceptible, including the King Edward. The most important example of private initiative, and one that helps demonstrate that plants are technical levers by which different forms of governance can be applied, is that of the Ormskirk Poor Law Institution (workhouse), whose trials were eventually taken over by the BoA, expanded and reconstituted as the responsibility of a new Ormskirk Potato Testing Station. In this section plants are technologies because they are material points upon which communities, experts and the state can exert pressure, in the process redistributing power and renegotiating the environment.

The fullest available account of the origins and early work on wart at Ormskirk is an article published in 1919 by George C. Gough in the *Journal of the Royal Horticultural Society*. Some 10 years earlier Gough had visited Ormskirk as an Inspector of the Board, having been made responsible for conducting a national survey of potato wart disease. On reporting to the BoA that the disease was indeed widespread and particularly prevalent in areas such as Lancashire and Cheshire, the BoA sponsored some official trials.[20] These included variety trials, as it had already been recognised by farmers and breeders that certain varieties were immune to the disease. Indeed Gough acknowledges the role played by local farmers and breeders in organising early investigations into wart disease, as they had done at Ormskirk in partnership with the Lancashire Farmers' Association. The workhouse setting is also significant as the inmates (some ill and temporarily admitted, others itinerant and passing through) were required to perform duties around the building and grounds, including working on the farm. The farm supervisor and gardener, Preece, had around 8 acres to manage including field strips used for potato testing. For the inmates, working on the farm was just one of a number of ways in which they were expected to pay for the food and lodging they received.[21]

Cooperation between growers, breeders and representatives of the Board was characteristic of the governance of potato diseases at the outset. This locally focused and voluntarist approach would continue to matter greatly even as it came under increased pressure from two interrelated influences: the threat and eventual outbreak of the Great

War and the institutionalisation of genetics, addressed in the second section of this chapter. Building on these initial trials, steps were taken towards greater levels of state intervention regarding safe and proper potato growth, though it was immediately recognised that such legislation could be highly controversial. Decisions as to how best to deal with wart disease incorporated decisions as to proper farming, where authority might lie between individuals, County Councils and the BoA, the administration of different countries (England, Wales, Scotland and Ireland) as part of the empire or something otherwise, and ultimately of course what kinds of plant were acceptable for inclusion in the Great British landscape. Such questions of governance are already explored in historical and social scientific research into science, and can motivate historical investigation of the environment, looking at changes in understandings and valuations of nature as embodied in legislation and any attempted ameliorative measures.[22]

The first Wart Disease of Potatoes Order was issued in 1912, and would be regularly reissued or updated throughout the period to the end of the Second World War. In the first instance it required that notices be placed throughout potato growing regions through the public press, sent to allotment holders and smaller growers, explaining that it was their responsibility to report all cases of wart. With the help of BoA Inspectors, and in some cases the police, potato growing districts were subject to inspection. As the BoA described, 'This led to the discovery of a very large number of cases in allotments and cottage gardens.' They go on to write that: 'In two instances only was it necessary to take legal proceedings. In one case a workman at Perry Barr, Staffordshire, was convicted and ordered to pay costs. In another an occupier of a garden in Worcestershire was fined ten shillings and costs.'[23] That so much time and attention was directed to small growers working in their home gardens might evidence what were considered to be the necessary minimum steps to ensure wart did not spread outward from infected regions, but also perhaps the extent to which the Board sought to demonstrate to farmers that they were not the only ones who this legislation inconvenienced. Further measures then focused on ensuring that within infected regions only varieties that had been determined immune (by tests verified at Ormskirk, Harper Adams and in some cases the laboratories at Kew Gardens) were planted.[24] Here we can see how plant technologies are used to consolidate or redistribute state powers and institutional responsibilities.

In this new governmental arrangement proven immune varieties would rapidly increase in popularity in those zones scheduled as wart infected, and the process of deciding which varieties were immune – including those novel varieties annually introduced to the marketplace – ceased to be of only local interest, but became an obviously powerful gatekeeping process, one that could determine whose varieties had access to these captive consumers. From 1913 onwards then, the breeders and farmers at Ormskirk, in collaboration with a Mr John Snell (BoA Inspector stationed in Lancashire), took on the responsibility of trialling and issuing reports on immunity, organising an annual varietal showcase. We can better understand what the latter was like through a detailed newspaper report published in the *Preston Guardian* in 1920.

> This week Ormskirk has been the Mecca of men from all parts of Britain interested in the scientific as well as the commercial aspects of potato culture ... to see the Ormskirk Potato Society's annual exhibition, which is held under the joint auspices of the Lancashire and South Westmorland Farmers' Association (who have supported this work from the inception of the trials), the Ministry of Agriculture, and the National Institute of Agricultural Botany. The show is not only increasing in magnitude, but arousing wider interests, which occasions no surprise in view of the recurrent outbreaks of wart disease and the significant remark of Sir Arthur Griffith-Boscawen, Parliamentary Secretary to the Ministry, at the opening ceremony, on Wednesday, that before many years had elapsed nothing but immune varieties would be grown in the country. There were over 300 entries of immune varieties sent for competition in the 27 classes, but, as in past years, the great educational feature of the exhibition was the display of over 500 varieties grown in the Ministry of Agriculture's trials at Ormskirk showing new seedlings or old varieties not previously tested which have matured in the infected soil and been lifted without a blemish, and indicating very plainly by the offensive-looking fungoid growths the 'susceptibles' that will have no legal right to be found in the scheduled areas.[25]

That the significance of Griffith-Boscawen's prediction was seized upon is indicative of the close watch given to the potential for government intervention. It is also clear from the general description of the event, and the accompanying photograph of organisers (Figure 9.1), representing trade, farming, government and scientific institutions, that Ormskirk's success as a recognised and reliable site for varietal trialling was dependent on maintaining cooperation across all social and industrial levels of potato interest. Less clear is the extent to which any of these actors interpreted zoning legislation in terms we might recognise as environmental. Zoning

Figure 9.1 Ormskirk show organisers, 1920. Group photograph
of the Ormskirk show organisers for the year 1920, published in
the *Preston Guardian*, 30 October 1920. Some of those identified
include: front row, seated (left to right): Messrs J. Wood, Chief
Inspector of Scottish Board of Agriculture (Ormskirk judge), S.T.
Rosbotham, JP, CC (Chairman of the Ormskirk Potato Society), Sir
Arthur Griffith-Boscawen, MP, Parliamentary Secretary of the Ministry
of Agriculture, Alderman W. Fitzherbert-Brockholmes, CBE (Chairman
of the Lancashire Agricultural Committee and the Lancashire and
South Westmorland Farmers' Association) and Mr F.J. Chittenden,
Director of the Royal Horticultural Society's Gardens, Wisley (Ormskirk
judge). Back row, standing (left to right): Miss N. Whitehead (hon.
show secretary), Messrs H. Bryan, BSc, Director of the Ministry of
Agriculture's immunity trials, W. Parker, director of the National
Institute of Agricultural Botany, Dr Salaman, MD, W. Cuthbertson, JP
(Messrs Dobbie & Co.), A.C. Cole, General Inspector in charge of the
exhibition department.
Source: reproduced by kind permission of the Syndics of Cambridge
University Library. Box 18, MS Add. 8171, Salaman Archive.

measures remain to be placed in context of the longer history of environmental regulation.[26]

Pictured among the organisers of that year's show (Figure 9.1)
are four people whose significance will be dealt with in the next section
thanks to what they can tell us about the potato as a technical object. In
truth I will be focusing mainly on the first of these, R.N. Salaman (back

row, fourth from left), thanks to a wealth of archival material associated with him. The others, W.H. Parker (back row, third from left), H. Bryan (back row, second from left), and N. Whitehead (back row, first on the left), have the potential to be all the more revealing, though they will be dealt with only briefly. Incidentally, these four people are all standing together for a reason.

Potato knowing

Along with changes in governance, the early twentieth century was characterised by newly invigorated debates as to what constituted a proper potato variety. As we see in Mat Paskins' argument regarding modern British forestry (Chapter 12), different imaginaries of the state were exhibited in different practices of knowledge and calculation, occurring at both local and national levels. In this section, rather than beginning with institutions for the poor, we instead find a Mendelian working at home in his own private gardens in Barley, Hertfordshire. There, Dr Redcliffe Nathan Salaman, MD, began in the early 1900s (at almost exactly the same time as Ormskirk began to pursue its first variety trials) investigations into potato varieties, diseases and breeding. He had no botanical training, initially having trained as a medical doctor. Soon after earning this qualification, however, he contracted tuberculosis, moving out of London to the countryside for the benefit of the country air (see Jennifer Wallis, Chapter 5, for the history of the garden in the medical machine). In the process he switched to investigations of breeding. What Salaman might tell us about the early history of genetics and the environment has not yet been explored, though some time ago Paolo Palladino noted the uniqueness of his 1949 *The History and Social Influence of the Potato*, which contains a number of different arguments and theories as to the relations between humans, food supply and their surroundings (including theories about the potato as a means for keeping Ireland in a state of peasantry).[27] He is a prime candidate for examination in histories of UK agriculture thanks to his prodigious output, the positions that he came to adopt in key agricultural science institutes (including the National Institute of Agricultural Botany and the Potato Virus Research Station), his arguments on race and eugenicism, and because through him we can reach agricultural development and its meanings across the British Empire and at other crucial sites, particularly the Middle East and the state of Israel at its founding. None of this can be covered here. It is simply enough to say that thanks to his work on the potato Salaman entered into correspondence with people invested in different kinds of agricultural improvement all around the world, and took on particular responsibilities in the Middle East through the Hebrew University, the laying of the foundation stone for which he personally witnessed as an enlisted man in the British 39th (Labour) Battalion – or 2nd Battalion of Judeans – shipped to Palestine in 1918. Techno-environmental histories of Britain can therefore also be of other places. In this section I instead focus on Salaman's broad interests in genetics. Aside from being among the first six people to be published in the new *Journal of Genetics* upon its establishment in 1910, Salaman was also its first author to write on human genetics, on a Mendelian view of Jewish heredity. Through claims to this kind of expertise over heredity and breeding – claims that were by no means exclusive to geneticists – the potato was remade as a newly technical object.

According to Salaman, he first contacted William Bateson around 1905 after a couple of years convalescing in Barley, 30 miles or so from Cambridge.[28] Initially Bateson recommended he use animals for breeding and genetic study, including mice, guinea-pigs and double-combed fowls, before eventually Salaman approached his own gardener, 'a man of stately mien who, looking down on me from his 6 ft 2 ins said that if a gentleman in my position *must* use his spare time in playing about with vegetables, he would advise the POTATO'. Some of his surviving breeding notebooks are available for 1910 onward. By putting these records in dialogue with stocks of potatoes grown at his house and in field sites across England, Scotland and Ireland, circulating these potatoes among those in the know, Salaman built his reputation as a potato expert.

Rather than moving towards any particular potato end it seems Salaman simply focused on the systematic production of novelty, perpetuating anything unusual that appeared in his garden and selecting multiple plants for cross-breeding. Of this work the best-known result is his discovery of potatoes resistant to blight.[29] Aspects of his method are encapsulated in Figure 9.2, including the dependence on potato varieties acquired from countries around the world, as seen in the 'Congo' potato included at bottom right. This is just one example of many dozens of photographs that are affixed to the pages of his notebooks, photographs that also provide an example of Agar's sixth and seventh tech-enviro forms in combination (Chapter 1). Photographs allowed Salaman to switch back and forth between the tabulated information about individual plants and the tubers they produced over time. In the tables he records their names, such as H19 (pictured near the centre of

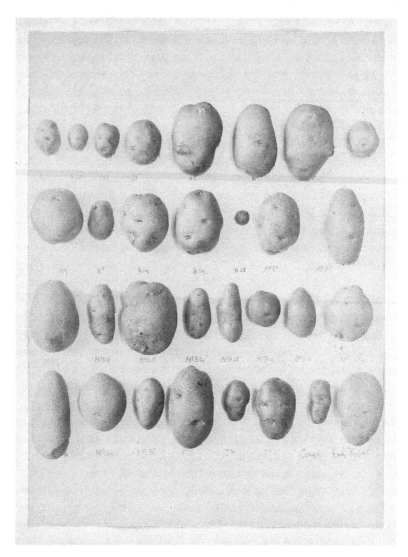

Figure 9.2   Example of a potato experiment photograph from Salaman's 1910 notebook. Photograph taken of potatoes grown by Salaman in his genetical research.
Source: reproduced by kind permission of the Syndics of Cambridge University Library, and thanks to Jane Miller and Nina Wedderburn for agreeing copyright permissions for its publication. 'Potato Harvest 1910', Brown Folder, Folders and Volumes 1, MS Add. 8171.

Figure 9.2), alongside the colour tuber produced (red, white, black), shape (long or round) and the number of eyes found on it. These are precisely the kinds of characteristics that were being investigated by eager Mendelians around the world at this time, and we can see evidence of Salaman thinking through his productions in Mendelian terms as he evaluates them. Against H24, for example, he records the shape as round but places a question mark alongside it, noting 'Shapes are fairly close rounds but not typical. The no. of eyes is against round.' Here 'against' meant that the high number of eyes (in this case 11) would suggest the hereditary unit characteristics were those of something other than a round variety. He is playing with the notion that a high number of eyes is correlated with long-shaped tubers.

Over the years, the extent of his variation making and the complexity of his crosses expanded. He also sought out opportunities to share his productions with interested parties, including owners of farming estates, such as Walter Wooll West of Needham Hall, Wisbech and proprietors of the UK's most influential seed businesses, such as Martin Sutton of Suttons Seeds, along with official representatives of state-managed trialling stations, such as M. Caffrey of the Cereal Station in Ballinacurra, Ireland. Remarkably, given the relationship between trialling and convalescence described earlier at the Ormskirk workhouse, a second though vastly more lavish medical site played an important role in Salaman's own work. Presumably thanks to connections made through his medical training, Salaman had convinced Dr Charles Easterbrook, Physician Superintendent at the Crichton Royal Institution in Dumfries, to grow on stocks of his varieties in Scotland. The Crichton was a world-famous asylum for lunacy, run according to new principles described by medical historians as 'moral treatment, under medical control', with large gardens and considerable air.[30] Salaman had around three hundred distinct varieties grown on and under observation at the Crichton, making that hospital a resource of considerable importance. Through places like the Ormskirk workhouse and the Crichton Royal our account of plant technologies bolsters themes covered elsewhere in this volume while drawing in starkly different social contexts. The differences between these contexts will come to matter in the third section of this chapter.

A quick look at Salaman's correspondence reveals how he circulated potato crosses among influential agriculturalists, potatoes that were named and organised according to his own record-keeping schemes, at the same time retaining control (or at least asserting control) over the uses to which they could be put. In this respect his technical knowledge of the plant was at one and the same time a means of establishing intellectual property and organising social relations.[31] In a letter to Sutton he writes:

> I think on the whole, the plan you suggest would be best, that I should let you have 7lbs of each of the varieties to test and then, if you think anything of any of them you will be able to say so and offer a price next year. At the same time they will be grown in Scotland under measured conditions, and I shall, of course, naturally retain the right of showing them to anybody else.[32]

If we look at the list of materials that Salaman then arranged to have sent to Sutton, these included the products of crosses such as 'M5B32 x fb2', 'H2 x ER2' and 'M5B18 x Wat.24'.[33] Even if Salaman had sent his complete records it would have been a job of work to understand what any of these crosses meant or the varieties from which they descended. Throughout the 1910s he continued to organise and manage these disparate stocks by letter, all the while becoming ever more embroiled in war work, initially as a medical officer signing off men as fit to fight before eventually joining the Royal Army Medical Corps. In 1917 he wrote to West of Wisbech that the varieties then being sent to him 'are the result of eleven years experimental breeding during which time I have raised well over a quarter of a million seedlings'.[34] Of course not all of these seedlings had been raised by Salaman directly; he had had considerable help from Will at the Crichton, and his own assistants at home.

Either side of the Great War, which had placed increased significance on domestic agricultural production, the credit that Salaman and other geneticists like him had accrued through their research and promises of agricultural improvement was rewarded with an expansion of funds available for the institutionalisation of agricultural science and genetics, primarily accomplished through grants from the Development Commission (DC).[35] This brings us to the other three persons highlighted in the Ormskirk show photograph: Parker, Bryan and Whitehead.

All were stood together with Salaman thanks to their being involved with Cambridge-based agricultural botany: Salaman through his early work with Cambridge University geneticists and eventually, from 1927, as Director of the Potato Virus Research Station established in Cambridge on DC funds; Wilfred H. Parker as the first Director of the National Institute of Agricultural Botany (NIAB), established in 1919 in Cambridge on DC funds; Harold Bryan, the Superintendent who took over from John Snell when the Ormskirk trials were taken over by NIAB in 1919 resulting in the establishment of the Ormskirk Potato Testing Station; and Nora Whitehead, assistant to Bryan at Ormskirk and a significant potato authority who would later author (among other things) the widely used and repeatedly reprinted *Key for the Identification of Commercial Potato Varieties and Rogues in the Field*.[36] It is important to stress how significant the initial research and commercial trialling begun at Ormskirk had been in gathering support for such nationally funded agricultural science institutes in the first place. What I argued and explained above for Salaman can be extrapolated and expanded many times over for the collective community of breeders (some Mendelian, many not), who each through the manipulation, analysis, circulation and sale of plants had built a number of different overlapping communities of expertise, each ostensibly dedicated to plant improvement.[37] Recognising how linked their expertise was to control of physical plant material is one result of exploring plants as technical artefacts embodying knowledge. If one were to trace the working practices of persons such as Bryan, who died shortly before the Potato Testing Station was closed in 1940, or even more importantly that of Whitehead, one of a number of women who made successful scientific careers through agricultural plants, we would get a sense of how and in what ways the understanding and management of the British environment has developed in techno-environmentalist mode.[38]

Potato culture

Having established the range of persons and interests in the potato, both as a means of governance and as a technical artefact, this section turns to demonstrate that being a technology does not place an object outside society and culture. Again, this will not come as a surprise to historians of technology, though few of them have extended their arguments thus far to encompass biological things (see my note 6 for a list of exceptions). The work of Barbara Hahn in *Making Tobacco Bright* is an important exception, explaining how different kinds of production methods, and different kinds of division of labour throughout tobacco growing, harvesting, preparing, packaging and trading – often depending on the use of different kinds of technology and technique – were constitutive of the plant product. Different kinds of society and culture are made through and with tobacco plants and potatoes, just as commodity historians might focus on how different kinds of tobacco plants and potatoes make up societies and cultures. We have already glimpsed this with regard to people being fined for not growing the correct potatoes, making them criminals, and people building analytical schemes and reference systems around the potato, making them experts. This can all be brought together by attending to the relations between our two most important sites, Ormskirk and Cambridge, and the societies and cultures of breeding that they represented.

From the outset, Ormskirk was considered a problematic kind of research site by geneticists in Cambridge, including Rowland Biffen of the PBI, Salaman and towards the end of the interwar period, even its new owner NIAB. In part these complaints were about geographies of power, and the desire to have as much as possible centralised around the various different headquarters in Cambridge. In addition, Ormskirk was not considered the most sophisticated research site (it was equipped only with an 'Elsan closet' toilet that was not replaced until the 1960s).[39] Most importantly, complaints were also about how potato breeding should be organised, according to whose expertise and on what principles the potato growing industry should be taken forward. In pulling this story together it becomes impossible to distinguish between when the potato is technology, environment or culture.

The earliest evidence that not all was well with relations between Cambridge and Ormskirk comes from a letter from Rowland Biffen to Lawrence Weaver, two people central to the founding of NIAB.[40] Biffen mentions 'I must manage Ormskirk somehow or other this time – if only to carry wart disease spores back on my boots & so start a fresh centre here.'[41] Where this is said in passing, much more pointed was Salaman in a letter to Sutton, explaining that he finds the role played by Ormskirk to be galling.

> I will certainly keep H2 x E2 B2 for a fortnight till you come back and I don't want to deal with the man I spoke to you of till I know who he is. If I seemed to be disappointed it really is not so much that, as I felt a little irritated about this seedling. I do not think there is any doubt – and it was the general opinion at Ormskirk that it is probably one of the very best Earlies that have ever been seen, it has been going nine years now, not only you have seen it but Matthew Wallace and heaps of practical people have seen it in Scotland year after year, said it is nice, etc. and it is only when it receives official recognition at Ormskirk and is incidentally declared to be susceptible, that its real virtues are recognised. As a mere scientific worker I recognised its merits seven or eight years ago.[42]

In Salaman's own self-representation there is a clear distinction between scientific work and whatever goes on at Ormskirk, with implications for the who and how of potato assessment. Ormskirk coming under the operation of NIAB in 1919 did not necessarily settle Salaman's difficult relations with the rest of the trade as represented here, or indeed with NIAB's own enterprises, as the following two episodes reveal.

One of the first challenges that Salaman set himself once he had the full resources of NIAB at his disposal was what he considered to be the problem of synonyms. As much as it may have perplexed the likes of Salaman, the problem of synonymity has been of considerable use to historians. Synonymity is the identifying or selling of a plant of one variety under the name of another variety. The practice was common and represents aspects of a culture of innovation that recognised multiple ways in which value could be added, often collectively, to a variety.[43] For Salaman though, the problem of synonymity was an ideal way in which to assert his status as a potato-identifying centre of power, and also the primacy of a genetic interpretation of plants. Establishing the Potato Synonym Committee at NIAB, Salaman set about creating the infrastructure for the annual assessment of potato varieties and their comparison with one another, to determine (at least to his own satisfaction) when a novel variety sent to NIAB was truly novel and when it was simply the same as another. While no legislation followed to intervene, Salaman did arrange for the state-funded and *National* Institute of Agricultural Botany to circulate notices in national newspapers when seed traders and breeders were stocking what his Committee had declared to be a synonym. These activities were not received well by established growers, the evidence for which is both remarkable and enjoyable.

In a series of articles and letters sent to the *Nurseryman and Seedsman* correspondents and authors complained, got angry and mocked Salaman for his intervention on behalf of true potatoes. Their displeasure became acute following the 1922 Ormskirk show, when Salaman was given the podium. 'It is all very well for Mr R.N. Salaman, JP, MD, to get up on his legs at Ormskirk', wrote the author of the *Nurseryman* 'By the Way' segment, 'and lecture the seed trade upon its "deplorable and dishonest practices", but if he were in the seed trade and not in the medical profession, he would very probably sing quite a different song.'[44] In the following issue another author, W. Cuthbertson, lamented how Salaman's whole approach undermined the cooperative foundations of Ormskirk. He reminded readers that it was traders and farmers who began the early trials with John Snell, which when later taken over by NIAB, continued to be assisted by the trade who lent the

Station their finest potato experts.[45] In the next issue Salaman had a poem dedicated to him, 'Do Tell', which made fun of his performance at Ormskirk and ended by again stressing his merely medical credentials. The culmination though came in the Christmas issue, in the form of a one-act play titled 'Synonyms'. In the play a gardener, Will Orange, is down the pub boasting about his line of work and his knowledge of potatoes. After a little back and forth about synonyms, his companions, the grocer, landlord, butcher and tailor, entice him to explain:

'A Potato is a Potato, and a synonym is another one morpho-, er- wait a bit while I look that up. Yes, here it is – morphologically and physiologically identical with it. That's what a synonym is, and as nearly everybody grows synonyms, it stands to reason that they are much cheaper than the real thing'.

'But how do you know which is the real thing?' said Harry.

'Ah! That's where art comes in. It's an art that's a cut above the ordinary – in fact, I would venture to say there are only two people who can speak with authority on the subject. One's a celebrated London doctor, and the other's myself. The doctor's made a lifelong study of Potatoes, and he knows so much more about them than any-body else, that there's nobody can tell him when he's right or wrong. Consequently he's always right'.

'And how about yourself?' said Dicky.

'Well, I know so much more about the subject than anybody else in this locality, that what I says here abouts "goes" too. See!' responded Will.[46]

These sources evidence an internally coherent society of breeding with its own standards of best practice and expectations as to what kind of relationship with fellow breeders and traders is necessary in order to imbue a person or institution with authority. They also capture that these arrangements as they had been understood were now becoming subject to change. This play even evidences the remaking of national or localised centres of authority, as between 'London' and what 'goes' in any given place. Lastly, their incredulity at how one could even tell the difference between potatoes that look the same in every respect (beyond the skills of any normally trained farmer or breeder) also highlights the importance of the technical component of the argument being made by geneticists. Salaman, approaching the issue of synonyms away from the kinds of society that invested in Ormskirk, and working from the perspective of the new Mendelians based in Cambridge, had intruded upon that culture on behalf of his own potato technologies and culture.

Lest potato identification and assessment be taken as too classic an example of an emerging rational modernity imposing itself on a pre-existing 'natural' order, my final example demonstrates that Salaman did, on at least one occasion, also throw over that same rationalising process either out of carelessness, or in pursuit of an even larger demonstration of his authority and technical skill, or perhaps out of ignorance of the social aspects of technical knowledge. At some time during the early years of the first Ormskirk trials, while John Snell was still involved in their coordination, the Lord Derby Gold Medal competition was begun. Awarding of the Medal became part of the annual show. It was organised by a local committee of breeders and farmers, though eventually also became the responsibility of NIAB. In 1928 Salaman caused a major embarrassment. That year the Gold Medal Committee decided to award the medal to '520', a variety bred by Donald MacKelvie, famed breeder of the Arran varieties. Unbeknownst to the rest of the Committee (of which Salaman was a member), Salaman had a few weeks earlier written to MacKelvie telling him to drop the variety. On hearing about Salaman's letter Bryan wrote to the NIAB's secretary F.C. Hawkes. 'To my mind it is an incredible happening,' Bryan wrote. 'Salaman is a member of the Gold Medal Committee, he is Chairman of the Committee responsible for the Institute's potato work and he is also Chairman of the Institute itself: you may therefore imagine the importance MacKelvie attaches to any letters from Salaman.' It was not only, or even mainly, Salaman's authority that agitated Bryan, rather how it all looked to MacKelvie who he called 'the most important man in potato world'. He closed his letter by stating:

> The reputation of the Institute, even for sanity, and the future conduct of the trials are surely involved. In view of the gravity (unless my judgement is completely at fault) of this note, I should be very glad to know what Parker [NIAB Director] intends to do as soon as possible. If the award goes through, then MacKelvie must have some explanation of the extraordinary position … The more I think of this the more mad it seems. The Gold Medal Committee bestow on 520 a Gold Medal for outstanding merit, and the Chairman of the Institute's Potato Committee advises the owner to scrap it.[47]

Technical knowledge has to be *shown* to work, and is shown to work through social organisation. Bryan feared a collapse. Salaman's potatoes were very different from MacKelvie's, circulating in two very different social worlds, each producing different kinds of agricultural governance (Salaman's more technocratic and individualist, as seen in the newspaper notices shaming synonym growers, MacKelvie's and other breeders' more voluntarist and communal) and different kinds of expertise. Salaman and other geneticist breeders were also different from the rest of the 'potato world' thanks to their broader biological interests. Salaman did, for instance, consider knowledge of potato heredity as of a piece with human heredity in ways that commercial breeders might have speculated on but rarely turned into the subject of writing. Recognising this offers one final insight as to how analysis of plants as technologies helps produce histories of the techno-environment.

As with the majority of the first geneticists, Salaman saw in genetics answers to societal problems well beyond the agricultural. As I have explored elsewhere, emergent views of heredity in the early twentieth century sat extremely comfortably with – and fed off – a range of other goals and values, whether they be in architecture, design, industrial manufacture or public health.[48] Eugenics was part and parcel of British culture and in this much Salaman was no different, giving talks on the subjects of eugenics and public health and writing for publications such as the *Eugenics Review*. This observation is worth including so that we know such themes will permeate histories of British environmental change, either through policy and governance, through the biological objects that make up the environment and industry, or through social worlds of agriculture and environmental management, right up to the present day.

Conclusion

I bring this chapter to a close with a passage that brings us full circle to the argument that plants are technologies. While farmers and breeders had had to make do with the pages of *Gardeners' Chronicle*, in 1930 Salaman was given the full weight of the BBC. Broadcast on 14 January, radio listeners heard some of the earliest material from what would later become his full-length historical treatment of the potato, here titled 'The History and Economic Influence of the Potato'.

> The chipped flint, the potter's sherd, the invaluable evidence of mass progress before the days of the sculptured stone or the inscribed word, are not the only records of our ancestors' victory over their environment. Man has stamped the impress of his genius no less on the living than on the inanimate. The records are scanty, the gaps

therein are many and great, but the end results – the bending of the plant and animal world to his own uses – are the outstanding achievement of prehistoric man.[49]

This is by no means the earliest example of this kind of statement, though it is doubtless among the first to be broadcast on the BBC. It has contributed to a formula of representation and understanding that is today widespread and seemingly inescapable. How much of the meaning of his talk as understood by his audience in 1930 would match that of audiences today? Is it disquieting that scientists and other experts have been saying much the same thing for nearly a century? Does bending nature to man's will, making biology into a technology, carry all that much significance in and of itself? Instead of allowing the suggestion of parity between the animate and inanimate to shock, excite or perplex us, perhaps we should shrug off claims to novelty and control, and get on with negotiating the who, when and why of potato growing. From a study of Salaman's career we can at least recognise that the extent of any bending of plants and animals to human will has always first and foremost required the bending of persons.

There are many ways that one might tell environmental history, and ways in which environments can be historicised with and through technology. In this chapter I have addressed the techno-environments of spore-filled fields in Ormskirk, of well-recorded gardens in Barley and of circulating potatoes themselves. Breaking out of analytical tropes regarding the natural historical and the synthetic, the biological and technological, does not have to be done in response to contemporary biological engineering, though I have written this chapter with such an audience directly in mind. Here I have attempted to provide part of the platform for new techno-environmental histories of Britain in the form of suggestions and recommendations regarding how to conceptualise organisms in fields, streams, forests and the air, by refusing to look for only certain kinds of qualities in plants and animals and other kinds of qualities in fences, bridges, paths and pylons.[50] All of the unparseable organic and technological are full to the brim with social and cultural meaning, the sum total of which will make up techno-environmental history.

Acknowledgements

This research was supported by the European Research Council through Consolidator Grant (616510-ENLIFE) and a Research Incentive Grant from The Carnegie Trust for the Universities of Scotland (70589). The latter provided for the time of a Research Assistant, Dr Thokozani Kamwendo, whose contribution in data collection and discussion of the research was invaluable. My thanks also to Pablo Schyfter for making me think about plants as technologies, Gill Haddow for encouraging its exploration, David S. Ingram for sharing so much enthusiasm for these topics and hours of botanical chat, and Jane Calvert, Deborah Scott, Berris Charnley, Mat Paskins and Sarah Wilmot for providing feedback and much needed criticism of earlier drafts. In this same respect my thanks go to Ralph Harrington, Jacob Ward, Jon Agar and all the participants of UCL's Technology and Environment in Modern Britain workshop. Images reproduced by kind permission of the Syndics of Cambridge University Library.

Notes

1   Hugh S. Gorman and Betsy Mendelsohn, 'Where Does Nature End and Culture Begin?', in *The Illusory Boundary: Environment and Technology in History*, ed. Martin Reuss and Stephen H. Sutcliffe (Charlottesville: University of Virginia Press, 2010), 277.

2   Martin Reuss, 'Afterword', in *The Illusory Boundary: Environment and Technology in History*, ed. Martin Reuss and Stephen H. Sutcliffe (Charlottesville: University of Virginia Press, 2010), 298.

3   'Are Animals Technology?' – archive of messages posted on Envirotech, Envirotech@lists. Stanford.EDU, 2001. Last accessed 16 August 2016. http://envirotechweb.org/wp-content/ uploads/2007/05/animaltech.pdf

4   Though not addressing organic material directly, Timothy Mitchell's formulation of agency in relation to environmental features such as the Aswan Dam have greatly influenced the work presented here. Timothy Mitchell, *Rule of Experts: Egypt, Techno-Politics, Modernity* (Berkeley and Los Angeles: University of California Press, 2002). My sincere thanks to Deborah Scott for recommending Mitchell to me.

5   Many examples could be cited, but the following are useful for their breadth of origins and intended audiences: Robert H. Carlson, *Biology is Technology: The Promise, Peril and New Business of Engineering Life* (Cambridge, MA: Harvard University Press, 2011); Richard Dawkins, *The Selfish Gene* (Oxford: Oxford University Press, 1976); Nuffield Council on Bioethics, *Genome Editing: An Ethical Review* (London: Nuffield Council on Bioethics, 2016), 4; William Hoffman and Leo Furcht, *The Biologist's Imagination: Innovation in the Biosciences* (Oxford: Oxford University Press, 2014), 83.

6   Helen Curry, *Evolution Made to Order: Plant Breeding and Technological Innovation in Twentieth-Century America* (Chicago: Chicago University Press, 2016); Barbara Hahn, *Making Tobacco Bright: Creating an American Commodity, 1617–1937* (Baltimore: Johns Hopkins University Press, 2011); Dolly Jørgensen, 'Not By Human Hands: Five Technological Tenets for Environmental History in the Anthropocene', *Environment and History* 20 (2014): 4; Edmund Russell, 'Can Organisms Be Technology?', in *The Illusory Boundary: Environment and Technology in History*, ed. Martin Reuss and Stephen H. Sutcliffe (Charlottesville: University of Virginia Press, 2010), 249–62; Susan R. Schrepfer and Phillip Scranton (eds), *Industrializing Organisms: Introducing Evolutionary History* (New York: Routledge, 2004); Tiago Saraiva, *Fascist Pigs: Technoscientific Organisms and the History of Fascism* (Cambridge, MA: MIT Press, 2016).

7   For a historiographical overview, with references to articles that explore and offer definitions, see David Edgerton, 'Innovation, Technology, or History: What is the Historiography of Technology About?' *Technology and Culture* 51, no. 3 (2010): 680–97.

8 In referring to these as 'nature's technologies' I am inspired by the category of 'nature's experiment' as defined and developed by Mary Morgan, 'Nature's Experiments and Natural Experiments in the Social Sciences', *Philosophy of the Social Sciences* 43 (2013): 3.

9 Edmund Russell, James Allison, Thomas Finger, John K. Brown, Brian Balogh and W. Bernard Carlson. 'The Nature of Power: Synthesizing the History of Technology and Environmental History', *Technology and Culture* 52 (2011): 2; Barbara Orland, 'Turbo-Cows: Producing a Competitive Animal in the Nineteenth and Early Twentieth Centuries', in *Industrializing Organisms: Introducing Evolutionary History* (New York: Routledge, 2004).

10 Jon Agar, 'Historiography and Intersections', Chapter 1 this volume.

11 The literature on synthetic biology, particularly from within science and technology studies, is extensive. Starting points include Andrew S. Balmer, Katie Bulpin and Susan Molyneux-Hodgson. *Synthetic Biology: A Sociology of Changing Practices* (Basingstoke: Palgrave Macmillan, 2016); Alexandra Daisy Ginsberg et al., *Synthetic Aesthetics: Investigating Synthetic Biology's Designs on Nature* (Cambridge, MA: MIT Press, 2014); Evelyn Fox Keller, 'What Does Synthetic Biology Have to Do with Biology?', *BioSocieties* 4 (2009); Markus Schmidt et al., *Synthetic Biology: The Technoscience and its Societal Consequences* (Dordrecht: Springer, 2009); Technology Strategy Board, *A Synthetic Biology Roadmap for the UK* (Swindon: Technology Strategy Board, 2012).

12 My thanks to Mat Paskins for drawing this implication out in a discussion of how our chapters relate.

13 Andy Stirling, '"Opening Up" and "Closing Down": Power, Participation, and Pluralism in the Social Appraisal of Technology', *Science, Technology, & Human Values* 33 (2008): 2.

14 *Journal of the Board of Agriculture* (September 1894): 25–8, 14–16, 54.

15 *Journal of the Board of Agriculture* (December 1898): 354–7. A set of potato seed is 'cut' when, as one might guess, the seed is cut and planted as two or more independent seeds.

16 Stewart Richards, '"Masters of Arts and Bachelors of Barley": The Struggle for Agricultural Education in Mid-Nineteenth-Century Britain', *History of Education* 12 (1983): 3; Stewart Richards, 'The South-Eastern Agricultural College and Public Support for Technical Education, 1894–1914', *Agricultural History Review* 36 (1988): 2; Paul Brassley, 'Output and Technical Change in Twentieth-Century British Agriculture', *The Agricultural History Review* 48 (2000): 60–84.

17 George Cooke (ed.), *Agricultural Research, 1931–81* (London: Agricultural Research Council, 1981).

18 Figures for England and Wales: Edward J.T. Collins and Joan Thirsk, eds, *The Agrarian History of England and Wales, Vol. 7, 1850–1914* (Cambridge: Cambridge University Press, 2000), 1772. Combined figures for Scotland and Wales in Christabel S. Orwin and Edith H. Whetham, *History of British Agriculture 1846–1914* (Newton Abbot: David & Charles, 1971), 350–1.

19 For the longer history of such measures, though with a focus on animals, see Abigail Woods, 'Partnership in Action: Contagious Abortion and the Governance of Livestock Disease in Britain, 1885–1921', *Minerva* 47 (2009): 2; Abigail Woods, 'A Historical Synopsis of Farm Animal Disease and Public Policy in Twentieth Century Britain', *Philosophical Transactions of the Royal Society, B* 366 (2011): 1943–4.

20 George C. Gough, 'Wart Disease of Potatoes', *Journal of the Royal Horticultural Society* 45 (1919): 301–12.

21 Betty Underwood, *Ormskirk Workhouse, Two World Wars, NHS Hospital* (Accrington: Nayler Group, 2007).

22 Sheila Jasanoff, *Designs on Nature: Science and Democracy in Europe and the United States* (Princeton: Princeton University Press, 2007); Don Leggett and Charlotte Sleigh, eds, *Scientific Governance in Britain, 1914–79* (Manchester: Manchester University Press, 2016).

23 Board of Agriculture and Fisheries (1914), 35.

24 Board of Agriculture and Fisheries (1914), 36.

25 Anon., 'The Ormskirk Potato Show', *Preston Guardian* (30 October 1920).

26 Peter Brimblecombe, *The Big Smoke: A History of Air Pollution in London since Medieval Times* (London: Methuen, 1988); John Sheail, 'Government and the Perception of Reservoir Development in Britain: An Historical Perspective', *Planning Perspectives* 1 (1986): 1; John Sheail, 'River Regulation in the United Kingdom: An Historical Perspective', *River Research and Applications* 2 (1988): 3. My thanks to the reviewers for suggesting these works as providing a valuable foundation to histories of environmental regulation.

27 Paolo Palladino, *Plants, Patients and the Historian* (Manchester: Manchester University Press, 2002).

28 Redcliffe N. Salaman, 'Half a Century of Potato Research', *Proceedings of the Second Conference on Potato Virus Diseases Lisse-Wageningen, 25–29 June* (1954).

29 Kenneth M. Smith, 'Redcliffe Nathan Salaman. 1874–1955', *Biographical Memoirs of Fellows of the Royal Society* 1 (1955).

30 Michael A. Finn, 'The West Riding Lunatic Asylum and the Making of the Modern Brain Sciences in the Nineteenth Century' (PhD diss., University of Leeds, 2012), 42.

31 Christine MacLeod and Gregory Radick, 'Claiming Ownership in the Technosciences: Patents, Priority and Productivity', *Studies in History and Philosophy of Science Part A* 44, no. 2 (2013): 188–201; Berris Charnley and Gregory Radick, 'Intellectual Property, Plant Breeding and the Making of Mendelian Genetics', *Studies in History and Philosophy of Science Part A* 44, no. 2 (2013): 222–33.

32 Letter to Martin H. Sutton from R.N. Salaman, 9 November 1914, MS Add. 8171, Box 3, University Library, University of Cambridge.

33 Letter to Will (Gardener at the Crichton) from R.N. Salaman, 10 November 2014, MS Add. 8171, Box 3, University Library, University of Cambridge.

34 Letter to Walter Wooll West from R.N. Salaman, 13 March 1917, MS Add. 8171, Box 3, University Library, University of Cambridge.

35 Robert Olby, 'Social Imperialism and State Support for Agricultural Research in Edwardian Britain', *Annals of Science* 48, no. 6 (1991): 509–26. For a recent account unpacking the history of UK genetics in terms of its role in the governance of agriculture, see Berris Charnley, 'Geneticists on the Farm: Agriculture and the All-English Loaf', in *Scientific Governance in Britain, 1914–79*, ed. Don Legget and Charlotte Sleigh (Manchester: Manchester University Press, 2016), 181–98.

36 After marriage Nora McDermott. McDermott (1948).

37 For a recent survey of the historiography of genetics in agricultural plant breeding, demonstrating the breadth of responses to Mendelism and the bounds placed on its usefulness for breeding, see Jonathan Harwood, 'Did Mendelism Transform Plant Breeding? Genetic Theory and Breeding Practice, 1900–1945', in *New Perspectives on the History of Life Sciences and Agriculture*, ed. Denise Phillips and Sharon Kingsland (Heidelberg, New York, Dordrecht, London: Springer, 2015), 345–70.

38 Another excellent case that should be pursued is that of Mary D. Glynne, MSc, who while working at Rothamsted Experimental Station developed tests for susceptibility to wart that could be conducted indoors rather than relying on a field space. Mary D. Glynne, 'Infection Experiments with Wart Diseases of Potatoes. Synchytrium Endobioticum (Schilb) Perc', *Annals of Applied Biology* 12, no. 1 (1924): 34–60. Her 'pot experiment' was soon adapted by McDermott at Ormskirk, though reported on by Bryan who records his thanks to McDermott as being 'largely responsible for the technique employed'. Harold Bryan, 'Wart Disease Infection Tests', *The Journal of Agricultural Science* 18, no. 3 (1928): 507–14. On women and scientific careers in genetics, see Marsha Richmond, 'Women in the Early History of Genetics: William Bateson and the Newnham College Mendelians, 1900–1910', *Isis* 92, no. 1 (2001): 55–90.

39 Minutes of 164th meeting of the Executive Committee, 14 June 1963, E-5.9, NIAB Archive, Huntingdon Road, Cambridge.

40 The founding and first 50 years of NIAB's history are the subject of Dominic J. Berry, 'Genetics, Statistics, and Regulation at the National Institute of Agricultural Botany 1919–1969' (PhD diss., University of Leeds, 2014). Much greater attention to the role of Rowland Biffen in the growth of UK genetics more broadly can be found in Berris Charnley, 'Agricultural Science, Plant Breeding and the Emergence of a Mendelian System in Britain, 1880–1930' (PhD diss., University of Leeds, 2011); Berris Charnley, 'Experiments in Empire-Building: Mendelian Genetics as a National, Imperial, and Global Agricultural Enterprise', *Studies in History and Philosophy of Science Part A* 44, no. 2 (2013): 292–300. The later years of NIAB from 1970 onwards are the subject of Matthew Holme's PhD thesis, and Matthew Holmes, 'Crops in a Machine: Industrialising Barley Breeding in Twentieth-Century Britain', Chapter 8 this volume.

41 Letter from R.H. Biffen to R.N. Salaman, 11 October 1918, C-5.10 Donations and Subscriptions 1917–1921, NIAB Archive, Huntingdon Road, Cambridge.

42 Letter to Martin H. Sutton from R.N. Salaman, 25 August 1919, MS Add. 8171, Box 3, University Library, University of Cambridge.

43 Dominic J. Berry, 'The Plant Breeding Industry after Pure Line Theory: Lessons from the National Institute of Agricultural Botany', *Studies in History and Philosophy of Science Part C: Studies in History and Philosophy of Biological and Biomedical Sciences* 46 (2014): 25–37.

44 Anon., 'By the Way', *Nurseryman and Seedsman* (16 November 1922): 23.

45 W. Cuthbertson, *Nurseryman and Seedsman* (23 November 1922): 25.

46 Anon., 'Synonyms', *Nurseryman and Seedsman* (21 December 1922): 27.

47 I am deeply grateful to Tricia Cullimore for finding this letter during her own archival research and passing it on to me. I unfortunately cannot confirm the location of the letter in this archive. From its contents, it would likely be included in T-3.19, Potato Committee Papers and Ormskirk Committee Papers, NIAB archive, Huntingdon Road, Cambridge. A PDF can be supplied.

48 Dominic J. Berry, 'Agricultural Modernity as a Product of the Great War: The Founding of the Official Seed Testing Station for England and Wales, 1917–1921', *War & Society* 34, no. 2 (2015): 121–39.

49 'The History and Economic Influence of the Potato', MS Add. 8171, Box 13, University Library, University of Cambridge.

50 My reference to pylons here picks up on a line of research undertaken by the historian and literary theorist James Purdon, who has looked at artistic responses to the arrival of the electricity network across Britain in the interwar period. Thus far he has only published a small section of this research: James Purdon, 'Landscapes of Power', *Apollo Magazine* (21 December 2012). Last accessed 5 April 2017. https://web.archive.org/web/20130121141715/www.apollo-magazine.com/features/7920953/landscapes-of-power.thtml. James Purdon, 'Electric Cinema, Pylon Poetry', *Amodern* (October 2013).

10

Oceanscapes and spacescapes in North Atlantic communications

Jacob Ward

The decades following the Second World War saw significant advances for international telecommunications. In 1956, TAT-1, the first ocean-spanning telephone cable, crossed the Atlantic, linking the USA, Canada and Britain. Six years later, in 1962, Telstar, one of the first active communications satellites, was launched by NASA, linking earth stations in the USA, Britain and France. The decades that followed saw the pro-liferation of two global telecommunication infrastructures: telephone cables in the ocean and communication satellites in orbit. In this chapter I focus on these two environmentally situated infrastructures through the activities of several organisations: AT&T, the British Post Office, British Telecom and INTELSAT. AT&T and the British Post Office, as their states' respective telecommunication monopolies, collaborated on TAT-1 and Telstar in the 1950s and 1960s. This collaboration continued with undersea cables through the 1970s and into the 1980s, when the British telephone service was transferred from the Post Office to British Telecom (BT) in 1981. AT&T and BT collaborated on TAT-8, the first fibre-optic transoceanic cable, laid in 1988. Meanwhile, communication satellites after Telstar were largely handled by one organisation: INTELSAT, the International Telecommunications Satellite Organisation. Formed in 1964 as an international consortium owned by member nations, INTELSAT sought to create a single global satellite system, and rapidly expanded through the 1970s, competing with submarine cables for inter-national communications market share.

I address the relationship between these two infrastructures in this chapter. I show how these infrastructures were environmentally situated and shaped by their environments, and, moreover, how the landscapes of these infrastructures – the oceanscape and spacescape – were deployed to publicise international communication links. Infrastructure and the environment has increasingly gathered historical attention, although perhaps not as much as it ought. Jeffrey Stine and Joel Tarr's landmark essay on technology and the environment notably neglects infrastructure, addressing it only in the context of urban environmental history and elsewhere implicitly relegating it to a secondary role, such as pipelines in the oil industry.[1] However, others have explicitly addressed the relationship between infrastructure and environment. Per Högselius, Arne Kaijser and Erik van der Vleuten have expansively demonstrated the imbrication of nature and infrastructure in nineteenth- and twentieth-century Europe.[2] Of direct relevance to this chapter, Nicole Starosielski has recently drawn attention to the undersea lives of telecommunication cables, and Lisa Ruth Rand has highlighted environmental fears about satellites in the 1970s as falling 'space junk' hazards.[3] Starosielski and Rand draw attention to environments usually neglected in predominantly terrestrial histories of infrastructure; Steve Pyne attributes this neglect to the near-total absence of human life from these 'extreme' environments and calls for greater historical attention to extreme environments.[4] Helen Rozwadowski and Roger Launius take up the calls for ocean and space history respectively.[5] Rozwadowski accepts the broader category of extreme history but maintains that the ocean's distinctive physical characteristics merit distinctive treatment, and, indeed, ocean history has had a distinctive niche carved out within environmental history.[6] On the other hand, as Launius notes, space historians have largely neglected the space environment altogether and so he echoes Pyne's call.

This chapter thus draws attention to the extreme environments of ocean depths and orbital space, and shows that they are not wildernesses into which cables and satellites intrude, but landscapes that shape and are shaped by technology, and are presented to the public as fusions of nature and technology. John Brinckerhoff Jackson defines landscape as 'a composition of man-made spaces on the land', and David Nye underscores that landscapes are 'inseparable from the technologies that people have used to shape the land and their vision'.[7] Coincidental to this chapter, Jackson points out that even extreme environments such as the moon and the ocean floor have been referred to as landscapes, and this has been expanded upon elsewhere: Rozwadowski uses the term 'ocean-scape' in her study of mid-nineteenth-century Atlantic hydrography, while Alice Gorman has called for a cultural landscape approach to recognise the mutual constitution of technology, human activity and the space environment.[8] Karl

Benediktsson has also argued for greater attention to the political significance of landscape, contending that the aesthetic values attributed to landscape are crucial to the political processes through which the environment is constructed and transformed.[9] In an age where global communications are marketed as instantaneous, dematerialised and omnipresent, I think it thus important to address not only the ocean and space environments of North Atlantic communications, but the oceanscapes and spacescapes, from TAT-1 in 1956 and Telstar in 1962, through the proliferation of transatlantic telephone cables and development of the INTELSAT satellite system in the 1960s and 1970s, to the laying of the first fibre-optic transatlantic cable, TAT-8, in 1988.

The categorisation provided by Jon Agar in this volume's opening chapter helps to lend a more synthetic terminology to this analysis.[10] Agar outlines eight categories of technology–environment combination, and several of those are relevant to the oceanscapes and spacescapes of North Atlantic communications. The second combination – environment as something natural made into (a component within) a technological system – is a natural starting point, as I shall demonstrate with the ways that cablelayers sought to use the seabed itself to protect cables and came into territorial conflict with the trawling industry over dominion of the ocean floor. The fourth combination – environment as something alongside the natural world – is another obvious category, as I explore the lengths that cable and satellite manufacturers went to in order to divorce technology from its environment, both practically and rhetorically. However, there are other, less obvious combinations that are also relevant. The fifth combination – environment as something untouched by artifice – features in 1950s Cold War discourses about the Atlantic, while the sixth combination – environment as something represented through technology – appears in various promotional films and adverts made about transatlantic communications. Finally, the seventh combination – environmental knowledge registered through technology – reveals itself as a crucial element through the ways cable operators searched for new techniques and technologies to mediate and access cables and their environments. A broader point that all of these combinations touch upon is the way in which these environments are transformed or mediated at great distances from the institutions responsible for these combinations; technology here extends far beyond the conventional borders of the nation, and so, as other chapters in this book also touch upon, these combinations of technology and the environment highlight how the spatial borders, extensions and projections of modern Britain extend far beyond the British Isles.[11]

These combinations shall be explored in three sections: first, I explore the laying of telephone cables in the post-war period, starting with TAT-1 in 1956. Second, I address the launching of communication satellites, starting with Telstar in 1962 and continuing with INTELSAT, and highlight the influence of satellites and spacescape on the inauguration of TAT-8, the first transoceanic fibre-optic cable. Finally, I conclude by addressing the place of business history at the intersection of technology and the environment.

Laying cables

TAT-1, the first transoceanic telephone cable, jointly developed by AT&T in the USA and the Post Office in the UK, opened on 25 September 1956. The cable was the first wired telephonic link between Europe and America and involved multiple transformations of technology and the environment. These transformations occurred along two lines: protecting the cable and routing the cable.

The protections for TAT-1 showcase the ways in which a boundary was created between the natural and the technological, and yet also highlight the contradictions of such boundary-making. The protection of undersea cables has a long ecological history: John Tully has shown how gutta-percha, a natural plastic 'gum' used to protect undersea telegraph cables and sourced from tropical plantations in the British Empire, was over-extracted to the point of 'ecological disaster' in the nineteenth century.[12] However, in contrast to the naturally sourced gutta-percha coatings of Victorian telegraph cables, TAT-1 was the first transoceanic cable to use a new synthetic coating, polyethylene.[13] However, despite polyethylene's synthetic nature, its status as a plastic, and thus an oil derivative, surfaces the material as, like gutta-percha, another artificial transformation of the environment. Unlike previous polyvinyl chloride coatings and gutta-percha, which – despite acting as an effective insulator – had proved attractive to groups of molluscs and insects known as 'submarine borers', polyethylene was impervious to attack from marine bacteria.[14] The use of polyethylene thus highlights a progressive separation of technology from the environment in protecting submarine cables, from the insulation from water alone to insulation from both water and organisms.

TAT-1's repeater design, on the other hand, shows how environment was also used as an input. The repeaters, which amplified and extended telephone signals, were the crucial innovation that had enabled TAT-1's creation, but to be viable, they needed three features: reliable operation at the bottom of the Atlantic, resistance to physical damage and the immense pressure at the ocean floor, and flexible construction so that they could be spooled up with the cable length on the cable ships. The flexible repeater units were composed of several sub-sections, arranged in a series, that were then attached by helical springs, allowing them to be coiled up. The repeater was then enclosed in two layers of steel rings, surrounded by a copper drum. The crucial feature here is how the extremity of the Atlantic ocean floor environment was used as an input: as the cable was laid out by ship, it would be bent by the laying process, which would separate out the steel rings and stretch the copper beyond its elastic limit. However, the hydraulic pressure at the bottom of the Atlantic would then return the steel rings together and reform the copper tube into its original state. Rather than purely separating technology from its environment, the repeater cladding used the Atlantic as an input to work as intended.[15]

The routing of TAT-1 highlights the ways different sciences and techniques were drawn upon by the Post Office and AT&T as registers of environmental knowledge. There is historical precedent here: Rozwadowski has shown how the first transatlantic telegraph cable spurred oceanographic inquiry, and how hydrographers' apparent discovery of a mid-Atlantic 'Telegraph Plateau' transformed the Atlantic floor from a hazardous environment to a peaceful zone for cable-laying. The irony to modern eyes is that the 'Telegraph Plateau' never existed, and Victorian hydrography had missed the mountainous mid-Atlantic ridge.[16] By the time of TAT-1, cable-layers at AT&T and the Post Office were well aware of the dangers of the Atlantic floor – in 1929 the Grand Banks region of the Atlantic, near Newfoundland, had experienced an earthquake. The earthquake caused a turbidity current – a sediment-laden flow of water along the ocean floor – which had snapped 12 telegraph cables. Contrary to the Victorian hydrographers' view of a peaceful 'Telegraph Plateau' Atlantic, by the time of TAT-1, cable-layers were aware of the ocean floor's hazards, and as I will show later, this played a role both in the cable-laying and in AT&T and the Post Office's corporate discourses about transatlantic communications.[17]

The topography and dangers of the Atlantic floor were not the only factors that influenced the routing of TAT-1: the location of undersea telegraph cables also played a role. Prior to TAT-1, 15 cables had been laid directly across the North Atlantic, and a further 5 via the Azores islands. This proved problematic: the most attractive route, from Cornwall to Newfoundland, was so congested that the cables were hazards to one another. This meant that a longer route, from Oban, Scotland, to Newfoundland was selected instead. The route selection shows how these cables came with their own environmental territories, which were important in planning new trans-atlantic communication infrastructures, but it should also be noted that there were political considerations as well: one proposed route, between Newfoundland and Ireland, was rejected because of the risks in laying an onwards connection across Ireland to London; three further routes via the Azores were all also rejected because of the issues involved in staffing cable stations in foreign territory.[18]

However, considering the various combinations of technology and environment involved in the protection and routing of TAT-1, it becomes clear that the Atlantic was not only something distinct from TAT-1, nor was it only a space to be registered through oceanographic science and technique: it was also a natural space made into a component within the TAT-1 system. The Atlantic environment was unavoidable, from the threat of biological and physical attack to the topology and pressures of the ocean floor. These features were transformed into components most visibly in the way that water pressure was used to reform the repeater housings into effective cladding, but it is apparent that, even in the form of resistance, avoidance and separation, Post Office and AT&T engineers incorporated many other features of the Atlantic environment into the design of TAT-1.

The Atlantic environment was not only incorporated into the TAT-1 technological system, but also into corporate discourses about trans-atlantic communications. However, rather than present the Atlantic oceanscape as a fusion of environment and technology, AT&T and the Post Office instead presented it as untamed wilderness that technology had to merely survive. Adverts in American popular science and youth magazines such as *Science World* and *Boys' Life* told of how the sea 'could make a "meal" of telephone cables' and publicised AT&T's 'experimental ocean', used to test cable specimens.[19] Adverts in business publications, including *American Banker*, *The Wall Street Journal* and *Fortune*, explained how the 'stormproof' TAT-1 would enable the expansion of American business interests into Europe.[20] A militaristic rhetoric also appeared in many of these features: one series of adverts portrayed TAT-1 and AT&T as 'conquering the Atlantic', while others spoke of TAT-1's 'far-reaching value in national defense' and compared the cable to US Cold War defence projects such as the Distant Early Warning Line and the Ballistic Missile Early Warning System.[21]

An AT&T publicity film for TAT-2, TAT-1's successor, made in 1959, captures this discourse while also demonstrating how technological mediations of the environment – in this case one of the first films made about transoceanic cable-laying – played a crucial role in constructing the struggle between transatlantic cables and the Atlantic.[22] The film, *Cable to the Continent*, opens with scenes of powerful waves battering shorelines and emphasises the cable's victories over the 'many-mooded sea'; icebergs, which posed a threat to the cable ships and the cables, are called 'great white deceivers'. The film goes on to conclude that the cable 'should do much to bring many nations together, both politically and economically, and contribute significantly to the defence needs of the free world'. The film captures how, for a discourse emphasising the importance of extending America's military and economic influence into Europe, the environment was depicted not as a partner or feature of the transatlantic cable system, but as an unruly subject.

The Post Office presented the Atlantic as a wilderness to emphasise British contributions to the project. At the cable's opening ceremony, Charles Hill, the Postmaster General, highlighted the British research that made the cable possible, while the Post Office's official souvenir booklet emphasised British oceanographic knowledge.[23] A talk for the BBC Home Service radio show 'Science Survey' by Sir Gordon Radley, Director General of the Post Office, portrayed the Atlantic oceanscape as an alien environment into which the cable had encroached. Radley described the cable resting in 'the perpetual darkness and ooze of the sea-bed', a turn of phrase that evokes Rudyard Kipling's poem *The Deep-Sea Cables*: 'There is no sound, no echo of sound, in the deserts of the deep / Or the great grey level plains of ooze where the shell-burred cables creep.'[24] However, where Kipling breathed life into the seabed, describing the 'shell-burred' cable that itself 'creeps' along the ocean floor, which resonated with his broader theme of telegraph cables drawing humankind together, Radley instead emphasised the extremity and lifelessness of the Atlantic environment.

TAT-1 was followed by many more transatlantic telephone cables: TAT-2 was laid in 1959, and TAT-3 through to TAT-7 were laid in 1963, 1965, 1970, 1976, and 1978. This criss-crossing of the Atlantic necessitated further strategies of prophylaxis and maintenance for this transatlantic infrastructure. I shall first deal with strategies of prophylaxis, taken to prevent damage to cables, and then strategies of mainten-ance, which were utilised when prevention failed.

One of the most significant risks to cables was the intersection of their environmental territory with the trawling industry. This had long been a problematic area: the first international submarine communications cable, laid between England and France in 1850, was severed by a French fisherman after only three days, and from 1965 to 1966, TAT-3 and TAT-4 were severed on four separate occasions by trawlers scouring a newly discovered bed of scallops off the New Jersey coast.[25] The primary strategy taken to prevent such damage was to bury the cable using a 'sea plough' developed by AT&T, which by 1969 had buried nearly 400 miles of cable in the North Atlantic and Mediterranean. The sea plough was a tractor-like vehicle, remotely controlled from a cable ship, which proceeded along the ocean floor using caterpillar tracks, digging a trench and burying cable in the trench. The sea plough proved hugely successful and the Post Office soon undertook its own trials using a modified sugar-cane plough in the North Sea. However, the sea plough not only transformed the environment physically, but also symbolically: the agricultural envirotechnological system of tractors and ploughs served as inspiration for the transformation of the oceanscape, surfaced in metaphors such as 'sea plough' and 'tilling the ocean floor'.[26] Here, there are parallels with the bulldozer in Ralph Harrington's chapter earlier in this book, in which the bulldozer not only transformed the environment, but took on a symbolic status in the process.[27]

Alongside the sea plough, efforts were also made to educate and deter the trawling industry from the cables' territories. The UN's Food and Agriculture Organisation (FAO) released a report in 1970, *Trawling and Submarine Cables*, which acknowledged the long history and territorial claims of submarine communication cables, as well as the significant lengths and expenses that cable companies like AT&T and the Post Office had gone to in order to protect these cables. This included not only the sea plough, but also the re-drawing of hydrographic and fishing charts to include cables, the commissioning of multi-language radio announcements by the BBC, adverts in fishing magazines and even a research programme by the Post Office to develop new, safer, trawling gear for fishermen.[28] That the UN FAO circulated this report to all government fishery departments highlights the significance that communication cables – not ordinarily under their purview – were seen to have. AT&T and the Post Office also undertook cable patrols in the North Atlantic, using their cable ships to patrol the cable territories and warn off trawlers.[29] These strategies further underscore how swathes of the Atlantic environment were rendered into disputed territories within competing industrial and technological systems.

If cables were damaged, maintenance procedures were needed to repair them, and these procedures highlight the ways in which the extreme Atlantic environment was registered and transformed to make it more accessible. Two strategies that were explored and practised, but never formalised, were diving units and underwater habitats. Diving units had been used in the early years of the sea plough in case it jammed, and so in 1970 the Post Office commissioned a report on the viability of also using divers to repair cables on the seabed itself.[30] The Post Office also collaborated with the United Kingdom Atomic Energy Authority's Marine Technology Support Unit on developing underwater habitats into which cables could be pulled for repairs.[31] The diving unit and submersible habitats were not deployed, in part because of two more successful alternatives for accessing the deep Atlantic: mini-subs and grapnels.

A mini-submarine, like divers, had also proved effective in early sea-plough trials, where it was used to dive to the seabed and locate the cable. In the 1970s, the Post Office used mini-submarines called 'Pisces' to bury and repair cable and later in the 1970s a consortium of North Atlantic telecommunications companies, the North Atlantic Cable Maintenance Agreement, pooled funds to purchase two remote-controlled unmanned submersibles, called SCARABS, to undertake submarine maintenance work.[32] Grapnel development was also a significant area of research for the Post Office, and by 1979, the Post Office had developed the 'cut-and-hold grapnel', which could simultaneously cut a cable and lift it for repairs. The grapnel also innovatively mediated the Atlantic, using a sonar surveillance system that could be used to locate the damaged cable. The 'cut-and-hold' grapnel reduced grappling time by a third and total repair time by a fifth, and was marketed and sold internationally by the Post Office to other cable companies.[33] These developments were about removing the need for humans to enter the ocean environment, further extending the division between nature and technology; as Agar notes in the opening chapter to this volume, maintenance is an area in which the boundary between nature and technology is heavily policed. However, as some of these developments show, in extreme environments such as the ocean floor, methods like sonar and remote control are also needed to mediate those environments that humans can rarely access. It is another of those environments, and the communications technologies that were situated within it, to which I now turn, as I explore the combinations of technology and environment involved in the installation of satellites in the remote and difficult-to-access orbital space of Earth.

Launching satellites

In the early 1960s, space-based communications took off as an alternative transmission method to submarine telephony. Telstar, the first satellite to relay telephony and television across the Atlantic, was designed by AT&T and launched in July 1962. Telstar relayed signals between AT&T's earth station in Andover, Maine, and the British and French earth stations in Goonhilly, Cornwall and Pleumeur-Bodou, Brittany. After Telstar, international satellite communications were reorganised into a single entity through the creation of INTELSAT, the international satellite consortium, in 1964. In this section, I will draw attention to combinations of technology and the environment in satellite communications in two distinct phases: first, with Telstar, in which the extremities of the space environment were visibly surfaced, and second, with INTELSAT, in which the extreme space environment was obscured in comparison with the environment of submarine communications.

Telstar's combinations of technology and environment highlight it as an artefact that was not only a communications satellite, but also a way of registering environmental knowledge. AT&T publicity explained how Telstar was a 'space laboratory', 'operating in the unknown environment of hostile radiation and micrometeorite dust'.[34] Telstar was used by Bell Labs and NASA to gain a greater understanding of the Van Allen radiation belt surrounding Earth, which had been discovered in 1958 by James Van Allen at the University of Iowa, using data from the Explorer 1 and Explorer 3 satellites.[35] Telstar thus fits into a history not only of satellites as communications technologies, but also of satellites as methods for registering environmental knowledge, as was the case with early satellites, like Explorer 1, and as Hogselius, Kaijser and van der Vleuten have shown, with later satellites used for meteorology and land surveying.[36]

Telstar also shared a common feature with practically every other satellite: the use of environment as an input, through solar panels, alongside a contradictory division of technology and environment, through radiation shielding. Telstar was covered in sixty solar cells used to power the satellite, and radiation shielding was achieved by packaging Telstar's electronics into an inner, hermetically sealed, container within the satellite structure. The hermetically sealed container contained normal Earth atmosphere, in which the components were known to work reliably. This was not unusual for satellites of the time, but is uncommon in modern satellites, which usually operate in the vacuum of space, and so, while Telstar itself was situated in the space environment, its electronics can, in a sense, be thought of as working in a small pocket of Earth taken up into space by satellite.[37]

This radiation shielding, however, was insufficient, and in November 1962, Telstar fell silent due to radiation damage to a transistor. Telstar's silence was not merely evidence of the hazardous space environment, but also of human damage to the space environment. The day before Telstar was launched, the USA detonated Starfish Prime, the largest man-made nuclear explosion in outer space, and part of a series of high-altitude nuclear weapons tests conducted by the USA called Operation Fishbowl. This detonation energised the Van Allen belt, and the additional radiation damaged Telstar, causing it to fail. The failure of Telstar and seven other satellites, including Ariel I, Britain's first satellite, caused by Starfish Prime, fed into what Lisa Ruth Rand has labelled a 'proto-environmentalist' moment in the Cold War, which interlinks international communications, the space environment and the militarisation of space.[38]

The relationship between the Cold War and the growth of envir-onmentalism in the 1960s and 1970s has been recognised previously. John McNeill and Corinna Unger call environmentalism 'among other things, a child of the Cold War', noting the link between antinuclear protests in the early Cold War and the development of environmen-talism.[39] Jacob Hamblin explores a similar link, attributing 'catastrophic environmentalism' in the latter half of the twentieth century to a growing environmental awareness during the Cold War and the parallel growth of modern science.[40] Toshihiro Higuchi particularly draws attention to nuclear testing and Operation Fishbowl as a stimulus for environmental insecurity in the Cold War.[41] These tests, however, did not unintentionally damage the space environment as a side-effect, but rather had the envir-onment firmly in their sights: Hamblin highlights the American physi-cist Edward Teller's remarks about nuclear weapons testing in space: 'We know how we can modify the ionosphere. We have already done it.'[42] Telstar's failure was linked in newspaper articles to Operation Fishbowl, showing that Telstar was not only a communications satellite or register of environmental knowledge, but also one of the first friendly-fire victims in the militarisation of the space environment.[43] Telstar's failure, and the public attention to the possible consequences of Starfish Prime, highlight how space communications in the early 1960s was entangled with the concept of orbital space as a natural environment.

It is also important to note that the space environment was not the only environment of satellite communications, which consists not only of satellites in space, but also a vast, material, terrestrial infrastructure composed of, by now, hundreds of earth stations around the world. Earth stations are important nodes in communications satellite infrastructure, yet relatively neglected in satellite histories, which tend to focus on the cosmic and not the terrestrial.[44] I shall focus on the British Post Office's earth station, Goonhilly, to explore the combinations of technology and environment at work in earth stations.

The Post Office pursued a unique design direction for Goonhilly, which surfaces the interlinkage of national prestige with different approaches to technology and the environment. Goonhilly, built by the Post Office in 1962 to communicate with Telstar, was situated on an isolated, elevated plateau on the Cornish peninsula, which gave good sightlines for satellite tracking and horizon-to-horizon communications. Goonhilly's first antenna, Antenna One, was the world's first satellite communication antenna with a parabolic design (i.e. a satellite 'dish'), and here was influenced by British precedent in antenna design: Charles Husband, the engineer responsible for Antenna One, had also designed Jodrell Bank's Lovell Telescope, which had earned prestige through its unique, world-first, 'dish' design.[45] The Post Office proclaimed this parabolic design as a uniquely British design concept, which did not need to be protected from the environment. It was explicitly contrasted with AT&T's earth station at Andover, Maine, which utilised a 'horn' antenna that was protected from the elements – wind, rain and snow – by a distinctive 'golf ball' protective radome. Goonhilly's dish was mobilised as part of Post Office publicity, highlighting the technological sophistication of the Post Office, which would later proudly tout how the British parabolic design became the template for subsequent earth stations around the world.[46]

Goonhilly thus shows an interesting combination of technology and the environment, where, in contrast to other earth stations with radomes, it was not the physical separation of environment from technology that was deemed significant, but rather the capability of the earth station to embrace 'wilderness' without being changed by it. More broadly, however, Goonhilly for the Post Office and Telstar for AT&T show how, in the early years of space communications, combinations of technology and the environment were significant features of corporate discourses about satellite communications.

After Telstar, a new satellite order was created in the form of INTELSAT. INTELSAT surfaced the space environment in a very different manner to AT&T and the Post Office, and I argue that this resulted from INTELSAT's different views, compared with organisations like AT&T and the Post Office, on the relationship between satellite and submarine communications. INTELSAT was spearheaded by the USA as a satellite system which would be globally accessible, but also had clear Cold War objectives: transmitting television and propaganda from the USA through a global satellite system was seen as a powerful weapon of American foreign policy. The USA started negotiations with foreign governments in the early 1960s to gather a consortium of nations to invest in and support the global system. Nigel Wright and Hugh Slotten have both previously pointed out the tensions between the commitments of potential European partners – particularly Britain and France – to their existing submarine cable networks and the US desire for a single satellite system to realise foreign policy objectives.[47] In the end, the INTELSAT vision won out, although caveats were made for interworking between satellites and submarine cables. In 1964, INTELSAT was created, with 61 per cent owned by the USA via COMSAT, its domestic satellite corporation, and the remainder owned by partners from around the world.

The tension between satellites and submarine cables informed publicity about INTELSAT, which presented satellites and cables as differing significantly on technical and environmental grounds. The greater communications capacity of satellites was heralded as a breakthrough in international communications. A brochure entitled *New Communications Era* created by COMSAT, the US partner in INTELSAT, explained how 'archaic' cable systems were no longer necessary, and that INTELSAT's first satellite, nicknamed Early Bird, had almost double the capacity of a transatlantic cable at under a fifth of the cost.[48] In 1971, an article for *Popular Science* by Werhner von Braun, the famed German-American aerospace engineer, touted the superior capacity of INTELSAT's fourth series of satellites, INTELSAT IV, compared with the 'puny' capacity of even the 'most sophisticated transatlantic cable'.[49]

The fragility of undersea cables was also emphasised: in June 1965, COMSAT publicised Early Bird's rescue of transatlantic communications after it offered a temporary replacement service for the failure of CANTAT, the Anglo-Canadian transatlantic cable.[50] A similar episode occurred three years later, when COMSAT publicised how INTELSAT satellites had carried their heaviest ever load of Atlantic traffic after two transatlantic cables had been damaged.[51] A particular feature of this rhetoric was the notion that satellites were able to escape the environment in a way that terrestrial communications could not: INTELSAT publicity presented communication satellites as surpassing the 'inherent limitations' of terrestrial communications, unlike the intrinsic, fragile, materiality of cables.[52] Another INTELSAT publicity feature, in *TV Guide* magazine, collapsed these technical and environmental differences into one with the rhetorical question, 'Meet that demand with undersea cables? They'd drown in an ocean of words. But satellites can handle it.'[53] The INTELSAT discourse submerged cables, literally and figuratively, while presenting satellites as environmentally transcendent.

However, this rhetoric was not quite accurate, and from the early days of the INTELSAT system, the British Post Office had researched and monitored the proficiency of both satellites and submarine cables as international links, including the various combinations of technology and environment that might influence performance.[54] Post Office research drew attention to the susceptibility of satellite communications to rain and atmospheric conditions that could cause bursts of noise, while other research addressed the environmental vulnerability of earth stations, noting that a satellite TV broadcast from Germany had been cancelled because the German earth station's radome had been covered in snow, and that radome repairs had also been responsible for putting Andover and Pleumeur-Bodou out of action for extended periods. Satellites were seen as less susceptible to malicious and electrical interference, although it was also pointed out that, in the event of damage, cables could be repaired, whereas satellites could not. Overall, the Post Office's conclusions on submarine cables and satellites were clear: the reports emphasised that cables held numerous advantages over satellites, which included the more adept handling of the interfaces between environment and technology outlined above.

INTELSAT's articulation of the spacescape, however, was clearly victorious over the oceanscape of submarine cables. In 1988, the first transoceanic fibre-optic cable, TAT-8, was laid across the Atlantic. TAT-8 was a leap forward in transatlantic communications capacity, carrying almost 10 times more calls than its predecessor, TAT-7, and over three times more calls than its satellite competitor, INTELSAT V.[55] However, rather than draw attention to this significant expansion in capacity, AT&T and British Telecom, the privatised successor to the Post Office, instead chose to use satellites and the spacescape for their marketing. A 1988 BT advert – the same year TAT-8 was laid – opens with a shot of an empty business office.[56] A businessman appears, picking up his address book and dialling his telephone. Echoing the famous documentary shorts *Powers of Ten* by Charles and Ray Eames, the camera slowly zooms out, revealing that the shot is coming from outside the office window, and carries on pulling out: London, the UK, Europe, until, finally, Earth, floating in space. A communications satellite orbits the Earth, before the camera quickly zooms back in to an office in New York City. Another businessman picks up his phone and answers the call. In 1989, AT&T ran their own advert about international communications: in contrast to BT's, this advert explicitly mentioned TAT-8, but using similar imagery, the advert opened and closed with shots of the Earth from space and went on to deploy space-age motifs, from satellites and earth stations to NASA-style communications and telemetry control centres, in support of AT&T's

'worldwide intelligent network'.[57] Another AT&T advert, in 1991, particularly highlights the choice of spacescape over oceanscape to market international communications. The advert bore the slogan 'We'd like to be the first to say hello', and depicted a message in a bottle (see Figure 10.1). However, highlighting the influence that the spacescape

Figure 10.1 'We'd like to be the first to say hello.' AT&T/N.W. Ayer. Source: courtesy of the National Museum of American History Archives Center.

had over communications discourses, the bottle was not afloat on the ocean, but instead in space.[58]

Conclusion

Paul Edwards points out that constructing infrastructure also constructs a specific nature in which it is situated.[59] In this chapter I have outlined numerous combinations of technology and the environment in the construction of submarine and satellite communications infrastructure: TAT-1 was separated from the Atlantic and yet also used it as an input; cable protection and maintenance strategies transformed, mediated and registered the Atlantic in various ways; Telstar registered knowledge about outer space while surfacing space as a vulnerable environment; and INTELSAT articulated a countervailing view that downplayed the extremity of the space environment. It is thus also clear that the construction of these infrastructures, which were not purely technological but fusions of technology and the environment, was used by their corporate builders to articulate specific visions of these extreme environments. AT&T and the British Post Office composed TAT-1's oceanscape as an untamed Atlantic environment, which was used by AT&T to emphasise the cable's contribution to US economic and military interests and by the Post Office to highlight British scientific and engineering skill. Telstar's spacescape was an unknown and potentially hazardous space environment, which was unwelcomely surfaced by US nuclear weapons testing, while Goonhilly's landscape was a wild opponent, again used by the Post Office to emphasise national engineering ingenuity. INTELSAT's spacescape was environmentally transcendent, while INTELSAT also further emphasised the hazardous oceanscape of undersea cables to promote satellite communications. The dematerialised, instantaneous aesthetic of modern telecommunications may thus owe this image to the commercial contests between satellite operators and cable-layers, and associated marketing of spacescape and oceanscape, from the 1950s to the 1980s.

The interplay between national and business histories is thus also a significant influence on this history of transatlantic communications. Modern British history is particularly apparent in the Post Office's discourse about TAT-1 and Goonhilly, which could be well captured by Robert Bud's concept of 'defiant modernism': Bud draws attention to the British pursuit of high-profile 'world-first' projects after the Second World War, such as Comet, the first civilian jet airliner, and Calder Hall, the first commercial nuclear power station.[60] TAT-1 and Goonhilly, as combinations of technology and environment, were deployed by the Post Office in a similar vein. Likewise, AT&T's presentation of TAT-1 and INTELSAT's rhetoric of satellites over cables could be positioned in John Krige's historical concept of 'consensual hegemony': the USA's use of scientific and technological projects in the Cold War to aid European reconstruction and also serve US Cold War interests.[61] However, while national history is important, the fact that it was businesses that reified and articulated these combinations of technology and the environment should not be overlooked. Business history and history of technology have long gone hand in hand, but stronger links could perhaps be forged between environmental and business history. Christine Meisner Rosen has been a staunch advocate here, and there are further signs of change.[62] David Nye has described landscape as 'a process embedded in narrative',[63] and businesses, as organisations with both national and transnational contexts, are thus crucial narrators of the landscapes of technology and environment. The oceanscapes and spacescapes of international communications thus owe as much to modern national histories as they do to the histories of AT&T, the British Post Office, British Telecom and INTELSAT.

Acknowledgements

Thanks to Jon Agar, Jessica van Horssen and Thomas Turnbull for discussion on this chapter and responses to an earlier draft and to audiences at the National Museum of American History and the National Air and Space Museum for their comments.

Notes

1   Jeffrey K. Stine and Joel A. Tarr, 'At the Intersection of Histories: Technology and the Environment', *Technology and Culture* 39, no. 4 (1998): 630.
2   Per Högselius, Arne Kaijser and Erik van der Vleuten, *Europe's Infrastructure Transition: Economy, War, Nature* (New York: Palgrave Macmillan, 2016).
3   Nicole Starosielski, *The Undersea Network* (Durham, NC: Duke University Press, 2015); Lisa Ruth Rand, 'Orbital Decay: Space Junk and the Environmental History of Earth's Planetary Borderlands' (PhD diss., University of Pennsylvania, 2016).
4   Steve Pyne, 'Extreme Environments', *Environmental History* 15, no. 3 (2010): 509–13.
5   Helen M. Rozwadowski, 'Ocean's Depths', *Environmental History* 15, no. 3 (2010): 520–5; Roger D. Launius, 'Writing the History of Space's Extreme Environment', *Environmental History* 15, no. 3 (2010): 526–32.
6   Helen M. Rozwadowski and David K. Van Keuren, *The Machine in Neptune's Garden: Historical Perspectives on Technology and the Marine Environment* (Sagamore Beach, MA:

Science History Publications, 2004); Jacob Darwin Hamblin, *Oceanographers and the Cold War: Disciples of Marine Science* (Seattle: University of Washington Press, 2005); Helen M. Rozwadowski, *Fathoming the Ocean: The Discovery and Exploration of the Deep Sea* (Cambridge, MA and London: Belknap, 2005); Jennifer M. Hubbard, *A Science on the Scales: The Rise of Canadian Atlantic Fisheries Biology, 1898–1939* (Toronto: University of Toronto Press, 2006); Jacob Darwin Hamblin, *Poison in the Well: Radioactive Waste in the Oceans at the Dawn of the Nuclear Age* (New Brunswick and London: Rutgers University Press, 2008).

7   John Brinckerhoff Jackson, *Discovering the Vernacular Landscape* (New Haven and London: Yale University Press, 1984), 7; David E. Nye, 'Technologies of Landscape', in *Technologies of Landscape: From Reaping to Recycling*, ed. David E. Nye (Amherst: University of Massachusetts Press, 1999), 6.

8   Helen M. Rozwadowski, 'Technology and Ocean-Scape: Defining the Deep Sea in Mid-Nineteenth Century', *History and Technology* 17, no. 3 (2001): 217–47; Alice Gorman, 'Cultural Landscape of Space', in *Handbook of Space Engineering, Archaeology, and Heritage*, ed. Ann Garrison Darrin and Beth Laura O'Leary (Boca Raton: CRC Press, 2009), 335–46.

9   Karl Benediktsson, '"Scenophobia", Geography and the Aesthetic Politics of Landscape', *Geografiska Annaler: Series B, Human Geography* 89, no. 3 (2007): 203–17.

10 Jon Agar, 'Historiography and Intersections', Chapter 1 this volume.

11 Esa Ruuskanen, 'Encroaching Irish Bogland Frontiers: Science, Policy and Aspirations from the 1770s to the 1840s', Chapter 2 this volume; Simone Turchetti, 'HMG's Environmentalism: Britain, NATO and the Origins of Environmental Diplomacy', Chapter 13 this volume.

12 John Tully, 'A Victorian Ecological Disaster: Imperialism, the Telegraph, and Gutta-Percha', *Journal of World History* 20, no. 4 (2009): 559–79.

13 L.R. Snoke, 'Resistance of Organic Materials and Cable Structures to Marine Biological Attack', *The Bell System Technical Journal* 36, no. 5 (September 1957): 1095–27.

14 Charles Bright, *Submarine Telegraphs: Their History, Construction and Working* (Cambridge: Cambridge University Press, 1898/2014), 381.

15 T.F. Gleichmann et al., 'Repeater Design for the North Atlantic Link', *The Bell System Technical Journal* 36, no. 1 (January 1957): 91.

16 Rozwadowski, 'Technology and Ocean-Scape', 243.

17 C.H. Elmendorf and B.C. Heezen, 'Oceanographic Information for Engineering Submarine Cable Systems', *The Bell System Technical Journal* 36, no. 5 (September 1957): 1035–94.

18 J.S. Jack, C.W.H. Leech and H.A. Lewis, 'Route Selection and Cable Laying for the Transatlantic Cable System', *The Bell System Technical Journal* 36, no. 1 (January 1957): 293–326.

19 AT&T and N.W. Ayer, 'The Sea Could Make A "Meal" of Telephone Cables' 1958, Folder 1, Box 32, Series 3, Collection 59: N.W. Ayer Advertising Agency Records, National Museum of American History Archives Center [hereafter NMAH]; AT&T and N.W. Ayer, 'Why We Have Our Own "Ocean"' 1959, Folder 1, Box 32, Series 3, Collection 59: N.W. Ayer Advertising Agency Records, NMAH.

20 AT&T and N.W. Ayer, 'Beneath the Broad Atlantic' 1954, Folder 3, Box 13, Series 3, Collection 59: N.W. Ayer Advertising Agency Records, NMAH; AT&T and N.W. Ayer, 'You Can Telephone Britain Over New Stormproof Cables' 1956, Folder 2, Box 24, Series 3, Collection 59: N.W. Ayer Advertising Agency Records, NMAH.

21 AT&T and N.W. Ayer, 'Tele-Facts: Laying the First Atlantic Telephone Cable' 1955, Folder 2, Box 32, Series 3, Collection 59: N.W. Ayer Advertising Agency Records, NMAH; Cleo F. Craig, 'Equipping Ourselves for Today's Responsibilities', *Bell Telephone Magazine* 34, no. 4 (Winter 1955): 217–24; C.C. Duncan, 'Communications and Defense', *Bell Telephone Magazine* 37, no. 1 (Spring 1958): 15–24.

22 *Cable to the Continent* (AT&T, 1959).

23 'Speeches at the Opening of the Transatlantic Telephone Cable' 1956, HIC W04/03 Network/Cable/Submarine Cables/Trans-Atlantic, BT Archives [hereafter BTA].

24 'The Transatlantic Telephone Cable', *Science Survey* (BBC Home Service, 27 September 1956), HIC W04/03 Network/Cable/Submarine Cables/Trans-Atlantic, BTA; Rudyard Kipling, 'The Deep Sea Cables', in *The Seven Seas* (London: Methuen & Co., 1896).

25 'Tilling the Ocean Floor', *Post Office Telecommunications Journal* 20, no. 2 (1968): 12–15.

26 'Tilling the Ocean Floor'; Captain O. Bates, 'Ploughing the Seabed', *Post Office Telecommunications Journal* 22, no. 4 (1970): 2–4.

27 Ralph Harrington, 'Landscape with Bulldozer: Machines, Modernity and Environment in Post-War Britain', Chapter 3 this volume.

28 'Trawling and Submarine Cables', *FAO Fisheries Circular* (Rome: Food and Agriculture Organisation of the United Nations, November 1970).

29 'Ice-Cold in Alert', *Post Office Telecommunications Journal* 20, no. 1 (1968): 17.

30 'The State of the Art of Diving at the Present Time and in the Foreseeable Future, with Particular Reference to the Repair of Buried Submarine Cable' 1970, TCB 711/30/5, BTA.

31 I.R. Finlayson to F.A. Hough, 'Proposed Organisation of R&D by the Underwater Engineering Group, CIRIA', 19 November 1969, TCB 711/30/5, BTA.

32 Edward Fennessy, 'Submarine Recovery' 13 September 1973, TCB 711/30/5, BTA; 'Purchase of Submersible Vehicles for North Atlantic Cable Maintenance' 28 January 1975, TCB 711/30/5, BTA.

33 The British Post Office, 'Cut & Hold Grapnel' 1979, TCB 711/35/3, BT Archives; J.E.H. Cosier, 'Getting to Grips with Undersea Cables', *Post Office Telecommunications Journal* 30, no. 1 (Spring 1978): 7–8.

34 'Facts from Space via Telstar', *Bell Telephone Magazine* 41, no. 3 (Autumn 1962): 38–39.

35 A.B. Crawford et al., 'The Research Background of the Telstar Experiment', *The Bell System Technical Journal* 42, no. 4 (July 1963): 747–51; AT&T, 'Project Telstar' 1962, HIC W04/02 Network/Transmission Systems/Satellites 2, BTA.

36 Högselius, Kaijser and van der Vleuten, *Europe's Infrastructure Transition*, 241, 319.

37 R.H. Shennum and P.T. Haury, 'A General Description of the Telstar Spacecraft', *The Bell System Technical Journal* 42, no. 4 (July 1963): 820–1.

38 Rand, 'Orbital Decay', 127.

39 J.R. McNeill and Corinna R. Unger, 'Introduction: The Big Picture', in *Environmental Histories of the Cold War*, ed. J.R. McNeill and Corinna R. Unger (Cambridge: Cambridge University Press, 2010), 11.

40 Jacob Darwin Hamblin, *Arming Mother Nature: The Birth of Catastrophic Environmentalism* (Oxford and New York: Oxford University Press, 2013), 8.

41 Toshihiro Higuchi, 'Atmospheric Nuclear Weapons Testing and the Debate on Risk Knowledge in Cold War America, 1945–1963', in *Environmental Histories of the Cold War*, ed. J.R. McNeill and Corinna R. Unger (Cambridge: Cambridge University Press, 2010), 319.

42 Hamblin, *Arming Mother Nature*, 12.

43 'AT&T Sets Telstar II Launch for Spring; Lessening of Radiation Damage a Main Goal', *Wall Street Journal*, 2 January 1963.

44 An earth station-oriented history can be found in Craig B. Waff, 'Project Echo, Goldstone, and Holmdel: Satellite Communications as Viewed from the Ground Station', in *Beyond the Ionosphere: Fifty Years of Satellite Communication*, ed. Andrew J. Butrica (Washington, DC: NASA History Office, 1997), 41–50.

45 Jon Agar, 'Making a Meal of the Big Dish: The Construction of the Jodrell Bank Mark 1 Radio Telescope as a Stable Edifice, 1946–57', *The British Journal for the History of Science* 27, no. 1 (1994): 3–21.

46 'Ten Years of Technological Progress' 1972, HIC W04/02 Network/Transmission Systems/Satellites 2, BTA; 'Progress' 1962, TCB 420/IRP (PR) 1, BTA.

47 Nigel Wright, 'The Formulation of British and European Policy Toward an International Satellite Telecommunications System: The Role of the British Foreign Office', in *Beyond the Ionosphere: Fifty Years of Satellite Communication*, ed. Andrew J. Butrica (Washington, DC: NASA History Office, 1997), 157–70; Hugh Richard Slotten, 'Satellite Communications, Globalization, and the Cold War', *Technology and Culture* 43, no. 2 (2002): 315–50; Hugh Richard Slotten, 'International Governance, Organizational Standards, and the First Global Satellite Communication System', *Journal of Policy History* 27, no. 3 (2015): 521–49.

48 COMSAT, 'New Communications Era' 1967, Folder 5, Box 58, Acc. No. 1987-0125, National Air and Space Museum Archives Center [hereafter NASM].

49 Wernher von Braun, 'Now at Your Service – The World's Most Talkative Satellite', *Popular Science*, May 1971, 56–57, 138.

50 COMSAT, 'Early Bird to Provide Emergency Service after Transatlantic Cable Break' 18 June 1965, OE-037000-03, NASM.

51 COMSAT, 'Interruption of Service on Two Transatlantic Cables' 15 February 1968, OI-435000-01, NASM.

52 TRW, 'INTELSAT III Press Handbook' 1969, OI-435100-01, NASM.

53 'Switchboard in Orbit', *TV Guide* (4 December 1971).

54 'Long Distance Cable and Satellite Systems' (6 May 1968) TCB 711/2/5, BTA.

55 Roger Smith, 'BT Bites Back ... New Cable More than a Match for "Jaws"', *BT Journal* 8, no. 2 (1987): 38–41; 'INTELSAT V Facts' 1981, OI-435400-01, NASM.

56 *British Telecom International ... It's You We Answer to* (BT, 1988).

57 *AT&T's Universal Telephone Service* (AT&T, 1989).

58 AT&T and N.W. Ayer, 'We'd Like to Be the First to Say Hello' 1991, Folder 13, Box 33, Series 4, Collection 59: N.W. Ayer Advertising Agency Records, NMAH.

59 Paul N. Edwards, 'Infrastructure and Modernity: Force, Time, and Social Organisation in the History of Sociotechnical Systems', in *Modernity and Technology*, ed. Thomas J. Misa, Philip Brey and Andrew Feenberg (Cambridge, MA and London: MIT Press, 2003), 189.

60 Robert Bud, 'Penicillin and the New Elizabethans', *The British Journal for the History of Science* 31, no. 3 (1998): 305–33.

61 John Krige, *American Hegemony and the Postwar Reconstruction of Science in Europe* (Cambridge, MA and London: MIT Press, 2008).

62 Christine Meisner Rosen, 'Industrial Ecology and the Greening of Business History', *Business and Economic History* 26, no. 1 (1997): 123–37; Christine Meisner Rosen and Christopher C. Sellers, 'The Nature of the Firm: Towards an Ecocultural History of Business', *The Business History Review* 73, no. 4 (1999): 577–600; Christine Meisner Rosen, 'The Business–Environment Connection', *Environmental History* 10, no. 1 (2005): 77–9; Hartmut Berghoff and Adam Romem (eds), *Green Capitalism?: Business and the Environment in the Twentieth Century* (Philadelphia: University of Pennsylvania Press, 2017).

63 Nye, 'Technologies of Landscape', 7.

11

The Thames Barrier: climate change, shipping and the transition to a new envirotechnical regime

Matthew Kelly

The process that led the British government to build the Thames Barrier resembled the trajectory of many infrastructural projects in post-war Britain. A generally accepted need generated proposals, counter-proposals and the concomitant weakening of vested interests, until under growing public pressure the government decided to push enabling legislation through the Parliament. This decision departed from this norm, however, because the government did not seek to improve living standards or economic effectiveness, but to counter the existential threat nature posed to London, as intimated by the devastating tidal flood of 1953.[1] As this chapter will show, by the 1950s the government accepted the scientific consensus that climate change was causing glacial ablation, or melt, and rising sea levels and increased storminess threatened all North Sea coastal terrains. London was made peculiarly vulnerable by the added effect of north–south 'tilt', by which the south-east of England was gradually sinking. The Waverley Report, the official response to the 1953 flood, identified the flood not as a unique weather event, despite the unusual concatenation of factors that caused it, but as an episode in a series whose ultimate origins were geological and climatic, against which human beings could only mount a defence. In this respect, to adapt the distinction Braudel drew between the *longue durée* and the 'conspicuous' events that usually preoccupy historians, the 1953 flood was a moment when the *longue durée* became conspicuous.[2]

Whitehall accepted Waverley's recommendation that improved coastal and estuarial flood defences be augmented with a retractable barrier across the Thames, and the long policy-making process that followed left little meaningful distinction between technological and political questions. Although the most decisive politicking occurred in private meetings between representatives of various public bodies, the public debate was periodically enlivened by the Thames barrage lobby, which hoped to transform the existing estuarial regime upstream of Gravesend or Woolwich into a slow-moving fresh water lake, 'liberating' the city from the tide. This revived a Victorian agenda but was promoted in terms of improving the river's 'amenity' value, one of the key governing concepts in post-war Britain. Although able to mobilise influential backers, the barrage lobby achieved little traction in Whitehall; the agenda seemed the stuff of outmoded visionaries, an extravagant alternative to the relatively simple technological fix represented by a retractable barrier. That this was so did not make the eventual decision to build the barrier or its location predetermined. To avoid 'naturalizing technological change', this chapter will consider the case made for the barrage and trace the fate of the different barrier schemes proposed.[3]

Moreover, the case made for a barrage threw into sharp relief how the barrier sought to protect London against a potentially devastating flood while preserving the existing estuarial regime. For much of the 1950s and 1960s, discussion concerned how the tidal energies of the Thames were harnessed by shipping, washed away the city's waste and prevented a build-up of silt. In the late 1960s, a new consideration entered the discussion. Improved sewage plants had begun to restore something of the river's pre-industrial condition and the return of migrating fish relied on levels of oxygenation and salinity partly created by tidal action. The failure of the barrage lobby should not, therefore, obscure the degree to which the River Thames was already, in William Cronon's influential formulation, 'second nature', or, to follow Thomas P. Hughes, an 'ecotechnological system', or, indeed, an example of Richard White's 'organic machine'.[4] Note, for example, how in the nineteenth century the upstream limit of the Thames tideway was lowered from Kingston-Upon-Thames by the construction of Teddington Lock, while Richmond Lock (1894), technically a half-tide lock and barrage, was necessary to maintain a navigable depth of water between the two locks following the dismantling of Old London Bridge. And this is to say nothing of the significant implications banking, flood defences and London's complex system of docks and harbours had for water flow and riverine ecology. Nonetheless, if the 'high modernist' desire of the barrage lobby to transform and control nature more closely resembles today's climate engineers than yesterday's pragmatic technocrats, the retractable barrier was contrastingly of tremendous anthropic importance but of relatively low environmental impact.

That said, in her history of the post-1945 development of the Rhône, Sara B. Pritchard argues that histories of technology tend to treat the environment as 'an unproblematic, ahistorical backdrop to studies of technological change, implying that nature and technology are entirely distinct, and that environmental and ecological process play no role in technological development'.[5] Whether this critique still holds is questionable, but certainly no meaningful history of the development of the Thames Barrier could sustain this distinction. Pritchard's argument that the Rhône is best understood as an 'envirotechnical regime' in which nature is an actant chimes with the argument made here. Given contemporary debates about climate change, it seems remarkable that during the protracted political process preceding the passage of the 1972 Act, the case made for greater flood protection attracted little opposition. Even the Port of London Authority, which contested specific proposals, did not fundamentally challenge official thinking. It can be supposed that this was because rising water levels were thought symptomatic of the interglacial cycle rather than human actions, but this explanation should not obscure how the politics of the Thames Barrier complicates the claim that 'climate change did not emerge as a *political* issue until the 1990s' and helps contextualise the responses of the British government to evidence of anthropogenic climate change in the early 1970s.[6]

It seems equally telling that the recommendations of the Waverley Report, which urged extensive improvements to downstream and coastal flood defences, including the development of an early warning system, retained authority within government over the whole period.[7] This reflected the quality of the scientific data underpinning the report and the dire prognosis that data heralded, but it also suggested much about the governing ethos of the time. A distinguished peer was given a responsibility – few were more distinguished than Sir John Anderson – and his recommendations behoved the government to respond accordingly, however ill-thought through that response proved in practice. As such, the progress of the issue was conditioned by the peculiar alchemy of deference, the authority of the Establishment – that nebulous but palpable presence in post-war Britain – and the confidence placed in state-led technological and engineering infrastructure projects. That the final decision was so delayed also makes the history of the barrier a suggestive case study in the history of post-war industrial design and engineering and, more particularly, London's historic decline as a port city. In the event, the construction of the Thames Barrier did not help sustain a threatened envirotechnical regime, but helped create a new one.

And when placed in its proper global context, the history of the Thames Barrier raises far-reaching questions about environmental justice in the context of anthropogenic climate change.

The Waverley Committee and climate science

The first of the Waverley committee's terms of reference was 'to examine the causes of the recent floods and the possibilities of a recurrence in Great Britain'.[8] The committee sought submissions from a range of scientific authorities, including sundry academics, the Council of the Institution of Water Engineers, the Hydraulic Research Station at Wallingford in Oxfordshire, the Observatory and Tidal Institute at Liverpool and the Royal Navy, the last a considerable source of expertise. The immediate causes were rather straightforward to establish. A relatively high tide combined with a surge, the latter caused by record-breaking northerly winds, channeled an atypically large quantity of water down the narrowing north–south axis of the North Sea to the bottleneck at the Straits of Dover. The rotation of the Earth ensured that the water was deflected to the west of the tidal currents, making the east coast of England south of Flamborough Head in Yorkshire one of the most vulnerable coastal regions in the North Sea. A significant quantity of the excess water was forced up the Thames Estuary.[9]

Expert opinion emphasised that the tide and the surge were distinct phenomena, the surge being the exceptional event. Neither was dependent on the other. Higher surges caused by fierce northerlies were on record but they had occurred in conjunction with low tides. On 1 January 1922, for example, a surge caused the sea level at Southend to rise 11 feet above the expected level, but because the peak occurred two hours after low water there were no serious consequences. By contrast, on the night of 6–7 January 1928, the surge had a height of only 5 feet at Southend but coincided with the high water of spring tides, causing serious flooding and loss of life in the Thames Estuary. To top London's flood defences, the peak of the surge had to occur within an hour or two of high water and within a day or two of spring tides.[10]

The effect of the 1953 surge could have been worse but for another factor. Rainfall had been below average, leading to low fluvial discharge into most east coast rivers; had the peak fluvial discharge of 1947 coincided with the tidal surge, the flood effect of the surge would have been greater.[11] Although the probability of a recurrence was low, the 1953 tidal surge was the greatest on record and consistent with a trajectory of increasingly threatening and unpredictable weather events. Diagrams prepared by London County Council showed that the highest recent tidal surges – depicted in relation to London's flood defences – had in each case been up to 9 feet higher than expected; in one exceptional case – the night of 7–8 April 1943 – that figure reached 18 feet.[12] The surges of November 1897, January 1928 and February 1938 confirmed the upward trend, while water levels of 11, 12 or 13 feet above Ordnance Datum Newlyn at Sheerness or Southend were becoming increasingly frequent.[13] The general trend seemed incontrovertible. Higher tides and stronger tidal surges were to be expected. As Waverley observed, however, the scientists did not argue that this was caused by 'any appreciable change in the tides themselves', but was 'due to a steady rise of mean sea level relative to the land along the coasts of southern and south-eastern England'.[14]

How could this be explained? Several factors were thought instrumental. First, rising sea levels were a consequence of glacial ablation or melt, a symptom of climate warming. In 1939, the research of the Dutch geologist François E. Matthes into glacier regrowth in the Sierra Nevada, California, following its melting away in the Hypsithermal of the early Holocene, led him to coin the phrase the 'Little Ice Age' to describe the period 1300–1850.[15] Although the phrase was not common currency, scientists considered the climate to have 'improved' over the previous century. Second, the phenomenon of tilt: the north-west and north of England was gradually rising and, correspondingly, the south-east was gradually sinking – or downwarping – a notion that had some cultural traction at the time, particularly in East Anglia.[16] Third, and this was less well understood, a shift in wind pattern meant that southerlies and sou'westerlies were becoming marginally less predominant and northerlies marginally more common, making the North Sea stormier and tidal surges more likely.

Research conducted in the 1920s and 1930s on the extent of Norwegian glaciers by H.W. Ahlmann showed that glacier ablation had occurred at a rapid rate. These results echoed the findings of the Leningrad Arctic Institute with respect to the North-East Passage and observations of the limits of the ice edge between Denmark Strait and Novaya Zemlya made during the war by British Coastal Command. Arctic fauna had followed suit, both fish and fowl now found further north than was previously the case. This gave, as Ahlmann put it, 'proofs of climatic improvement', noting that R. Scherhag had suggested that this trend was of such significance that it could be termed 'the warming of the Arctic'.[17] A survey of the existing literature published in 1940 confirmed that glacial ablation had occurred concurrently throughout the world since the middle of the nineteenth century, leading to an increase in sea levels of about 0.05cm per year, approximately 5cm a century. The evidence did not suggest that the interior parts of Greenland or the Antarctic were melting, though scientists recognised the danger this would pose.[18] As L.C.W. Bonacina and E.L. Hawkes observed in 1947, 'If the polar inland ice-sheet were to melt as rapidly as the glaciers, the rise in sea-level would become a far-reaching phenomenon of great practical importance.'[19] In its submission to the Waverley committee, the Hydraulics Research Station echoed these arguments, but offered more dramatic figures, suggesting that the sea level was rising 1–2mm per annum, a calculation confirmed by current thinking. They also cited an alarmist paper that suggested the rate of rise could be as much as 3.5mm per year.[20] Historical records, current measurement and predicted future trends thus informed Waverley's recommendations.

Tilt was also linked to warming. At the end of the last ice age, some 20,000 years ago, glaciation had reached as far south as the line from the Bristol Channel to the Wash. With the weight of the ice no longer acting on northern Britain, a correction – post-glacial isostatic uplift – had been long in train and this accentuated the effect of rising sea levels in the south-east. As Waverley explained, the academic evidence suggested – and he got quite exercised about this material – that in Roman times the Thames had been tidal only as far as London Bridge;[21] other evidence suggested the tide had overtopped Teddington Weir on 20 March 1874.[22] Dr Harry Godwin, University Reader on Quaternary Research at Cambridge, explained to Waverley that the evidence of the past 50 years demonstrated that in recent times the degree of tilt had been two feet per century, though that level of subsidence could not have been maintained since Roman times. Godwin thought they could probably count on between one and two feet of further subsidence in the south-east over the next 50 to 100 years. A decade or so later Dr Anthony Michaelis, prominent science journalist and friend of Dr Hermann Bondi, later significant to our story, claimed the south-east of England was sinking an inch every 10 years,[23] predictions that wildly exceed current thinking.

Scientific papers circulated in Whitehall in 1970 once Solly Zuckerman, the government's Chief Scientific Advisor, put his weight behind the barrier scheme did not offer a fundamentally different interpretation of overall geophysical trends. J.R. Rossiter of the Institute of Coastal Oceanography and Tides confirmed that sea levels were rising and that south-east England was probably continuing to sink, causing a relative rise in sea level of about a foot a century. Although wary of making firm predictions, Rossiter hypothesised that the river's increased depth should mean a larger surface area, smaller velocities and, crucially, less bed friction. This would lead to faster propagation of the tide between Southend and Tower Pier, which had already enjoyed a mean interval decrease of about 16 minutes over the course of the previous century. The effect of this in the upper reaches of the Thames was a rise in high water levels of two feet.[24]

A radical paper by Hubert Lamb, the meteorologist and pioneering climate change scientist, was excitedly received by Rossiter. Lamb attributed the increased frequency of great wave height observed by the German Navy in the North Sea and the Norwegian Sea to the higher incidence of northerlies or nor'westerlies with a long sea fetch. These conditions, originating in the Atlantic Ocean, funnelled large quantities of water into the North Sea, and were the cause of the abnormally high frequency of slow-moving cyclonic centres in the region 50–60°N 10°W–10°E responsible for the prolonged rains and flooding seen in southern Britain in 1968 and 1969. Lamb argued that these weather patterns were part of a long-term sequence that scientists were only just beginning to understand. Data stretching back to the medieval period, although imperfect, suggested a repeated oscillation with a period length of about two hundred years in which mean frequencies of sou'westerlies over London coincided with the thicknesses of annual snow layers at the South Pole. Consequently, the increased frequency of nor'westerlies in the North Sea was likely to persist for much of the next century, continuing to generate the slow-moving weather fronts that increased the likelihood of tidal surges.

Lamb's argument then developed a distinctly sceptical discussion of theories posited by American and Soviet meteorologists – and since exploited by anthropogenic climate change deniers – that the Earth had entered a period of cooling. Short-term temperature trends did indeed suggest this, and Lamb was apparently comfortable with the idea that the increased levels of CO_2 in the atmosphere that might have accentuated warming could now offset the increase of particulates and other pollutants in the atmosphere, but evidence of long-term weather patterns saw him ultimately reject the cooling thesis. Either way, he insisted, whether the trend was towards warming or cooling, a continued increase in North Sea nor'westerlies or northerlies was likely, leading to continuing storminess.[25]

Whitehall deliberations: technocrats, civil engineers and the PLA

From the perspective of Whitehall, the scientific consensus and the broad case for a barrier was clear. The positioning of the barrier was more contentious. The more upstream the barrier's position, the greater the cost of improving downstream flood defences; the more downstream the barrier's position, the greater the engineering challenge and cost associated with the design and construction of the barrier as the river widened. These cost–benefit calculations had implications for London's spatial politics: riverine communities upstream of the barrier would be relatively privileged by the consequent need for lower-impact flood defences, whereas the possible effect of a 'reflected wave' caused by the barrier on downstream estuarial communities and infrastructure might necessitate more substantial flood defences than those already planned, an issue repeatedly raised by Kent and Essex county councils. For much of the process, however, the Port of London Authority proved the most influential voice. Since the 1908 Thames Act gave the PLA sole statutory authority for the management of the Thames tideway it had resisted any intervention that might undermine the navigational capacity of the river or the shipping interest.[26] In the 1950s and 1960s, the PLA enjoyed an Indian Summer as the weight of shipping and employment in London's docks reached historic highs, but the creation of the Greater London Council in February 1968 and the rapid decline of London as a port city in the 1970s radically transformed the play of power with respect to the governance of the Thames. Ultimately, the positioning of the barrier reflected this fundamental change in what London was.

In June 1955, the Ministry of Housing and Local Government (HLG), the lead ministry on the barrier project, published an internal report arguing that the middle of Long Reach, a section of the Thames 20 miles downstream of London Bridge, was the most suitable site for a barrier. As an easily navigable long straight stretch of river relatively free of dense industrial or residential development, landward approaches were easy and the river bottom – gravel overlying chalk – offered a hard substratum, relatively resistant to scour, with a high load-bearing capacity. Should the engineering solution require a mid-stream pier, the needs of shipping could be met by allowing clear waterways of 500 feet either side and vertical clearance of 200 feet above Newlyn Ordnance Datum. Side spans of 250 feet with vertical clearance of 50 feet would be needed to allow barges and light river craft to pass unhindered.[27] Once these requirements were agreed in principle, the Treasury granted permission to appoint consultant engineers, though it insisted this did not constitute a commitment to finance the project and the riparian authorities were told that at best the Treasury might partner other authorities.[28]

The Institution of Civil Engineers recommended Messrs Rendel, Palmer and Tritton and Sir Bruce White, Wolfe Barry & Partners. Both firms were long established with SW1 addresses, had close contacts in government and much experience of major civil engineering projects. Brigadier Bruce White had overseen the construction of the 'Mulberry' piers used in the D-Day landings, while Rendel, Palmer and Tritton's significant works included the Royal Albert Docks (1880), the West India and Millwall Docks Improvements (1929) and the Tilbury Docks Improvements (1930), as well as similar projects elsewhere in Britain and a host of imperial projects, including major undertakings in India and several significant commissions related to the development of the oil industry in the Middle East and Britain.[29] Firms like these helped the British Empire function. The ministry intended that the two firms would work in friendly competition before coming together to produce a final plan; the Hydraulic Research Board and the Geological Survey of Great Britain would provide free advice and the PLA grudgingly agreed to hire out its large model of the Thames.[30]

Friendly competition did not see the two firms arrive at the best solution. They differed over the best time to close the barrier and the volume of water that should pass through it – controlling the level of flow rather than entirely blocking a tidal surge was the aim. Further research resolved these questions, but both firms proved obstinate with respect to their preferred engineering responses. Each proposed a structure that would lower gates into the river. Rendel et al. favoured the 'vertical lifting type', comprising steel spans raised and lowered between towers built in the river, with hinged frames attached to the underside of the spans that carried vertical lifting gates. When the barrier was open, it would constitute a huge rectangular arch over the river. By contrast, White et al. advocated the 'horizontal swing type', comprising two 670 foot piers lying longitudinally in the river mounted with long arms carrying vertical lifting gates. Closing the barrier would mean swinging the arms into place across the river and then lowering the gates. Rendel et al. maintained that their proposals were preferable because the barrier could be closed by gravity in the event of a power failure, plus the 'vertical lifting type' would be significantly more expensive to build and maintain. A retractable barrier, whereby a huge cantilever girder would be housed in dry dock on either side of the river and then launched along tracks across the riverbed, was judged by both firms to be outside of engineering experience and, as such, too risky to pursue.[31] Still, attractive drawings of the scheme were produced, evidence of the speculative nature of the proposals at this stage.[32]

The Ministry of Transport was concerned that the plans suggested the horizontal clearances proposed for the centre and side spans were narrower than the terms of reference had stipulated and it objected to the placement of 'an obstruction in tidal waters in the main navigable channel of the greatest port in the world and of vital economic importance to the country'.[33] Despite this, HLG published 'Technical Possibilities of a Thames Barrier' as a Blue Book in March 1960 (Cmnd. 956). Local authorities downstream of the proposed barrier immediately raised some concerns. They asked about the effect massive piers would have on river flow, the risk of an oil spillage should a tanker strike a pier and whether a reflected wave caused by the barrier might worsen the downstream effect of a tidal surge, threatening civilian populations and industry alike. As Essex River Authority observed, there was 'no exact community of interest between those above and below the proposed barrier' and the downstream consensus was that the consequential costs of improving flood defence downstream of the barrier should be met by those upstream.[34] Still, a conference of the Thames estuary authorities that December approved the plans on the assumption that substantial funding would be forthcoming from central government. The only significant objection was made by a PLA engineer who explained that the plans raised fundamental questions about how the Thames was navigated. Ships sailed in on the flood tide and out on the ebb tide, but the piers would create turbulence and narrow the river, increasing the speed of the tide. Necessary speed restrictions during the construction period – and possibly thereafter – would be impossible to meet without reversing the pattern of navigation, so that 'ships were brought up against the stream'.[35]

In February, alarmed officials were reassured by Sir Leslie Ford of the PLA that the authority's position was not as hostile as the engineer had suggested, but it was troubled by estimates that during construction the tideway would be reduced to a 500-foot gap and the tidal speed increased from 3.5 to 6 knots. That April, Commander Parminter made the PLA's difficulty plain:

> the principal difficulty was that of taking a big ship through the gap between the piers and the difficulties in handling a big ship at slow speed would be enhanced by the effects of the eddies which were to be expected near the piers. Big ships always went up river on a

flood tide but if the barrier were constructed it would be necessary for them to proceed up the River on Ebb tide and down on the Flood tide with consequent delay which would result in missing favourable tides elsewhere.[36]

Moreover, when the barrier was closed advance warning would be needed so anchorages could be found for ships, while placing the barrier in the middle of Long Reach would not only bisect an industrial area of growing importance but also halve the last significant stretch of safe deep water anchorage for inward-bound ships. HLG acceded to PLA demands: it accepted there could be no permanent structure in the navigable channel, that a minimum gap of 1,400 feet was required when the barrier was open, and that it should be located at the western end of Long Reach. Cutting back the bank (and dredging) would be necessary to improve alignment and visibility. This meant a recourse to a retractable barrier housed in dry docks, costed at £20m.[37]

HLG's report to the Home Affairs Committee in July 1962 reiterated Waverley's assessment of the threat, placed the likely cost of a barrier at £30m, and requested £50,000 to finance further investigations. With its usual ritualised display of reluctance, the Treasury released £56,000, having sanctioned an additional £6,000 for on-site borehole testing; the original engineers began fresh hydraulic investigations and a complete redesign. What had been considered outside the realm of engineering possibility in the 1950s was now the basis for the new designs. Rendel proposed a 'high-level' type, in which the arms would be cantilevered out from dry docks and the gates lowered into the river; White went with a 'low-level' design, whereby the arms would be launched along a sill constructed across the river bed. Rendel argued silting might have an adverse effect on the 'low-level' design, and at £30m, it was some £9m more than the estimated cost of the 'high-level' design; the construction time for both was estimated to be six or seven years, a factor that would grow in importance as the process lengthened.[38]

In the meantime, consultations by HLG on the land requirements unearthed a difficulty peculiar to the time. The War Office intended to put Purfleet Camp and Magazine, a portion of which overlapped with the proposed barrier's northern site, up for auction in March 1963. The War Office insisted the process could not be reversed, for undertakings had been made to an Italian oil company that the land was available for purchase; although the War Office allowed that the PLA could also bid, it suggested the barrier be shifted 200 feet westward, allowing the two installations to exist side by side. This was a characteristic move by a ministry still struggling to come to terms with the reality of civilian power and priorities.[39] Notwithstanding the ill-advised observation by a War Office official that the negotiations on behalf of the Italians had been handled by the stepson of Peter Thorneycroft, former Chancellor of the Exchequer, HLG made it clear that the wants of a private company would not be placed above the public interest. War Office pressure on London County Council to buy the land also went unheeded. In 1964–5, stroppy queries from the War Office and the Ministry of Defence were brushed off by HLG: they would just have to wait, though the land could be offered for sale to the newly formed Greater London Council.[40]

Orthodoxies challenged: the case for a barrage and the weakening of the Port of London Authority

In the mid-1960s, ministerial attention was roused by the revival of the old question of whether the Thames needed not a barrier but a barrage. Tom Driberg, Labour MP, raised the question in response to the award of the President's Prize of the Royal Institution of Chartered Surveyors to Michael Wand, a constituent, for his paper 'A Town in the Thames – the New South-East Centre?' Characteristic of the futurism that could capture the public's imagination in the 1960s, it was but the latest attempt to imagine a re-engineered Thames, which, as Richard Crossman admitted, had never had a full public airing.[41] Herbert Spencer had promoted the agenda in the mid-nineteenth century and it was revived in 1903 when a group of parliamentarians were inspired by the decision to barrage the Charles River in Boston. Frustrated by the refusal of the Royal Commission on the Port of London to hear evidence in favour of a barrage, they formed an informal parliamentary committee, commissioned a series of expert studies and promoted three bills to draw attention to the question. The case was made in *The Port of London and the Thames Barrage: A Series of Expert Studies and Reports* (1907), which scrutinised the conversion of 'the river from a highway into a dock' and criticised legislative attempts to render the Thames more commercially efficient for giving the Port Authority monopolistic control over the estuary. Among the technical factors assessed were the suitability of the geological strata and the vexed question of silting and pollution, but the fundamental case was made in terms of improving the navigable capacity of the Thames and enhancing its 'popular use'. To maintain upstream water permanently at high tide with a barrage and eight locks at Gravesend would provide sufficient depth for ships of increased tonnage, end hazardous tide-waiting, and reduce the dues paid at dock entrances and the cost of barging, pilotage and labour. To make the full width of the river always available would ease congestion and allow obstructive floating piers to be dismantled and landing places to be built closer to the shores. To slow the water flow would allow reliable passenger services to be developed and more use by pleasure boats. Aesthetic improvements would follow too: supposedly ugly mudflats would not be exposed twice a day, fresh water would make for a more pleasant environment and the sewage system would be re-engineered so effluent was pumped out below Gravesend, making for a more fragrant river.[42]

Barrage enthusiasts like Lord Desborough, for 32 years the Chair of Thames Conservancy, were frustrated by what they took to be the PLA's failure to make an objective assessment of the case. As J.H.O. Bunge explained in *Tideless Thames in Future London* (1944), the 1928 flooding put the idea back on to the agenda among those who recognised that the 'only radical solution' was to keep the tides 'out of London altogether' by locating 'a barrage well below the floodable area'. The decision of a public enquiry in July 1934 that the tide made river buses unworkable on the Thames prompted the formation of the Thames Barrier Association in 1935.[43] When the barrage question was debated in the House of Lords in May 1937, Lord Dudley directed his fire at the 'complete dictatorship' the 1908 Act had given the 'pig-headed' Port of London Authority over the whole estuary below Teddington Lock. Under more enlightened direction, the Thames could become 'a slow-moving lake' and 'the playground and the pathway of London's citizens', becoming, according to Lord Jessel, like the Charles River, where 'there is a full river, which provides many amenities for the citizens – yachting clubs and every sort of thing of that kind'.[44] Lord Richie, Chair of the PLA, emphasised the negative effect a barrage would have on sewage and shipping and insisted the PLA had considered the idea, just as it had the less transformative and the more fantastical schemes that came its way. Ritchie cannily suggested a barrage would be vulnerable to bombing in the event of war.[45] Leslie Burgin, the Minister of Transport, when asked if he intended to establish a committee of inquiry, responded that this was a question for the PLA.[46] Under considerable political pressure, the PLA conceded an enquiry, but on its eve, 28 March 1938, Burgin urgently requested that the PLA halt its proceedings because the Committee of Imperial Defence feared that to publicly air these issues would compromise national security.[47] Bunge lamented the preparation done and remained convinced the PLA, knowing it would lose the argument, had made its influence felt in Whitehall.[48]

When the idea resurfaced in the post-war decades it was often considered in conjunction with the plan to link Britain and France with an underwater tunnel between Dover and Calais. A channel tunnel seemed imminently realisable and visionary planners re-imagined Britain's north–south transport links as a new road and rail network firmly to the east of London and integrated into a barrage. Despite lobbying activity, the view in Whitehall remained resolutely sceptical. In 1955, when the consulting engineers were contracted for the first time, the Ministry of Transport made it clear to HLG that they should not be asked to investigate a barrage. It was simply too expensive and its implications were too significant. To remove the 'scouring action of the tides' would completely alter 'the regime of the river', leading to greater siltation and more pollution, while a permanent rise in ground water level risked flooding basements in low-lying riverside areas and interfering with drainage and underground services, including London's underground railway. Engineering solutions could be found, of course, but at greater cost than the barrage itself.[49]

The barrage also got an outing in Professor Hermann Bondi's catalysing report on the need for a barrier, though the report was mainly significant for questioning the underlying assumptions that had steered discussion so far. On Zuckerman's recommendation, Bondi had been commissioned by Richard Crossman and Lord Kennet who, given the threatening water levels of 10 December 1965 and the engineering problems raised by the latest designs, agreed that a fresh look at the problem was needed by a scientist of 'Nobel quality'. Bondi, then professor at King's College London, clearly relished his task. He shared Waverley's assessment of the risk and wrote in melodramatic terms of the threat a major tidal surge posed London and, particularly, the underground railway system, which he thought could be put out of action for a year. Bondi did not attempt to calculate the probability of a major flood, but simply argued that it was foreseeable and so the government must act.[50] Although he was keen on a barrage, suggesting that at high tide the north bank of the Thames made London aesthetically superior to Paris and the equal of Leningrad, he was convinced a barrage would lead to greater siltation, making increased dredging a permanent charge of the PLA. Ultimately, Bondi believed a barrage should be pursued only as part of a fundamental rethink of the southeast, which must include a willingness to diminish the utility of the docks in the upper part of the river, but the apparent absence of any enthusiasm for the idea among the public made the cause a dead letter.[51]

Bondi found it absurd that the mariners had been allowed to dictate the barrier's requirements, and no attempt had been made to quantify the cost of different approaches. If the 1400-foot requirement was determined by the location, asked Bondi, would not a change of location make for more cost-effective or technologically viable solutions? How would the problem be approached if, for example, questions of cost meant the opening was limited to 800 feet? The wide opening needed at Long Reach reflected the large swing big ships needed to come alongside various jetties and wharfs, but to position the barrier upstream of the Royal Docks at the Isle of Dogs where big ships did not go would make a smaller opening serviceable without disrupting shipping. The money saved could be spent on improving downstream flood defences. Alternatively, placing the barrier at Woolwich was possible, though it would be necessary to 'sterilise' the banks by buying up existing berths used by large ships needing big swings, which would then allow openings as small as 350 feet. Protecting Plumstead Marshes, site of the new Thamesmead residential development (first stage due to begin on 1 March 1967), would mean additional costs. Other possible locations were just below either the Ford Motor Works or Dagenham Dock, where Bondi thought a minimum opening of 750 feet was needed.[52]

The significance of Bondi's intervention stemmed from both his proposals, which proved influential, and his approach to the problem. When Anthony Greenwood, Crossman's successor at HLG, took the report to the Cabinet's Home Affairs Committee, it was clear that Bondi had galvanised ministers. The Treasury, irritated by the suggestion it would meet two-thirds of the eventual cost, conceded that Bondi had raised questions needing answers and it took HLG to task for not recognising that a barrage would have to be looked at again.[53] Kennet, junior minister at HLG, Chair of the Flood Protection Policy Committee and barrage enthusiast, became increasingly vocal on the question.[54] More significant was the shift in responsibility away from HLG, which retained ultimate oversight for the project, to the newly constituted Greater London Council (the GLC) in February 1968. The Treasury agreed HLG should increase its share of these research costs from a third to a half and it allowed the overall cost to reach £539,000 by 1970.[55]

The transfer of responsibility affected the balance of power. It was less that London County Council had simply deferred to the PLA, than that it simply did not have the GLC's broad responsibilities or its broad perspective on how the Thames might serve London's citizens. A GLC working party examined the possibility of a barrage at Limehouse Reach, commissioning the Hydraulics Research Station to investigate silting; with respect to a barrier, it examined the dangers of a reflected wave. Initially, the prospect of improving amenity loomed large in the GLC's thinking, though this was tempered by its consciousness of how Battersea and Bankside power stations were dependent, respectively, on coal and oil supplied by coastal shipping.[56] The PLA's plaintive response to Bondi was that it had not been asked to consider sites other than Long Reach and it accepted that an 800-feet opening could be sufficient at the two Dagenham sites, though it considered Woolwich already too built-up. The authority privately admitted that Tilbury Docks, which had been adapted to containerisation, a development that loomed large in Bondi's projections, would in time take an increasing proportion of shipping, relieving pressure on the Royal Docks, though there was not yet much talk about their future, despite the closure of Surrey Docks in 1969.[57]

In the event, the GLC came down against a barrage. Its report of January 1970 confirmed the risk to London identified by Waverley, duly recognised the advantages a barrage would confer on London, but equally recognised the challenge it would pose to shipping, the problem of increased siltation, the ecological effect of lower levels of oxygen saturation caused by impounding the water, and the potential problem caused by the location of London's sewages outlets. Much would be determined by where the barrage was positioned and the GLC feared it could take another 10 years to make that decision: 'It is scarcely likely therefore that a decision now to build a barrage would result in London being protected against a surge tide before the 1980s.'[58] The irony of the statement cannot be missed. As for the alternative of raising the walls along the Thames to the six feet needed, the cost would be little different to a barrier but obscuring pedestrian views of the Thames would significantly damage the river's amenity value.[59] Moreover, the Hydraulic Research Centre had established that if the barrier was closed for six hours, it was hard to conceive of the circumstances when the fresh water flow would cause the water level to rise more than three feet; as for the risk of a reflected wave to land downstream of the barrier, tests demonstrated that if the barrier was closed at low tide and the sluices closed gradually, allowing some water through, the reflected wave could be reduced to no more than a few inches at Canvey Island or Southend.[60] A barrier, then, it had to be, and the GLC expressed its preference for the low-level type, arguing that research showed it to be within the bounds of current engineering capability, a claim that could not be confidently made of the more complicated high-level type.[61]

This, then, seemed clear enough, but that April Kennet felt bound to write to the prime minister, short circuiting 'all proper channels', to express his concern at government inaction. 'I have tried all I can think of', he wrote, 'and have now concluded it is only right to let you know personally and directly that this situation is now extremely grave.' In a somewhat humiliating letter, Harold Wilson replied that a HLG paper was due to be discussed at the highest levels of government, though the general election that June meant responsibility passed to the incoming Conservative government under Edward Heath.[62] By July, the government had accepted the GLC's recommendation that the barrier be located at Silvertown in Woolwich and was prepared to take on the vested interests of the PLA, which it had come to consider opposed a barrier *tout court*. Tasked to prepare a chair's brief ahead of the crucial meeting, S.K. Gilbert at HLG was determined that discussion would not be dominated by the PLA's 'hobby horses' of siltation and pollution or its increasingly dubious insistence that London's docklands would remain in full use for another 20 years: a push must be made towards the consideration of strictly practical questions.[63]

That dismissive reference to pollution is striking, for any tendency to dismiss the PLA as a purely reactionary force needs to be tempered by an awareness of its growing responsibility for managing pollution levels in the tideway and how these concerns were becoming part of the political mainstream. In an account presenting a largely positive picture of the PLA, Leslie Wood explained how since the 1940s it engaged in an extensive programme of research into the extent and causes of riverine pollution, which led to a major report in 1964 and the passage in 1968 of an Act giving the PLA the responsibility for pollution control in the Thames enjoyed by the other river authorities since 1951. If higher levels of oxygen saturation were achieved by treating sewage effluent more effectively and aerating the heated effluents released into the river from power stations and other industrial plant, migratory fish like salmon and trout would return to the Thames. A significant upgrade of the storm sewage tanks and the sewage treatment works meant that by 1980 the river's pollution load had been reduced by 90 per cent.[64]

In the event, the GLC's case against the barrage did indeed reflect the likely effect impounding the Thames would have on oxygen saturation and water temperature upstream. In October 1970, Zuckerman wrote to the prime minister advising that a barrier be built at Silvertown, financed at 50 per cent by central government, and investigations into all other sites be halted; in early November, the GLC was informed that Peter Walker, the first Secretary of State for the newly established Department of the Environment, agreed that a movable barrier was necessary and should cross the Thames at Silvertown.[65] With this the settled policy of both the government and the GLC, the type of barrier and the width of the openings once again became a source of contention. New solutions were prepared by the engineers, including a drum gate scheme which would involve deep excavation, until the idea of a 'rising gate type' was hit upon by Charles Draper. In this ingenious solution, the gates are attached to large wheel-like structures and sit on the bed of the river when open. Rotating the wheels lifts the gates into the closed position between the piers.[66]

Things moved swiftly at a succession of meetings in December. The choice put by the GLC was between a 'drop gate type' with a main opening of 450 feet and auxiliary openings of 150 feet or a 'rising gate type' with three 200-feet openings, which would give an estimated cost saving of £9–10 million. The PLA rejected the latter, explaining that tidal speeds and the angle of approach made it likely that large ships would hit the piers, but the response by the GLC's consulting engineers was that the drop gate type was not practical and, in any case, fewer than two ships of over 10,000 tons per day now passed Silvertown for upriver berths and their navigational difficulties could be resolved if Voight Schnieder tugs were introduced.[67] Allies of the PLA now gave way. The Department of Trade and Industry conceded that 450 feet was ideal, but 200 feet might work; the Chamber of Shipping and Trinity House insisted 450 feet was preferable, but an appropriate system of tugs would make 200 feet possible. Increasingly isolated, the PLA made a stand against a rising sector gate, insisting the risk of collision was high, but its offer to accept a 350-foot drop gate and two 200-foot rising gates was politely rejected by the Department of Environment. In a press release of 22 December, Peter Walker accepted the GLC's recommendation of a rising sector type and expressed his hope that construction would begin in 1973. According to a scribbled note on a draft letter, the 'clinching' moment came when Trinity House broke ranks with the PLA.[68]

Although the River Thames (Barrier and Flood Prevention) Bill 1972 had the backing of the Department of the Environment, the Ministry of Agriculture and the Department of Trade and Industry, it was, as befitted its statutory responsibility, a private bill promoted jointly by the GLC and the Kent and Essex river authorities. It was not, as such, a government bill, although in the end the Treasury footed most of the bill. In the months preceding the first reading, Whitehall was irritated by the PLA's continuing attempt to undermine the decision and the failure of the GLC to deal effectively with the political fallout, which seemed of a piece with its earlier assumption that the Department of Environment would make the difficult decisions on their behalf and force the PLA into submission.[69] Nigel Spearing, Labour MP for Acton, proved particularly awkward. During the parliamentary debate on the second reading he made it clear he did not wish to see the bill fail, but he questioned proceedings that had seen the navigational interest bullied into accepting the 200-foot opening at a 'murky' meeting in the Department of the Environment in late 1970.[70] There was some truth to this, but only the most partisan of observers could suggest that the PLA had been the victim of a process outlined by one Whitehall civil servant in 1968 as the 'History of the Thames Barrier Project. "The years the locusts ate."'[71]

A new envirotechnical regime

Waverley had urged the construction of a retractable barrier across the River Thames not simply as a response to the catastrophic events of 31 January 1953 but as a necessary defence against a set of geophysical and climatic developments that could not be controlled or overcome any other way. Of all the official documentation generated by the question, it was perhaps a note by the Treasury that best captured the meaning of this intervention. When the PLA claimed that the state should cover the costs of pilotage, comparing passage through the barrier to passage through Tower Bridge, the Treasury responded:

> It seems to us, however, to be dangerous to imply that the Barrier is in the same category as Tower Bridge. The Barrier is not simply an inconvenience: it is something which changed physical characteristics of the river itself have made necessary. In these circumstances our general line on compensation etc. should surely be negotiating the Barrier will become part of the true cost of using the Thames for shipping purposes in much the same way as natural reefs and obstructions.[72]

Foregrounding the need to respond to the 'changed characteristics of the river' identified a distinct category of state activity. The solution chosen, and the navigable costs associated with it, should not obscure the singularity of the problem: there was nothing whimsical about the proposed barrier. Consequently, it was plausible for the government to categorise its technological response to climate change as equivalent to 'natural reefs and obstructions': the specificity of the barrier made it no less inevitable. And by dissolving the distinction between technology and the natural world, the Treasury effectively categorised the Thames Estuary as 'second nature'.

But that 'second nature' had already been subject to change, as Leslie Wood's account of the PLA's successful attempt to restore the Thames tideway as a habitat for fishes made clear. This is part of a larger story that challenges the declensionist narrative that once dominated environmental history. De-industrialisation and improved sewage technologies have improved riverine habitats throughout the post-industrial world. To take one example, this is part of the story Leona Skelton tells in her history of the River Tyne.[73] Writing on the eve of the barrier's completion, Wood noted its environmental effects 'are likely to be insignificant' but he warned that if it were used as a half-tide barrier, as opportunistic barrage enthusiasts had proposed, it would create a localised thermal barrier that would interrupt fish migration and keep mudflats covered, thereby impeding wildfowl access to tubificid worms, which in turn would be affected by the size of particulate matter in the substrate caused by silting.[74] The ecological implications of the half-tide use of the barrier were understood in the early 1970s and the view in Whitehall appears to have been that this possible use of the barrier represented such a significant alteration to their intentions that it would need separate consideration by government and therefore need not be resolved in order for the primary legislation to go ahead.[75]

Although the Thames Barrier has not been used as a half-tide barrage, its meaning would nonetheless change, undermining the claim that its high-tech stainless steel structures were in some ontological sense 'natural'. Originally intended to protect London against the cyclical consequences of interglacial weather patterns, it is now regarded as the city's first defence against the threat posed by anthropogenic climate change, a shift in historical sensibility as significant as any that has gone before. A second Thames Barrier, much larger and further downstream, has been proposed, but the Environment Agency says there is no need before 2070: the existing barrier is used more frequently than was anticipated, but remains fit for purpose. Jon Agar asks if there is a British equivalent to the 'technological sublime' identified by David Nye with respect to the awe generated by the human-built component of the territories of the United States.[76] Does the Thames Barrier qualify? Its iconic design, particularly when the gates are raised, has inspired some magnificent photographs, but to visit its decaying visitor centre under more ordinary circumstances delivers a milder thrill. As a large-scale state project conceived at the height of post-war dirigisme but opened when under sustained ideological assault, the potential of the Thames

Barrier as spectacle has perhaps never been fully realised. The contrast with the highly commercialised redevelopment of London's docklands in the 1980s and 1990s, a symbol first of Thatcherite hubris and now of Britain's embrace of a post-industrial and neo-liberal economy, is striking. Aesthetically, the barrier is of a piece with docklands; politically, it is of a piece with Thamesmead. Either way, it was foundational to the transformation of an envirotechnical regime in a period of significant political change. In this respect, the barrier is comparable to the equally iconic *Intercity 125* locomotive (British Rail Class 43) of 1976, another engineering product of the mixed economy, which helped renew an equally significant envirotechnical regime and whose working life has also exceeded expectations: Class 43 was later privatised, a fate yet to befall the barrier.[77]

To observe that the barrier's existence, and more particularly its possible supplement, helps protect private as well as public interests raises profound questions about environmental justice. Londoners are safe from the potentially catastrophic consequences of climate change not because they are blessed by geographical good fortune but because they live in a wealthy state capable of mobilising sufficient resources to mount an effective defence. People facing periodic coastal flooding elsewhere, whether it is in peripheral parts of the developed world, which include the north or south-west of England, or on a more catastrophic scale in the developing world, do so not because they are peculiarly geographically vulnerable but because they are of low national priority or live in states incapable of developing the necessary defensive infrastructure. Historians of climate change, and particularly historians of the technological solutions intended to mitigate its effect, should not lose sight of the global inequalities that give structures like the Thames Barrier meaning. Faced with the global consequences of the carbon economy's long history, in no meaningful sense can it be said that the polluter pays.

Notes

1   For an account of the storm and its aftermath, see Peter J. Baxter, 'The East Coast Big Flood, 31 January–1 February 1953: A Summary of the Human Disaster', *Philosophical Transactions of the Royal Society of London A: Mathematical, Physical and Engineering Sciences* 363, no. 1831 (2005): 1293–312.

2   Quoted in Peter Coates, *Nature: Western Attitudes since Ancient Times* (Cambridge: Polity, 1998), 19.

3   Sara B. Pritchard, *Confluence: The Nature of Technology and the Remaking of the Rhône* (Cambridge, MA: Harvard University Press, 2011), 5.

4   William Cronon, *Nature's Metropolis: Chicago and the Great West* (New York: Norton, 1991),
    xvii; Thomas P. Hughes, *Human-Built World: How to Think about Technology and Culture*
    (Chicago: University of Chicago Press, 2004); Richard White, *Organic Machine. The Remaking
    of the Columbia River* (New York: Hill & Wang, 1996).

5 · Pritchard, *Confluence*, 12.

6   Daniel Bondansky, 'The History of the Global Climate Change Regime', in *International
    Relations and Global Climate Change*, ed. Urs Luterbacher and Detlef F. Sprinz (Cambridge,
    MA: MIT Press, 2001), 24; Jon Agar, '"Future Forecast – Changeable and Probably Getting
    Worse": The UK Government's Early Response to Anthropogenic Climate Change', *Twentieth-
    Century British History* 26, no. 4 (2015): 602–28.

7   On the flood warning system, see Alexander Hall, 'Risk, Blame, and Expertise: The
    Meteorological Office and Extreme Weather in Post-War Britain' (PhD diss., University of
    Manchester, 2012), 148–53.

8   Report of the Departmental Committee on Coastal Flooding (London: HMSO, 1954) [here-
    after Waverley], 1.

9   J.R. Rossiter, 'The North Sea Storm Surge of 31 January and 1 February 1953', *Phil. Trans.
    R. Soc. A* 246, no. 915 (1954): 371–400.

10 K.F. Bowden, 'Storm Surges in the North Sea', *Weather* 8, no. 3 (March 1953): 82.

11 HLG 50/2501 Memo. no. 3, 1953–4, HLG 50/2501, The National Archives [hereafter TNA].

12 The diagrams can be found at HLG 50/2502, 1953–4, TNA.

13 Ordnance Datum Newlym is the mean sea level at Newlyn in Cornwall, 1915–21, and is used
    by the Ordnance Survey as the basis for deriving altitude on maps; Waverley, 6.

14 Waverley, 6.

15 John A. Matthews and K.R. Briffa, 'The "Little Ice Age": Re-Evaluation of an Evolving Concept',
    Geografiska Annaler: Series A, Physical Geography 87, no. 1 (2005): 17.

16 David Matless, *In the Nature of Landscape: Cultural Geography on the Norfolk Broads*
    (Chichester: Wiley-Blackwell, 2014), 206–7.

17 Hans W. Ahlmann, 'Researches on Snow and Ice, 1918–40', *The Geographical Journal* 107, no.
    1/2 (1946): 11–25.

18 Sigurdar Thorarinsson, 'Present Glacier Shrinkage, and Eustatic Changes of Sea-level',
    Geografiska Annaler 22 (1940): 131–59.

19 L.C.W. Bonacina and E.C. Hawkes, *Climatic Change and the Retreat of Glaciers* (Royal
    Meteorological Society, 1947), 190.

20 Memo of evidence submitted by Hydraulics Research Station, 25 April 1953, HLG 50/
    2501, TNA.

21 Waverley, 46.

22 *The Port of London and the Thames Barrage: A Series of Expert Studies and Reports* (London: S.
    Sonnenschein & Co., 1907), 29.

23 *The Times* (November 1967) HLG 52/2020, TNA.

24 J.R. Rossiter, 'The Long-Term Stability of the Tidal Regime in the Thames' and 'Secular Trends
    in Levels at Thames Stations' enclosed to Zuckerman, 22 January 1970, CAB 168/259, TNA.

25 H.H. Lamb, 'On the Problem of High Waves in the North Sea and Neighbouring Waters and the
    Possible Future Trend of the Atmospheric Circulation', draft paper, CAB 168/259, TNA. For
    Lamb's pioneering but cautious work on climate change, see his *Climate History and the Future*
    (Princeton: Princeton University Press, 1977).

26 On the complexities of the Thames governance, see the meticulous work of Vanessa Taylor,
    especially 'Watershed Democracy or Ecological Hinterland? London and the Thames River
    Basin, 1857–1989', in *Rivers Lost, Rivers Regained: Rethinking City–River Relations*, ed.
    Martin Knoll, Uwe Lübken and Dieter Schott (Pittsburgh: University of Pittsburgh Press,
    2017), 63–81.

27 Thames Technical Panel Report and Chairman's Confidential Minute, 27 June 1955, HLG 50/
    2495, TNA.

28 Treasury to HLG, 25 January 1956, HLG 50/2495, TNA.

29 Rendel, Palmer and Tritton, *Yare Basin Flood Control Study* (London: Rendel, Palmer &
    Tritton, 1978).

30 See Minute by McNaughton, 15 May 1956, HLG 50/2495, TNA; Letter from Sir Claude Inglis
    of the Hydraulic Research Station, 29 May 1956, HLG 50/2495, TNA; Letter from W. J. Pugh
    of the Geological Survey of Great Britain, 28 May 1956, HLG 50/2495, TNA; the Treasury

approved the cost of hire at £50 per day for both firms, taking the overall cost of the consultation to £24,750, though this soon jumped another £1,000 by 25 June 1957, HLG 50/2496, TNA.

31 Rendel et al. to HLG, 15 August 1958, HLG 50/2497, TNA; White et al. to HLG, 19 August 1958, HLG 50/2497, TNA; Rendel et al. to Henry Brook, 24 September 1958, HLG 50/2497, TNA; Note on the Reports, 6 October 1958, HLG 50/2497, TNA.

32 Reproductions of these and later proposals can be found in Stuart Gilbert and Ray Horner, *The Thames Barrier* (London: Thomas Telford, 1984), although these are no match for the originals, which can be found in the above cited files of the National Archive, Kew.

33 O.F. Ginger, Ministry of Transport, to Pearce, HLG, 29 October 1959, HLG 50/2497, TNA.

34 Account of meetings of Thurrock and Canvey Island urban district councils, 1 December 1960, HLG 120/344, TNA.

35 Account of the meeting by W. O. Hart, clerk of LLC, to Evelyn Sharp, HLG, 19 December 1960, HLG 120/344, TNA.

36 Notes on meeting between HLG and PLA, 22 February 1961 and 11 April 1961, HLG 120/344, TNA.

37 Report of the Navigational Working Party under Capt. H. Menzies, RN, 25 January 1962, HLG 120/380, TNA; Report of Meeting of Interested Parties including the PLA, HLG & MAFF, 27 September 1962, HLG 120/380, TNA; Statement by Chamber of Shipping, 16 November 1962, HLG 120/380, TNA.

38 Thames Flood Barrier. Report of Steering Committee on Consulting Engineers' Reports, 6 July 1964, HLG 120/528, TNA; for correspondence on the boreholes investigation, see HLG 120/642, TNA.

39 See Matthew Kelly, *Quartz and Feldspar: Dartmoor. A British Landscape in Modern Times* (London: Jonathan Cape, 2015), 308ff.

40 Henry Butcher and Co., auctioneers, to Ass. Sec., HLG, 14 December 1962, HLG 120/381, TNA; War Office to HLG, 1 January 1963, HLG 120/381, TNA; H.H. Browne, HLG, to G.A. Wilson, PLA, 14 January 1963, HLG 120/381, TNA; Browne, HLG, to Hopkins, WO, 21 January 1963, HLG 120/381, TNA; Chief Engineer, PLA, to Browne, 21 January 1963, HLG 120/381, TNA; Fowler, HLG, to Hopkins, 25 January 1963, HLG 120/381, TNA; Corfield to Ramsden, 31 May 1963, HLG 120/381, TNA; Ramsden to Corfield, 15 May 1963, HLG 120/381, TNA; Peter Kir, WO, to Corfield, 31 December 1963, HLG 120/381, TNA; Corfield to Kir, 26 November 1963, HLG 120/381, TNA; MOD to HLG, 2 April 1964, HLG 120/381, TNA; HLG to MOD, 9 April & 19 October 1964, HLG 120/381, TNA; MOD to HLG, 7 January 1965, HLG 120/381, TNA; Fowler to Hopkins, 19 February 1965, HLG 120/381, TNA; Hopkins to Fowler, 29 March 1965, HLG 120/381, TNA; HLG to MOD, 9 April 1965, HLG 120/381, TNA.

41 Crossman to Tom Fraser, Ministry of Transport, 5 July 1965, JH 7/16, TNA.

42 Barber et al., *The Port of London and the Thames Barrage*, 4–5, 140.

43 J.H.O. Bunge, *Tideless Thames in Future London* (London: Thames Barrage Assn., 1944), 31, 45–6.

44 HL Deb., 26 May 1937, vol. 105, c235–7, 240, Hansard.

45 HL Deb., 26 May 1937, vol. 105, c243–8, Hansard.

46 HC Deb., 02 June 1937, vol. 324, c1012, Hansard.

47 *Daily Mail*, 29 March 1938.

48 Bunge, *Tideless Thames in Future London*, 48–50.

49 Watkinson, MT, to Sandys, HLG, 28 February 1956, HLG 50/2495, TNA. This exchange was promoted by the TBA's lobbying. Lord Brabazon, R.A. Allen, MP, A.W.J. Lewis, MP, and Bunge were fobbed off with a letter signed by the minister.

50 Bondi Report, 1966, HLG 145/105, TNA, 3–10.

51 Bondi Report, 1966, HLG 145/105, TNA, 11–17.

52 Bondi Report, 1966, HLG 145/105, TNA, 22–30.

53 P.F. Clifton, Treasury, to Lloyd-Davies, HLG, 13 June 1967, HLG 52/2020, TNA; Cousins, Treasury, to John Crocker, HLG, 15 September 1967, HLG 52/2020, TNA; Cousins to J. H. Street, HLG, 28 December 1967 and 4 January 1968, HLG 120/1381, TNA.

54 For example, see note by Kennet on Bondi, 14 November 1967, HLG 120/1381, TNA.

55 Home Affairs Committee and HLG press release on GLC, 20 February 1968, HLG 145/105, TNA; correspondence detailing what the Treasury described as 'rather unfortunate' increases can be followed in HLG 52/2020, TNA.

56 Note on meeting between HLG, MAFF and GLC, 2 May 1968, HLG 145/105, TNA; Draft Paper by Department of Economic Affairs, 11 June 1968, HLG 145/105, TNA.
57 Dudley Perkins, PLA, to Browne, 24 November 1967, HLG 120/1381, TNA.
58 GLC Report, 1970, HLG 120/1377, TNA, 10–13, quotation at 13.
59 GLC Report, 1970, HLG 120/1377, TNA, 14.
60 GLC Report, 1970, HLG 120/1377, TNA, 14–16.
61 GLC Report, 1970, HLG 120/1377, TNA, 44.
62 Wayland to Wilson, 10 April 1970, PREM 13/3261, TNA; Wilson to Wayland, 20 April 1970, PREM 13/3261, TNA.
63 S.K. Gilbert to B.W. Wood, 24 and 27 July 1970, HLG 120/1373, TNA.
64 In general, including data on the species of fish present in the tideway, see Leslie B. Wood, *The Restoration of the Tidal Thames* (Bristol: Institute of Physics Publishing, 1982), 87–125.
65 Zuckerman to Edward Heath, 14 October 1970, CAB 168/259, TNA; R.G. Adams, DoE, to Clerk, GLC, 4 November 1970, HLG 120/1373, TNA.
66 For a fuller account of the technical questions, see Gilbert and Horner, *The Thames Barrier*.
67 This recalled an earlier argument made by Gilbert that the PLA tended to deal in percentages as a means of obscuring the actual numbers of ships passing through.
68 Notes by S.K. Gilbert, 3 November 1970, 2 December 1970, HLG 120/1380, TNA; notes on meetings, 4 and 7 December 1970, HLG 120/1380, TNA; Adams to Beddoe, 9 December 1970, Beddoe to Edmonds, 11 December 1970, Simon to Griffiths, 16 December 1970, Griffiths to Simon, 22 December 1970, HLG 120/1380, TNA; Press Notice, 22 December 1970, HLG 120/1380, TNA; Walker to L. M. Potter, Trinity House, January 1971, HLG 120/1380, TNA.
69 Adams to Beddoe, 29 October 1971, HLG 120/1380, TNA.
70 HC Deb., 5 June 1972, vol. 838, c121–6, Hansard.
71 Note by S.K. Gilbert, 29 April 1968, HLG 120/1380, TNA.
72 Walley, Treasury, to Adams, DoE, 2 July 1971, HLG 120/1380, TNA.
73 Leona Skelton, *Tyne after Tyne: An Environmental History of a River's Battle for Protection 1529–2015* (Winwick: White Horse, 2017).
74 Wood, *Restoration of the Tidal Thames*, 146–8.
75 Adams to Beddoe, 15 November 1971, HLG 120/1380, TNA.
76 Jon Agar, 'Historiography and Intersections', Chapter 1 this volume.
77 Of Class 43, Simon Bradley observes that it is 'difficult to think of a more admirable product of 1970s Britain', in *The Railways: Nation, Network and People* (London: Profile, 2015), 231.

12
The woods for the state

Mat Paskins

Introduction

During the period and aftermath of the Napoleonic Wars longstanding fears about timber supply became an imminent risk throughout Europe as a result of an increased pace of ship-building for private as well as military purposes, increasing use of timber for non-naval purposes.[1] These years also saw a plethora of writing about trees in Britain that was strikingly literary, concerned with securing scientific knowledge, and highly critical of the state. In 1831, sixteen years after Napoleon had been defeated, the Scottish landowner Patrick Matthew published a book titled *On Naval Timber*, which discussed recent attempts by the government to plant timber trees in the Forest of Dean by depicting a scene of bucolic carnage. Matthew described how officials had 'planted and replanted trees, persevering even to the fifth time', but 'the seeds did not vegetate, and the plants refused to grow'. Then,

> the natural richness of the soil threw up such a flush of vegetation – of grass, and herbs, and shrubs, that most of these plants were buried under this luxuriance; and how the mice and the emmets, and other wayfarers, hearing, by the *bruit* of fame, of the wise men who had the governing of Dean, assembled from the uttermost ends of the island, expecting a millennium in the forest, and ate up almost every plant which had survived the smothering. Now, this is well; we rejoice over the natural justice of the native and legitimate inhabitants of the Royal Domain, the weeds mastering the invaders the plants, who, year after year, to the amount of many millions, made hostile entrance into the forest.[2]

This chapter is about the context that gave birth to Matthew's strange hybrid way of writing about trees and their meanings for the state. Its technologies are newly introduced species of tree, meant to be more productive than existing ones; new techniques meant to encourage the rapid growth of trees; systems of quantification that allow for the value of trees to be assessed; and the timber that was extracted, measured and tested. Talking about 'technologies' in the nineteenth century is somewhat anachronistic – while the word was coined in the eighteenth century, it was not widely used until the earlier twentieth. The eighteenth- and nineteenth-century term that did cover projects like tree planting or intro-duction of novel crops, as well as engineering projects, was 'improvement'. While it does not have quite the same connotations as 'technology' would gain during the twentieth century, improvement was a coverall term for enlightened and rationalising activity, natural and artificial.[3]

If we accept, however, the claim in the opening chapter of this volume that 'Environments, when cast as means towards ends, are technological in form', then technology can be treated as an analysts' rather than an actors' category.[4] In this sense woodlands and their products are technological in four main ways. First, they involve attempts to apply formal regimes of calculation to processes that had previously been going on without quantification of this kind. Second, they foreground processes and practices that were claimed to be artifi-cial – growing trees in soils that were thought not to suit them, introdu-cing new species, deliberately planting rather than allowing woodlands to grow of their own accord – against those that were claimed to be natural. Whether this division between the natural and the artificial really existed is not at issue: the point is that tree writers believed it did, and generally lined up on one side of the artifice/natural divide. In other words, the divide provided rhetorical resources through which different ideas about nature and artifice could be articulated. Third, timber played a major infrastructural role, as a major raw material of a wooden world.[5] Trees and timber were assessed for naval purposes, but also for pit props, firewood, hop-poles, fencing, building materials, furniture making and so on. These various uses were a major concern of tree writers during this period but have often been overlooked by historians who have been more concerned with the noise of naval con-troversies. Finally, this chapter emphasises the crooked roads that ran from woodlands to the use of timber: the technical activities of hacking, shaping, estimating, felling, squaring and hawing, which were involved in shaping timber out of trees.

Like other chapters in this book, in other words, this chapter is concerned with some of the ways in which plants might be considered as technologies. The organisms discussed here are not quite like the 'industrial plants' that Matthew Holmes discusses in Chapter 8 of this volume: although attempts were made to speed the growth of slow-growing timber trees, this was not entirely connected either to capitalist goals or to mechanical processes, and did not involve interventions from formalised science. And as we will see, there was also no neat fit between attempts to improve the production of timber and the enterprises that made use of them. Like the potato experiments that Dominic Berry describes in Ormskirk (Chapter 9), experimental planting efforts raised questions about the geography of technical improvement, with sites far from the beaten track presenting themselves as having made transformative discoveries.

We cannot, however, see technologised organisms and environments as operating against the static backdrop of an unchanging state. Because of trees' rich symbolic associations, woodland management has often been regarded as an allegory for state power. In James Scott's book *Seeing Like a State*, for example, the rationalising approaches of German Scientific Forestry are used to identify the ways in which centralising states abstract from local conditions – a problem that in Scott's view has run throughout schemes of modernisation in the nineteenth and twentieth centuries.[6] This is close to the story that Esa Ruuskanen tells in his chapter of this volume (Chapter 2), about how 'Imperial eyes' came to regard Irish boglands as potentially exploitable resources, neglecting existing customary uses. Although woodland historians have seriously contested the view of the relationship between woodlands and the state upon which Scott drew, arguing that it mistakes the aspirations and rhetoric of scientific foresters for their achievements, his view remains a powerful lens for thinking about how the state relates to its environments.[7]

As we will see, however, the British state's relation to timber was not centralising in the manner of Scott's scientific foresters. To write seriously about trees during this period, British authors had to re-assemble the varied users and producers of timber and their relations with the state. Natural historical information, political allegories, accounts of local customs and attempts to quantify different planting practices were inextricably linked in British works about forestry. The state appeared in a number of different aspects: as a guarantor for future security of the land; as a contractor with timber merchants; as a (bad) manager of its own lands; as an illegitimate incursion into the lives of the people and their rights to plant trees; as a collector of data and tester of materials. Rather than cohering into a central, domineering entity, the state proliferated in a huge variety of local manifestations. Writers on plantation could not, so to speak, see the woods for the state.

The linked representation of the state and woodland management was also conditioned by two significant absences. The first was a lack of clear information about what private reserves of timber Britain actually possessed. Surveys of private timber were attempted without success. The Admiralty distrusted the timber merchants who supplied timber to its dockyards but did not know how to do without them. There was little foundation for general theories of plantation and the information that did circulate was highly partial and particular.

The second absence was that only a relatively small proportion of timber for naval purposes was produced within Britain itself. Most was imported – from the Baltic, and increasingly from South Asia and North American colonial territories as well. These 'ghost acres' are of course a significant theme in British environmental history, because they allowed the kingdom to prosper on the basis of exploitation of resources from elsewhere.[8] But they had a strange effect on understandings of timber within Britain itself as well, intensifying the sense that woodlands might, potentially, be required in a situation of emergency, when imports were not available. Because of the long times that trees require to grow, this was a slow contingency, a delayed emergency. And as a result, most of the proposals for planting were intensely speculative, seeking purchase on unknown future conditions. And because they were speculative, considering future possibilities rather than absolutely immediate concerns, they could be used to imagine quite radical changes.

These absences, I argue, were at the heart of all British tree-writing during this period. For this reason, we need to be cautious in evaluating the different schemes for indigenous timber production that proliferated between the mid-eighteenth and early nineteenth centuries. While planters might offer potent and persuasive views about the relationship between woodlands and the state, advancing their arguments required entanglements with the state's existing ways of dealing with timber, and the two absences described above. For this reason, we should not see discussions of woodlands in Britain during this period as emerging from simple differences in attitude and policy between elite networks, as Fredrik Jonsson does in his book *Enlightenment's Frontier: The Scottish Highlands and the Origins of Environmentalism*. According to Jonsson, networks of improving landowners in the highlands of Scotland were divided between those who advocated a 'liberal ecology of commerce' based on free trade, and 'civic cameralists', who supported autarkic self-sufficiency and 'argued for the prudence of long-term manage-ment of forest resources by the government and eminent landowners'.[9] The division between these views was real enough, but in practice both points of view collided with existing state structures.

The second section of this chapter discusses the different aspects of the state that were relevant for woodlands, drawing on work by Martin Wilcox, Roger Knight and Joanna Innes. It then examines the state's failed attempts to conduct surveys of private timber resources and attempts to deal directly with timber merchants and the shift towards increasing imports. The third section of the chapter describes attempts to treat private reports of tree planting as experimental knowledge, focusing on the plantation reports received by the London-based Society for the Encouragement of Arts, Manufactures and Commerce. Although these reports could not provide the national overview for which their creators longed, they represented rich discussions of the perils and enthusiasms involved in plantation, and the connection between private planting and the activities of the state. The chapter then goes on to ana-lyse radical works on plantation of William Cobbett, William Withers, which reconfigured speculative hopes about plantation into a vision of resistance to the state and a revivified national landscape. Finally, the chapter examines the conclusions that can be drawn from these rich, hybrid works about trees as visions of the environment, technology and the state in modern Britain.

Woodland surveys and the contractor state

Historians since Robert Albion's 1926 book *Forests and Sea Power: The Timber Problem of the Royal Navy, 1652–1862* have taken pains to show that timber supply was never simply about trees.[10] Most timber for naval purposes was imported in any case; it was only for certain 'knee' timbers that British oak was deemed essential. Albion followed his sources in thinking that the real problem lay with corruption and mis-management in the Royal Dockyards; later histories have queried this conclusion by suggesting that the dockyards were more effective than Albion suggested, and that naval officials used the question of potential shortages to press for the reforms that they desired. Clive Wilkinson, for example, concludes that there was not a lot of neglect at the dockyards during the eighteenth century: at the end of 1770, the Royal Navy was able to use 12,177 loads of oak timber and 1,315 loads of oak plank, all of British origin.[11]

In dealing with timber and the question of its local availability, various agencies of the British state acted as what Martin Wilcox and Roger Knight term 'the contractor state'.[12] This was that aspect of state power devoted to dealing with private contractors that furnished the state with supplies: Knight and Wilcox's case is the Navy Victualling Board, which was responsible for providing provisions to the fleet.[13] They argue that working with a very large number of outside contractors offered four major advantages for the state. First, outsiders could buy in markets in which the government was unable to participate. Second, contractors had much greater knowledge of 'complex and sophisticated markets' – such as the international wheat market – than the government possessed. Third, the 'machinery of war was easier to dismantle', because the state did not have to put capital into projects serviced by contractors. Fourth, established contractors were able to 'expand when the state demand[ed] it with flexibility and with greater speed than [could] a state machine'. Wilcox and Knight conclude that 'it was the strong industrial base outside the state establishments which gave the British government the means to overcome its enemies'.[14]

In relation to trees, the contractor state appeared in the dealings of timber merchants with dockyards. Part of the relationship between the state and its contractors was the use of the Royal Forests. In 1706, Edward Wilcox – the surveyor general of the forests south of the Trent – wrote that the timber of the New Forest 'should not be cut except on extraordinary occasions, but should be preserved as a check upon the timber merchants, who, when it was gone, would impose what rates they pleased'.[15] Among state agencies, too, there were many spaces for suspicion of collusion with the merchants, by accepting poor quality timber at too-high prices. The dockyards were overseen by the Navy Board, whose actions were often challenged by the Admiralty. Suspicion fell upon the Navy Board, the dockyards and the merchants themselves; the Board and dockyards blamed the Admiralty for instituting policies that made it more difficult to obtain timber from the merchants. As we will see, many discussions of private timber production focused exclusively on the activities of landowners, but the contractor state was always present as well. As with victuals, there were aspects of dealing with timber that the state struggled to broker with landlords directly.

A second aspect of the state's activities that was important in dealing with timber was data collection and surveying exercises. The challenges that this involved can be seen from the county surveys of the Board of Agriculture, which were conducted from 1793 onwards.[16]

The Board was a hybrid of state agency and private concern. Its surveyors aimed to give a picture of the ways in which agriculture was conducted throughout the kingdom, and how it might best be improved. The Board's surveys were hesitant to talk about the policies of individual landowners: one surveyor, in a discussion of the management strategies of Joseph Banks, apologised as 'many of my readers will probably censure me, for entering thus widely into calculations, concerning the private property of an individual'.[17] Surveys of different counties varied tremendously as to the ways they collected data; where the Board's reports did discuss woodlands, they did so in very general terms. The report for Essex contained what the agricultural writer William Marshall described as 'an undigested mass of materials, giving the general idea that South Essex abounds in woodlands, but no estimate of aggregate extent'.[18] That for the North Riding of Yorkshire contained a section on the Disposal of Timber, which claimed that 'it is the practice in this Riding to sell the falls of wood to professional wood-buyers, who cut up the trees in the woods, according to the purposes for which they are best calculated, and the most valuable'.[19] Marshall thought this was nonsense, because timber merchants could not possibly possess expert knowledge of the shapes of timber which the navy would require:

> when he says … 'All the ship-timber grown in the Riding is thus cut up in the woods, into shapes ready for the builder to make use of,' he is certainly wrong. It is not probable that in any part of the Riding such a practice prevails: it being impossible for the woodman to know exactly the wants of the ship builder, unless the latter were to furnish him with molds.[20]

This was as close as any of the reports came to giving details of the interactions between the land on which timber grew and the people who dealt in it. The Board's surveys, as rich and locally oriented as they were, gave no serious impression of the workings of the contractor state, or the holdings of private landowners.

Other surveys tried to grapple with the absence of knowledge of timber supplies. Looking back from 1810, Henry Dundas described the 'alarm in the public mind' that had commenced in 1771, and how subsequent years had seen greatly 'increased consumption of oak timber for machinery for which fir cannot be substituted', including 'canals and wet docks, mill-work, engines, lighters, barges and other purposes'.[21] Dundas had been treasurer to the Admiralty from 1782 until 1800, when he resigned amid allegations of financial impropriety that finally led to his impeachment. He concluded that:

> if the prosperity of this country should continue, the consumption of oak timber, for its internal purposes, and for the shipping necessary for the whole of our Trade, including that of the East India Company, would, at no very distant period, furnish an ample demand for all that could be expected to be produced on private property in this kingdom.[22]

As a result, Dundas had decided in 1792 that 'the state of Naval Timber ... required an immediate and radical investigation', under the auspices of the Commission of Naval Revision. The investigation, Dundas claimed, was far-reaching:

> the enquiries ... took a most extensive range, so as to enable them to bring together and methodize a mass of useful information, collected from almost every part of the United Kingdom, from a great variety of different sources; and from persons whose interests and objects were not only different, but whose character and situation must place them above the suspicion of giving erroneous information.[23]

Informants included the 'most eminent land-surveyors and timber dealers in every part of the kingdom; the purveyors of His Majesty's Dock-yard, and the gentlemen assembled at the Quarter Sessions of every county in England', as well as 'various noblemen and gentlemen, whose extensive possessions, and knowledge of the resources and management of timber, gave them advantages of information which could not fail to render the communications from such authorities, highly useful and instructive'.[24] Despite the value of this material, however, its publication had been prohibited, a censorship that appalled Dundas as (he claimed) earlier moments of crisis had been averted 'by the publication of the state of timber in this kingdom and by looking at the danger of a scarcity boldly in the face'.[25]

At the time he wrote these observations on timber, Dundas was under investigation for corruption and his perspective was obviously partial. Heavily embroiled with the East India Company, he also argued for an increase in imports of teak as a substitute for oak. These other concerns guided his approach towards the question of local production: his 'appreciation of domestic forestry was undermined by his commitment to imperial rule in South Asia'.[26] Jonsson regards this as a controversy with autarkic policies of self-sufficiency on one side and the 'neo-mercantilist' commitment to trade and use of colonial imports on the other. The Admiralty commissioned a large number of ships built from teak, 'entrenching Dundas's vision in the naval budget, although the Asian timber never came into widespread use'.[27] In the longer term, British timber was increasingly imported from North America, and 'Pax Britannica underwrote the exploitation of ghost acres in the New World'.[28] So Dundas was perhaps being disingenuous about the difficulties in assessing local timber resources.

Whatever Dundas' intentions were, though, the fact remained at the beginning of the nineteenth century, naval authorities found themselves in an embarrassing situation in dealing with timber from British sources. Besides the lack of knowledge, one source of the embarrassment was an attempt by the state to work around the timber merchants and to deal with landowners directly. According to a very angry correspondence between the Admiralty and representatives of the Navy Board, supplies of oak timber had been in a state of uncertainty since 1802, when tightened restrictions in the dockyards had led to a significant reduction in the quantities of timber being accepted for naval use.[29] The Admiralty blamed the Navy Board for collusion with the merchants; the Board claimed that the Admiralty had refused to acknowledge the risk of reduced supplies if the terms for merchants were made less favourable. The Admiralty had since 1792 also been seeking timber from the Royal Forests, but here again there was a strong sense that stocks had been wildly overestimated. Sir William Rule, the chief naval surveyor, was dispatched to obtain timber directly from landlords; advertisements were published inviting merchants directly to tree auctions; many of the merchants rebuffed the offer, saying there wasn't enough time for them to make a fair estimation of the value of the timber, and the results were not the success for which the Admiralty had hoped. Where previously go-betweens had dealt with the processes of judging and transporting timber, authorities were now attempting to control the supply chain directly to undermine the merchants. These efforts had failed, however. As the Navy Board gloomily noted in 1804:

> Our late Advertisement, signifying the Navy Board was willing to treat for Oak Timber from the Growers, without the intervention of a Dealer, has not brought many offers; and the difficulty which attends the manufacturing of the Timber, namely, the felling, squaring and hawling to water-carriage, together with the sale of

the bark tops and lops is so great, that very few Gentlemen will give themselves the trouble to attend to it; and it is next to impossible for the Navy Board to undertake it, without running the risk of considerable depredation and loss, and employing Purveyors all over the kingdom. In those instances where we have tried it, the Timber has cost the Public much more than by the usual means; and I hold it to be next to impossible that 30,000 Loads of Timber is to be procured but by Timber-dealers.[30]

As a result of the immense difficulty of dealing directly with landowners, the Admiralty's increasing distrust of the timber merchants, and the increasing sense that underlying both was a genuine shortage of timber as a result of rising demand, the difficulties of dealing with indigenous sources appeared intractable.

Private accounts

If it was indelicate to publish calculations about landowners' property, would they nevertheless be willing to vouch for themselves? Many landowners were enthusiastic tree planters, committed to new species that might be useful substitutes for oak. Jonsson discusses the example of the Duke of Atholl, who supported the widespread introduction of larch, and even built a vessel from larch to demonstrate its timber's viability for naval construction.[31] Although Atholl's status enabled him to advance his claims further than others, however, his enthusiasm for larch was of a piece with excitable works published by other planters. Patrick Matthew regarded these enthusiasms with fond scorn:

> almost every author has his own particular mania, which few common readers have sufficient knowledge of the subject to discriminate from the saner matter: and as, from the nature of hobbies – from some shrewd enough guesses by the owner that they are his own undoubted property – and, perhaps, from some misgivings, that what he advances on these is not perfectly self-evident, he is thence the more disposed to expatiate upon them, and embellish. The credulous and inexperienced, partly from this, and partly from the fascination of the very improbability, rush at once into the snare; bring the speculations or assertions to practical test; get quickly disenchanted by realities, and ever after are disposed to treat all written directions on material science with contempt.[32]

Matthew was not dismissing all attempts to provide written directions for planting; only the tendency to focus on one approach over and against others. His metaphor of property for the planters' sense of their own hobbies captured the link between local experience and general enthusiasm: the tendency to treat successes in planting that had been achieved privately as though they should be enforced throughout the whole kingdom.

From the 1750s, the Society for the Encouragement of Arts, Manufactures and Commerce in London had offered money and medals for reports of tree planting. The Society was founded by a group of natural philosophers and philanthropists; early on its membership swelled to include numerous members of Parliament and landowners as well.[33] Initially intended to encourage planting on a small scale, the Society's scheme eventually rewarded correspondents from throughout the kingdom, who planted on scales that ranged from the great to the small.

In 1833 the Society's secretary Arthur Aikin noted that it had spent the past 80 years collecting testimonials of landowners' experiments on different ways of growing trees. What was needed, Aikin claimed, were 'histories of plantations sufficient in number and in their details to allow of a fair comparison to be made of different modes of management, modified by varieties of soil, of climate, and of exposure'.[34] The reason this had not happened so far, Aikin claimed, was because of the peculiar ways in which reports of tree planting were unlike descriptions of other agricultural experiments. Most experimental crops would grow within one year; in consequence 'all the particulars ... of an agricultural experiment, together with its final result, are easily observed and registered'.[35] Without difficulty, a single person could perform many experiments, 'with the reasonable hope of deciding in a few years the comparative advantages of different modes, either of general management, or of the culture of any particular crop'. But in general trees would not be ready for 50 years; oaks would take a century. As such, single experiments in planting could rarely be 'conducted from beginning to end by the same person'.[36] Moreover, memoranda recording effective practice were likely to be lost 'in the course of time, or by transfer of the property from one owner to another'. This was the source, Aikin argued, of the 'contrariety ... both in principle and in practice'[37] between tree planters.

Aikin suggested that accounts of planting should be made commensurate and comparable with each other. This did not happen, and may have been impossible. The accounts that the Society had received varied tremendously in their presentation and the amount of information they contained. This does not mean they contained no quantitative information: far from it. But the most quantitative accounts tended to include considerable speculative matter, projecting the likely growth of trees into the future; those planters who did draw on each other's accounts tried to base these future speculations on the experiments of others. Those experiments themselves, however, were often highly particularist and speculative. Mary Morgan has observed that the usual form for published agricultural experiments during this period was to present them in the form of profit and loss accounts. In this way the advantages of two different forms of management could be compared at a glance; with the self-evidence of a sum, one could be made to appear superior to another.[38]

When records of tree planting were given in the form of financial accounts, there was a neat fit between materiality and their accumulating value: trees were believed to grow in both size and value at a relatively steady rate, and it was tempting to present their growth as a form of materialised compound interest. Some planters noted that the rate at which trees grew declined as they aged, and a few used each other's accounts to calculate the optimal time for cutting by comparing likely rates of growth to likely increases of profit. As the natural philosopher and Bishop of Llandaff Richard Watson put it, 'if profit is considered, every tree of every kind ought to be cut down, and sold, when the annual increase in value of the tree, but its growth, is less than the annual increase of the money it would sell for'.[39] Watson used tables of tree growth that had been calculated by the landowner Robert Marsham and published in the *Philosophical Transactions of the Royal Society* during the 1750s; the Society subsequently rewarded other tables of tree growth as well.[40] Their use did not become standard, though. Marshall, for example, was critical of the serenity of these calculations: they were marvellous for those with existing plantations, and an 'eligible speculation' for those who could expect their oaks to reach maturity 'in ten, fifteen, or twenty years'. But for 'proprietors of younger timber, to play so high a game' entailed very significant risks, as he might hope for good return on the basis that the state might suddenly cut the price it was willing to pay.[41]

Young and Marshall argued about the meaning of interest in connection with forested land. Because land planted with slow-growing timber trees could not be let out for other purposes before the trees were mature, there was no hope of increasing rents on it on the basis of its increased productivity, as there might be for other improved land. Young argued that any understanding of 'interest' needed to interpret its value in national rather than private terms: an organisation like the

Society of Arts should consider how 'such interest is made every day, and every hour nationally, though not individually'.[42] That is, calculations of interest should not focus primarily on private profit. He contrasted timber trees, which produced nothing 'of their principal crop for 150 years', and corn, which

> instantly circulates, and produces a national compound interest; and the prodigious difference in the account at the end of 140 years, would be infinitely more valuable to the kingdom, where in the light of money, or profit, or of defence even than any thing to be derived in speculation from the possession of oak.[43]

That is, the ability to increase the power and happiness of the kingdom by feeding people was of more value than the shady hopes that were held out for oak. Moreover, the association with naval timber meant that even the private growth of oak was never innocent of the wider purposes of the state. Young alluded to:

> The trading system which makes the necessity of isles in the West manured with African blood; and conquests in the East acquired and kept, I will not say on what principles – creating the necessity of a navy to defend them – and, by re-action, rendering such possessions necessary to support the navy; a fine system, that has, in a single century, burthened us with a debt of 240 millions, and is in all its principles, bearings, and combinations, so abhorrent to what ought, and will sometimes or other be the feelings of country gentlemen, that they will plant plague, pestilence, and famine on their land, as readily as oak, the use of which they know, by cruel experience?[44]

Aikin hoped for an aggregation of local experience which could give a picture of the nation as the sum of richly different component parts. The private accounts of planters, however, saw the question of the relation between their woods and the state in different ways. For Young, woodlands connected landowners to the vicious system of imperial trade and sequestered the 'interest' of improvement in the present. For enthusiasts like Watson, young woods of mixed species symbolised rational inheritance; for others, they indicated the guarantees that the state would need to supply. And in their ardours these accounts were remote from the demands of timber merchants and the contractor state, instead imagining that timber was sold directly to supply the needs of the state.

Radical planters

By the 1820s, the lack of reliable information about private woodlands, criticism of state planting policy and particularist reports of local plantations formed a matrix that the journalist and provocateur William Cobbett, together with his friend William Withers, used to deploy a politically radical vision of tree planting. While Cobbett's advocacy of tree planting was as hobbyish as that of any of the authors who Matthew had condemned, it was different from what had come before because it deliberately challenged the presumptuous claims of large-scale landowners and the state. Through their publications and networks of correspondence, Cobbett and Withers constructed alliances that were intended to challenge the workings and inertia of the state.

The forum for these projects of planting was the explosion of print periodicals of the early nineteenth century, including the learned reviews and the specialist gardening press, as well as short, privately printed pamphlets and more expansive books.[45] The radical planters faced challenges in translating the rich combination of political symbolism and practical technique, their own ways of valuing timber, from their own networks to the world of print.

Part of Cobbett's authorial persona was that of a man who was always active in the landscape: who could always judge where a tree should be planted, and was always able to reckon the rational increase and compounding value of woodlands and other agricultural activities. This intuitive grasp of what rural life and the rural landscape required was contrasted with the rampant wickedness of the state, whose outrageous taxation appeared in the landscape in the form of unwanted fortifications, mindless road improvement schemes, fields lying fallow and the other activities of 'fund-holders, pensioners, soldiers, dead-weight, and other swarms of tax-eaters'.[46] As Cobbett put it, 'If I write grammars; if I write on agriculture; if I sow, plant, or deal in seeds; whatever I do, has *first* in view the destruction of those infamous tyrants.'[47]

Cobbett claimed that the difficulty of obtaining accurate information about timber production served as yet another example of the government's negligent corruption. He noted darkly: 'If I were a Member of Parliament, I *would* know what timber has been cut down, and what it has been sold for, since the year 1790.'[48] In the early 1820s, Cobbett started to lobby in favour of the timber of a tree he had observed in North America, the locust. In 1817, when Cobbett was living in Long Island, he had seen locust trees and products made from their timbers, and become excited about their possibilities. Returning to England in 1819, he brought a parcel of seeds, but (he claimed) had no means of sowing them until 1823. This he began on a very small scale, then 'sold the plants; and since that time [had] sold altogether more than a million of them!'[49]

Through his weekly political journal, *The Political Register*, in a book about woodlands and through a nursery behind his house in London, Cobbett claimed to have distributed hundreds of thousands of locust trees, whose timber – he claimed – would amply substitute for oak. To support his campaign, he sought testimonials from eminent friends, including Radnor's bailiff, Viscount Folkestone and various worthies he had known in Long Island. His book on woodlands incorporated multiple testimonials from these people, in which the durability of fence posts made of locust became something like relics, material proof of the tree's virtues. As Withers noted, 'Mr Cobbett proceeds to state his anxious desire to procure evidence, that might substantiate the facts he had witnessed: he adduces testimonials, signed by witnesses of credit, in proof of the durability of the timber in certain cases particularly related by him.'[50]

Cobbett's locust was a relatively familiar tree (the 'false acacia') made new and exciting by his advocacy on its behalf. Cobbett distributed trees from his farm in Barnes and the nursery at his house in Kensington, and '[a]lthough hundreds of the *Robinia pseudoacadia* stood unasked for in the British nurseries, the "locust plants", which everyone believed could only be had genuine from Mr Cobbett, could not be grown by him in sufficient quantities to supply the demand'.[51] So, it was alleged, Cobbett imported seed and 'procured others, as well as young plants, from the London nurseries, and passed them off as his own raising or impor-tation'.[52] For present purposes, the truth or falsity of these allegations is less important than the fact that people connected the locust with Cobbett's name and his personal custody of the precious trees.

Cobbett restricted his testimonials to his friends and admirers: he did not consult with timber merchants or dockyard officials, or try to locate locust trees within the existing networks of the contractor state. It was on this basis that his claims were challenged in print. A reviewer in the *Edinburgh Review* noted that advocates of the locust tree had 'incon-siderately proposed a multiplicity of uses to which they conceived it might be applicable, and have urged its extended cultivation with a zeal unwarranted by the test of experience'.[53] The influential horticultural writer J.C. Loudon was particularly dismissive of Cobbett's claims about the properties of the locust – because

> the uses which he has enumerated do not amount to a hundredth part of those to which timber is applied in this country. Hence,

were his predictions to be verified, and were the locust to become more prevalent than the oak, we should find its wood a miserable substitute, in the construction of ships and houses, for that of our ordinary timber trees.[54]

At the height of his locust-mania, Cobbett wrote:

> The time will come, and it will not be very distant, when the locust-tree will be more common in England than the oak; when a man would be thought mad if he used anything but locust in the making of sills, posts, gates, joists, feet for rick-stands, stocks and axletrees for wheels, hop-poles, pales, or for anything where there is liability to rot.[55]

Then he added: 'This time will not be distant, seeing that the locust grows so fast.' This symbolism was at least as important as the practical uses of the tree: Cobbett's version of patriotic sturdiness would not decay; its profits would always be realised; and the reasons this foreshadowed vision of the durable countryside had not come to pass was because of the physical and financial corruption of the state.

In 1842 Cobbett's friend William Withers, an attorney from Holt in Norfolk, also published a pamphlet on the locust tree, which collected information from French sources as well as Cobbett's allies. It was a sincere attempt to set Cobbett's claims on a sounder natural historical foundation, and to prolong the campaign beyond Cobbett's own death. During the 1820s, Withers had engaged in his own noisy campaign on behalf of a system of oak planting that required trees to be heavily manured during their early years of growth.

Withers opposed his techniques to what he called the 'Scottish system of planting', which was publicised in books and articles by the Scottish landowners Sir Henry Steuart and Sir Walter Scott. Steuart had argued that the properties of timber could only be judged by a landowner of long experience, and that increasing the rate of growth and moving trees from their natural place would injure the quality of the timber they produced, for 'whatever tends to increase the wood, in a greater degree than accords with the species when in its natural state, must injure the quality of the timber' and 'slowness of growth is essential to the closeness of texture and durability of all timber, but especially of the oak'.[56] This notion of 'phytological affinity' – that plants had certain soils that they favoured above others – was received wisdom among landowners.

The rhetorical strategy Withers adopted to support his claims resembled Cobbett's relentless, splenetic enthusiasm – with one major difference. Where Cobbett put his own actions and personal friendships front and centre of any account of the trees he supported, Withers corresponded with people from a number of different social situations and who worked in a range of official capacities. His correspondents included timber merchants, sawyers, surveyors, chemists, engineers and landowners. Withers incorporated their responses into his publications. Rather than simply raging against stockjobbers and tax-eaters, as Cobbett had, he emphasised how the specialisation of knowledge about the uses of timber in different parts of the kingdom had to be re-assembled for the real value of timber to be assessed. Where Cobbett had simply asserted the reliability of his allies, Withers attempted to insinuate his approach to planting and managing oak trees into the crevices of the contractor state. Withers quoted from timber merchants in East Grinstead, Lambeth and Uxbridge who all agreed that quick-grown oak timber could be as fine as any other timber provided. As one of his correspondents put it,

> the various forests of the government do not produce, on an average, more than one sixteenth of the oak timber consumed in his Majesty's yards; the remainder is collected in the various parts of the kingdom, without having regard as heretofore to the part of the kingdom in which it grows.[57]

This suggested that traditional prejudices in favour of oak from certain counties could be overcome.

Not all of the correspondence from which Withers quoted supported his claims about the indifference of timber to the conditions in which it was grown, however. Withers quoted from a letter he had received from Peter Barlow, an engineer at the Woolwich dockyard. At Woolwich, Barlow had pioneered new ways of testing the strengths of different materials. He reported there were significant differences between varieties of timber:

> I had much conversation with the different officers of the dock-yard here, on these and other practical points; and I know it was the decided opinion of those gentlemen, and I have proved it by experiment, that different specimens of English oak vary in their comparative strength in the proportion of five to four, and in some cases in the proportion of seven to five, and that they always considered the strongest and best to be the produce of the best soil. The grain was straighter, and more compact and dense, and of brighter colour.[58]

Barlow admitted, however, that he had not made 'particular enquiries as to the soil and circumstances of the growth of the trees', and so could not match his conclusions with actual conditions of growth. Nonetheless, Withers used Barlow's test as proof that it was possible that timbers' virtues did not ultimately depend upon the soils in which they were grown.

In botanical, agricultural and literary journals, Withers' system was reviewed much more respectfully than Cobbett's campaign for the locust tree had been, though there were serious criticisms. Several reviewers claimed that Withers had not travelled widely enough beyond Norfolk to know the soils and situations of the kingdom as a whole. The *Journal of Agriculture* in 1830 gave an extended review to Withers' work, from the perspective of a landed proprietor, which was nervous about the replacement of an existing system of plantation by one of greater cost. The reviewer censured Withers because '[h]is own experiments … seem to have been limited to a few acres',[59] contrasting such a limited outlook with that of a nobleman who had planted 'ten to fifteen thousand acres', and seen 'the wood of his own planting fashioned into ships of war'.[60]

Patrick Matthew, meanwhile, admired the boldness of Withers' political claims even if he was dubious about his system. He noted with evident pleasure that 'the discomfiture of the knights had been wrought by simple hands'.[61] Matthew also thought that Withers had insufficient experience of different soils and trees that had grown rapidly; he described a prodigal 'Celtic' oak on his own land. It was locally associated with 'miraculous virtue', but turned out to have '*soft tender*' wood; Matthew speculated that this might be because it fed on slops and 'like an animal in similar circumstances' was of 'soft flabby consistency'.[62] Matthew also conducted his own experiments into the strength of timber, and criticised Barlow's experiments on the pieces of wood that Withers had submitted, for not attending to the point of the tree from which the timber had been taken.[63]

In 1832, the Society for the Diffusion of Useful Knowledge published a book by George Sinclair on *Useful and Ornamental Planting*, which gave a sympathetic account of Withers' techniques. Even this, however, treated the question of the effect of early trenching and manuring on the growth of trees as unproved, dependent upon local circumstance and requiring more experimental observation. Like so many observers of experimental accounts of plantation before him, Sinclair noted that there were 'no satisfactory records of the comparative rate of increase of timber, or of solid vegetable fibre, after the first twenty or thirty years' growth of the different species of forest-trees, which had been planted on trenched and manured ground'.[64]

It is tempting to associate Cobbett's and Withers' campaigns with resistance to an ethos of conservation in favour of more rapid production. Cobbett's and Withers' radicalism sits somewhat askance to this opposition. In their enthusiastic ardour they were neither advocates for free trade nor supporters of the existing landed order. Instead, their works of natural history created a drama in which continuous human activity could produce a more fecund and secure nation; the virtue of such activity was directly materialised in techniques for planting and managing trees. Precisely because tree planting was so immediate and yet so intimately connected with future hopes and promises, and because it joined together the locality with the wider concerns of the state, it could perform the role of radical improvement for which Cobbett and Withers longed. And because timber trees were apparently tangible sources of value, about which rational expectations of profit could be entertained, planting stood in stark opposition to the vagaries and phantasmal nature of the state's productions. For Withers, investigations into timber also provided a way, from a marginal location, to acquire knowledge of what was happening elsewhere in the kingdom and how its activities fitted together; this correspondence also allowed him to contribute to public discussions of natural history. Making natural knowledge about trees was also making knowledge of the state.

Conclusion

Up until around 1850 timber prices remained high, and the traditional associations between timber production and state power persisted and timber continued to be used for a multitude of different purposes. In this context, the practical, political account of planting described in this chapter proliferated. Thereafter, sales of oak coppice declined in the face of chemical substitutes derived from non-timber sources; imports from Scandinavia replaced timber from British sources; ships were increasingly built of metal instead of wood; even hop-poles were replaced by string and wire.[65] Although the intoxicating and miscellaneous blend of pastoral parody, political commentary and practical advice declined after this period, problems about lack of information regarding the extent of private woodlands persisted into the later nineteenth century – and, indeed, into the present.[66]

What should we make of the hybrid works considered here with their bold speculations and varied representations of the state? I think

it is worth drawing two conclusions. The first is that the technologies –
meaning systems of management, tools, regimes of calculation and so
on – involved in surveying and extracting raw materials can be invested
with very potent meanings. This is true of all the technologies described
in this chapter: from surveys, to quantified financial accounts, to new
species that appeared to burst forth from the soil and new systems of
planting. The works of Henry Dundas, the planters recognised by the
Society of Arts and the radical planters considered here all mapped to
varying degrees on to the actual practice of timber extraction in Britain;
of all of them, Dundas' advocacy of teak was in the end by far the most significant in environmental and economic terms. Yet all these technologies
of planting, and the difficulties of securing knowledge about the existing
state of timber in the kingdom, created a discursive space in which the
local activity of planting trees and tending them could connect to much
larger-scale political concerns. They could play this role, in part, because
trees had to be imagined as growing for much longer time periods than
other crops. The question of whether other technologies open horizons of
temporal imagination in similar ways is beyond the scope of this chapter;
for present purposes, what is most important is that all the tree writers
considered here believed that thinking about the management of trees
was a way to think about the relation between the relatively proximate
and the relatively distant future.

From this follows a second conclusion: that the state is constructed
in different ways by different technical projects that are also invested in
shaping environments. For tree planters and naval officials in the period
considered here, the state was variously a collector of information, an
imposing enemy, a potential guarantor for maintaining plantations,
a surveyor and a client that dealt with timber merchants. Knowledge
about trees and timber was made in response to and in collision with
these different aspects of the state. In the introduction to this chapter
I contrasted this with both James Scott's panoptical view of the
centralising abstracting state and Fredrik Jonsson's view of the conflicts
between liberal trading ecologies of nature and 'cameralist' attitudes
towards self-sufficiency in timber production. While differences of policy
were very significant, discussions of trees in Britain during this period
were not abstract: they were embodied in particular trees, reports from
specific sites and the great difficulties of negotiating with the knowledge
and ignorance of the contractor state. Thus nineteenth-century tree
writers invite us to take all three parts of the title of this collection seriously: technology, the environment *and* modern Britain.

Notes

1   Paul Warde, 'Fear of Wood Shortage and the Reality of the Woodland in Europe, c.1450–1850', *History Workshop Journal* 62 (2006): 28–57.

2   Patrick Matthew, *On Naval Timber and Arboriculture* (London: Longman, 1831).

3   For a good introduction to eighteenth- and nineteenth-century ideas of improvement, see Sarah Tarlow, *The Archaeology of Improvement in Britain, 1750–1850* (Cambridge: Cambridge University Press, 2007). There are many examples of improvers in Esa Ruuskanen's 'Encroaching Irish Bogland Frontiers: Science, Policy and Aspirations from the 1770s to the 1840s', Chapter 2 this volume.

4   Jon Agar, 'Historiography and Intersections', Chapter 1 this volume.

5   Joachim Radkau, *Wood: A History* (Cambridge: Polity, 2011).

6   James Scott, *Seeing Like a State: How Certain Schemes to Improve the Human Condition Have Failed* (New Haven: Yale University Press, 1998).

7   K. Sivaramakrisnan, *Modern Forests: Statemaking and Environmental Change in Colonial Eastern India* (Redwood City: Stanford University Press, 1999); Brett M. Bennett, *Plantations and Protected Areas: A Global History* (Cambridge, MA: MIT Press, 2015).

8   Kenneth Pomeranz, *The Great Divergence: China, Europe, and the Making of the Modern World* (Princeton: Princeton University Press, 2000).

9   Fredrik Jonsson, *Enlightenment's Frontier: The Scottish Highlands and the Origins of Environmentalism* (New Haven: Yale University Press, 2013).

10   Robert Albion, *Forests and Seapower: The Royal Dockyards during the Revolutionary and Napoleonic Wars* (Cambridge, MA: Harvard University Press, 1926); Roger Knight, 'New England Forest and British Seapower: Albion Revisited', *American Neptune* 46 (1986): 221–9.

11   Clive Wilkinson, *The British Navy and the State in the Eighteenth Century* (Martlesham: Boydell & Brewer, 2004), 159.

12   Roger Knight and Martin Wilcox, *Sustaining the Fleet, 1793–1815: War, the British Navy and the Contractor State* (Martlesham: Boydell & Brewer, 2010).

13   Knight and Wilcox, *Sustaining the Fleet*, 9.

14   Knight and Wilcox, *Sustaining the Fleet*, 10.

15   Quoted in Sara Morrison, 'Forests of Masts and Seas of Trees: The English Royal Forests and the Restoration Navy', in *English Atlantics Revisited: Essays Honouring Ian K. Steele*, ed. Nancy L. Rhoden (Kingston, ON: McGill-Queen's University Press, 2014), 151.

16   Rosalind Mitchison, 'The Old Board of Agriculture (1793–1822)', *The English Historical Review* 74 (1959): 41–69.

17   William Marshall, *A Review of the Reports to the Board of Agriculture from the Eastern Department of England* (York: Thomas Wilson & Sons, 1811), 125.

18   Quoted in John Bailey and G. Culley, *A General View of the Agriculture of the County of Northumberland* (Newcastle: Sol Hodgson, 1797), 124.

19   John Tuke, *General View of the Agriculture of the North Riding of Yorkshire* (London: McMillan, 1794), 89.

20   William Marshall, *The Review and Abstract of the County Reports to the Board of Agriculture*, Vol. 1 (York: Thomas Wilson & Sons, 1818), 466.

21   Henry Dundas Melville, *A Letter from Lord Viscount Melville to the Right Hon. Spencer Perceval, on the Subject of Naval Timber* (London: S. Bagster, 1810), 7.

22   Melville, *A Letter*, 10.

23   Melville, *A Letter*, 18.

24   Melville, *A Letter*, 18–19.

25   Melville, *A Letter*, 22.

26   Jonsson, *Enlightenment's Frontier*.

27   Jonsson, *Enlightenment's Frontier*.

28   Jonsson, *Enlightenment's Frontier*.

29   See the correspondence reproduced in Charles Noel, *The Third Report of the Commissioners for Revising and Digesting the Civil Affairs of His Majesty's Navy*, House of Commons 19th-Century Sessional Papers, 312, 5.

30   A.S. Hamond, Letter 23 April 1802, reproduced in *Third Report of the Commissioners*, 63.

31   Jonsson, *Enlightenment's Frontier*.

32 Matthew, *Naval Timber*, 138.

33 Derek Hudson and Kenneth Luckhurst, *The Royal Society of Arts, 1754–1954* (London: John Murray, 1954); D.G.C. Allan, *RSA: A Chronological History of the Royal Society for the Encouragement of Arts, Manufactures, and Commerce: Founded 1754, Royal Charter 1847, Royal 'Prefix' 1908* (Private printing, 1998).

34 Arthur Aikin, 'Plantations', *Transactions of the Society for the Encouragement of Arts, Manufactures and Commerce for the Session 1831–1832* 49 (1832): 4.

35 Aikin, 'Plantations', 5.

36 Aikin, 'Plantations', 5.

37 Aikin, 'Plantations', 5.

38 Mary Morgan, *The World in the Model: How Economists Work and Think* (Cambridge: Cambridge University Press, 2012) 55.

39 Quoted in Marshall, *Review and Abstract*, 290.

40 Robert Marsham, 'Observations on the Growth of Trees', *Philosophical Transactions of the Royal Society* 51 (1759–60): 7–12; 'A Supplement to the Measures of Trees, Printed in the Philosophical Transactions for 1759', *Philosophical Transactions* (1759) 87, 128–32.

41 Marshall, *Review and Abstract*, 291.

42 Young, 'Transactions', 307.

43 Young, 'Transactions', 309.

44 Young, 'Transactions', 309.

45 Geoffrey Cantor and Sally Shuttleworth. *Science Serialized: Representations of the Sciences in Nineteenth-Century Periodicals* (Cambridge, MA: MIT Press, 2004).

46 Ian Dyck, *William Cobbett and Rural Popular Culture* (Cambridge: Cambridge University Press, 1992), 47.

47 Quoted in Robert Huish, *Memoirs of the Late William Cobbett Esq. MP for Oldham* (London: John Saunders, 1836), 260.

48 William Cobbett, *Rural Rides*, Vol. 1 (London: William Cobbett, 1830), 141.

49 William Cobbett, *The Woodlands, or, a Treatise on the Preparing of Ground for Planting, on the Planting, on the Cultivating, on the Pruning, and on the Cutting Down of Forest Trees and Underwoods* (London: William Cobbett, 1825), para. 326.

50 William Withers, *The Acacia Tree, Robinia Pseudo Acacia: Its Growth, Qualities, and Uses* (London and Holt: Longman, James Shalders, 1842), 210.

51 J.C. Loudon, *Arboretum et fruticetum Britannicum, or, The Trees and Shrubs of Britain*, Vol. II, (London: Longman, 1838), 616.

52 Loudon, *Arboretum*, 616.

53 Anon., 'Review of British Trees and Shrubs', *Edinburgh Review* 69 (1839): 210.

54 Loudon, *Arboretum*, 622.

55 Cobbett, *Woodlands*, para. 351.

56 Cobbett, *Woodlands*, para. 351.

57 William Withers, *A Letter to Sir Henry Steuart, Bart On the Improvement in the Quality of Timber, to Be Effected By the High Cultivation and Quick Growth of Forest-Trees* (London and Holt: Longman, James Shalders, 1829), 54.

58 Withers, *A letter*, 99.

59 Anon., 'On Economy in Planting', *The Journal of Agriculture* 2 (1831): 415.

60 Anon., 'On Economy', 416.

61 Matthew, *Naval Timber*, 211.

62 Matthew, *Naval Timber*, 211.

63 Matthew, *Naval Timber*, 212.

64 George Sinclair, *Useful and Ornamental Planting* (London: Baldwin & Craddock, 1832), 64.

65 Jan Oosthoek, *Conquering the Highlands: A History of the Afforestation of the Scottish Uplands* (Canberra: ANU Press, 2013), 29.

66 Anna Lawrence and Norman Dandy, 'Private Landowners' Approaches to Planting and Managing Forests in the UK: What's the Evidence?', *Land Use Policy* 36 (2014): 351–60.

13

The UK government's environmentalism: Britain, NATO and the origins of environmental diplomacy

Simone Turchetti

The combined administration of technological change and ecological degradation demands the introduction of far-reaching environmental protection laws. But it is only during the second half of the twentieth century that most international agreements set out to counter environmental threats have been introduced. Thus only in the past 60 years have many countries, including Britain, adjusted to the creation of an international space for negotiations on environmental provisions by setting up new government agencies and, through the work of qualified civil servants, setting the country's terms for complying with new treaties.[1]

One would hope that this critical commitment to the well-being of our planet will continue in the twenty-first century, although the recent transfer of the eight-year-old Department of Energy and Climate Change to one devoted to Business, Energy and Industrial Strategy reveals that environmental ambitions can be rather short-lived in the UK's corridors of power. As a number of civil servants relocate into a new government branch, one wonders if the government's stances on environmental protection are inextricably linked to the whims of its decision makers and bureaucrats. Certainly the transfer does not seem to equip Britain in the best possible way to fulfil the 2015 Paris agreement on climate change; nor does Brexit, which could lead the country to disengage from more stringent EU environmental regulations.[2]

But has the UK government's attitude towards international environmental affairs always been as aloof as in recent months? And at what point in the country's history did its representatives elaborate a convincing agenda to deal with environmental threats at home and internationally? An analysis of the current UK government's posture on environmental matters could definitely benefit from a historical reconstruction of past approaches, but there is a significant gap in the literature regarding Britain's environmental affairs. Scholars interested in finding out about the intersections between contemporary British diplomacy, technology and environment will struggle to retrieve enough on this subject. And although historians of science and technology like Bill Luckin have looked into environmental regulations at home, including those brought about by technological change, the twentieth century has never really come into focus.[3] There are important exceptions of course, including the works of John Sheail and others on the history of environmental policy-making, but international relations have been by and large overlooked.[4]

As this chapter shows, the British government's attitude in the late 1960s and early 1970s provides a compelling historical narrative. This was a period when, as Andrew Jamison and Stephen Bocking have pointed out, the emergence of protest and environmental movements re-defined the interactions between activists, scientists and decision makers, something that was decisive in framing environmental debates in Britain.[5] And talks on environmental affairs featured more in the international arena too, thus encouraging political leaders to elaborate a constructive approach. As Jacob Hamblin, Stephen Macekura and J. Brooks Flippen have shown, environmental diplomacy, intended as a space for negotiations on environmental sustainability, featured for the first time in the international arena at this juncture in the twentieth century, shaping relations of which the recent climate change talks can be seen as the main legacy.[6]

Of course, diplomats hardly ever embraced environmentalism solely to protect the environment. One transnational agency in particular, the North Atlantic Treaty Organisation (NATO), swiftly moved into the environmental diplomacy arena, in order to evade problems marring its cohesiveness. Since its establishment in 1949, this defence alliance has continued to be perceived, especially in Western Europe, as an emanation of the US administration, something that encouraged US officials to enter more frequently into the diplomacy arena in order to persuade allies to retain membership.[7] Furthermore, by the early 1960s NATO had already approached the study of the environment by exploring the possibilities of environmental warfare.[8] So by the end of that decade the alliance's integration problems, including France's exit from the alliance's coordinated military structure, and the allies' criticism over the US conflict in Vietnam, convinced president Richard Nixon to think innovatively about NATO's circumstances, which is the reason for his rather bizarre proposition that a defence alliance focus on environmental concerns.

This chapter posits that the emergence of environmental diplomacy at NATO paved the way to the elaboration of British stances on international environmental politics. Initially Nixon's initiative caught British officials off guard, but eventually it compelled them to pay greater attention to environmental matters. Yet, the archival documents discussing NATO's activities reveal that environmental concerns often represented minor indentations in the agenda of British negotiators, even in the presence of major environmental catastrophes hitting the country (such as the SS *Torrey Canyon* oil spill disaster). More generally, these officials displayed a tendency to consider environmental threats as second rank with respect to other business and industrial priorities, especially when protecting large-scale technological projects such as oil tankers and supersonic jets like the Concorde. Finally, their environmental propositions were rooted in diplomatic pragmatism; they were part and parcel of the specific diplomacy game that they were then busy playing.

What follows reveals the ambivalent position of the UK government towards NATO's environmental programme. Even when its officials warmed up to the alliance's plans to protect the environment, projects undermining economic interests in Britain continued to stir resistance.

A disaster or an opportunity? NATO's environmental programme

The origins of NATO's environmental diplomacy can be traced back to a discussion on the directions of the alliance's scientific research programme that started in the mid-1960s. By then a NATO Science Committee, comprising representatives from allied countries and elaborating schemes to assist scientists through fellowships and grants, had been in existence for 10 years. Its decline instigated a debate on a possible investment in environmental studies and these discussions informed the effort of making the alliance a champion of environmental actions.[9]

It was especially the UK's worst ever oil spill disaster that persuaded officials at the US State Department that this was a sound proposition. On 18 March 1967, the tanker SS *Torrey Canyon* ran aground on the Seven Stones reef between Cornwall and the Isles of Scilly and the oil it carried poured into the sea. The UK government's chief science adviser Solly Zuckerman set up a task force to deal with the disaster, but Operation Oil Buster was far from successful. The adviser called in RAF bombers to use napalm and chlorate bombs over the slick, which only helped the pollutant to mix with water.[10] The follow-up Operation Mop Up involved the spreading of chemical detergents to get rid of the oil hitting the Cornish coasts. It slayed its fauna and was equally unsuccessful. Both operations were the subject of intense media scrutiny as £250,000 per week was spent in the rescue and management of specific tasks remained 'awkwardly divided' among ministries.[11] Zuckerman also rebuffed an accusation that the amount of detergent used on beaches caused harm to marine life, naïvely claiming that 'unless people go on pouring the stuff needlessly on some beaches, we no longer expect anything significant'.[12]

The UK government's science adviser received numerous attacks because he appeared to be unprepared to deal with environmental disasters, and he presumably was not. 'Sir Solly' had built his career as adviser on the administration of defence research. A South African-born zoologist, at the outbreak of the Second World War, Zuckerman became involved in the scientific study of bombing campaigns (which might explain why bombing was the method of choice against the slick). Employed in a number of Whitehall's science policy departments in the post-war years, from 1960 he became the Ministry of Defence's science adviser. In 1964 the new PM Harold Wilson wanted Zuckerman to elaborate the UK government's science policy.[13]

Zuckerman was also the UK representative at the NATO Science Committee. And if the handling of the disaster damaged his reputation, it was actually viewed as an excellent opportunity at the US State Department. As one of its memoranda of January 1968 claimed, it demonstrated that more research was needed in the environmental field, something that its officers linked to the trajectory of the NATO committee. Set up in 1957, the committee had just entered a crisis phase. Its mission had partly failed as it had not captured, according to the memo, 'the real interest and concern of Delegations and Governments'.[14] The way forward was to 'take quick advantage of opportunities for constructive action presented by scientifically interesting events'. The *Torrey Canyon* incident, and its mismanagement, was mentioned as a key example of the lack of scientific knowledge on environmental disasters.[15]

That the environmental catastrophe could be an opportunity was openly stated three months later by the US representative at the Science Committee, the physicist Isidor Isaac Rabi, during the celebrations of the tenth anniversary of its foundation.[16] He spoke just after Zuckerman, who, still embittered by the polemic on *Torrey Canyon* at home, recalled only the committee's troubles. Instead his US counterpart put the environment at the centre of an optimistic portrayal. Rabi stressed that the problems of the alliance had grown increasingly complex: 'I refer to such matters as pollution of the environment: air, water, soil, noise, congestion.' 'Science and technology could offer solutions,' Rabi argued, especially if more funding was invested in finding out about environmental threats.[17]

The other delegates of the NATO committee continued debating Rabi's propositions for a few more years and in the early 1970s they agreed on the sponsorship of novel research on ecology and pollutants. By then, however, environmental protection had moved from the periphery to the centre of NATO's political debate partly thanks to the propositions of the newly elected US president Richard Nixon. In January 1969, when Eisenhower's former vice president moved into the White House, few thought that environmental protection would feature significantly in his programme. But Nixon understood that the environment now appealed more to the moderate US electorate and mobilised a growing number of activists, including the popular conservationist Russell Train.[18] A republican and lawyer by trade, Train had been involved in the nature conservation movement for more than a decade. Nixon thus wanted Train to lead his administration's environmental initiatives.

By taking environmentalism in NATO's stride, Nixon also hoped to instigate community-building, gain consensus for his (controversial) policies and especially allay criticism on the US conflict in Vietnam. On the occasion of the twentieth anniversary of NATO's foundation, the new president urged his allies 'to explore ways in which the experience and resources of the Western nations could most effectively be marshalled toward improving the quality of life'.[19] His proposition led to the creation of a new NATO Committee on the Challenges of Modern Society (CCMS). The underlying principles of its programme underscored the committee's practical goals.[20] In contrast with the Science Committee, the new one would not carry out research, but promote environmental actions. National delegations would voluntarily take responsibility for pilot projects on key environmental issues and seek the collaboration of other delegations that could, if they wished, feature as co-pilots. New legislation could also be elaborated and specific provisions put before the North Atlantic Council (NAC) for approval.[21]

Nixon's proposition caught the delegates of Western European countries unprepared and they wondered if a defence alliance was really equipped to deal with environmental protection. In particular, the proposition escalated a contrast between US and British officials, who had already collided on the directions and future agenda of the Atlantic alliance. Statements at NAC meetings and information relayed in secret US memos reiterated that UK delegates believed NATO to be an unsuitable candidate for promoting environmental actions and that other organisations should have taken the lead in environmental diplomacy.[22] This was a position that UK officials had developed much earlier on the basis of their understanding that NATO-led initiatives outside the realm of defence stifled East–West collaboration. UK criticism reiterated a stance popular at the Royal Society (and Whitehall departments) since a similar polemic had previously typified international scientific collaboration. For instance NATO work under the aegis of the Science Committee was viewed in the UK as disruptive to collaboration in the context of the International Council for the Exploration of the Sea (ICES), of which the Soviet Union was a member.[23] So the UK representative at NATO, Sir Bernard Burrows, rebuffed Nixon's initiative and warned that the OECD and the Council of Europe were equally apprehensive.[24] Britain's initial contribution to CCMS activities was thus construed as an effort to prove Nixon's initiative as ill-conceived. And it produced an embarrassing fiasco.

I'll prove you wrong! UK stances on NATO's environmental turn

Already in the summer of 1969, the US State Department had canvassed support for the CCMS among NATO delegations, also urging national representatives in private talks to take responsibility for pilot projects. Despite the initial scepticism of the allies, the diplomacy effort eventually paid off. Some delegates were persuaded especially when US representatives agreed to set up pioneering collaborative schemes. For instance, they were busy negotiating with West Germans and Turks a new pilot project focusing on air pollution of urban conglomerates. From 1970 the pollution in the cities of St Louis, Ankara and Frankfurt, which presented a similar degradation in the quality of air, was thus monitored in order to find out about the nature, volumes and circulation of air contaminants. A NATO list of environmental actions containing potential subjects for pilot projects was finally released in May 1969 with the view of promoting more collaboration.[25]

UK criticism of Nixon's initiative had by then already been reiterated at NAC meetings and in official correspondence. There was therefore little that could be done to restate it further. But UK officials feared that the lack of enthusiasm they had shown would make them look too negative in the eyes of other NATO delegations. The Science and Technology Department of the UK Foreign and Commonwealth Office (FCO) thus took responsibility for elaborating a new strategy. The option to 'give way to American pressure' and be more collaborative was now considered for the first time, but only, notably, on the condition of steering clear of problems of the physical environment.[26] Actually, since the NATO document sponsored CCMS interventions in the social environment too, they agreed that the UK would initially take responsibility *only* for actions in the social domain *exactly* to prove the Americans wrong in proposing NATO's environmental turn.[27]

So in June 1969 a number of government departments were hurriedly approached in order to find someone who could elaborate a UK pilot project *deliberately* ignoring environmental problems. The search proved successful when the Social Sciences Research Council (SSRC) appeared to be interested in carrying out an analysis of satisfaction and motivation in workplaces. This seemed to be promising as it could have been construed as a response to what a CCMS document had hazily indicated as a study on 'Individual and Group Motivation'.[28] The UK delegation thus proposed the study at the first CCMS plenary meeting of 8 December 1969. But since the SSRC had by then yet to commit, no further details were provided in Brussels and the UK delegate had to improvise ('obliged to try to make bricks without straw').[29] Unsurprisingly, the UK proposal failed to attract co-pilots.

If straw was lacking among UK diplomats, so was expertise on environmental issues. With Zuckerman about to retire from his role as UK Scientific Adviser, it was the metallurgist Alan Cottrell who agreed to take responsibility for upcoming CCMS business as he had done, on and off, for Science Committee affairs. Zuckerman's deputy at the Ministry of Defence, Cottrell shared with him the same lack of knowledge on how to tackle environmental threats, having previously dealt with defence research items, such as, for instance, UK's new reconnaissance aircraft.[30]

Moreover, in examining how to expedite a contribution to the new NATO scheme, Cottrell and FCO officials faced complications. Due to the looming general elections, the workers' conditions represented a particularly sensitive topic, especially given the growing mineworkers' influence in national affairs.[31] Proposing a NATO study of work satisfaction was thus rather unfortunate, as the FCO's senior minister Alun Arthur Gwynne (Lord Chalfont) pointed out, as it would have irked the trade unions and suggested a plan to involve NATO in industrial disputes.[32] The worst had yet to come though: SSRC managers now let the FCO know that they no longer intended to lead the pilot project. One of its civil servants now apprehensively remarked that championing a UK project that no UK agency was willing to carry out would make their NATO delegate 'look pretty silly'.[33]

Desperate not to look so in the eyes of allies as well as to retain the strategy of steering clear of environmental issues, Cottrell hurried to find one such organisation and, with time running out, he finally obtained a positive response from a Ministry of Defence retiring psychologist.[34] But aware of Lord Chalfont's remarks, Cottrell swiftly agreed to change the UK pilot project's title again, and gave instructions to avoid publicising its content.[35]

The now renamed 'Work and Its Satisfactions in a Technological Era' was presented for the first time at the CCMS meeting of 13/14 April 1970, but given the lack of preparation on the subject, Cottrell failed to impress. Particularly unconvinced was the US chief delegate Patrick Moynihan, who had led a 20-strong delegation comprising experts from a variety of US departments. Since the FCO sent Cottrell alone, Moynihan now made an official complaint about 'leaving our representation ... in the hands of a metallurgist from the Ministry of Defence who was not necessarily very well-versed in the problems of the environment'.[36]

Meanwhile, a question in the UK Parliament materialised Lord Chalfont's fears: a Conservative MP learnt about the unpublicised project and thus asked if NATO was to investigate workers' pay and conditions.[37] No longer happy to sponsor the scheme as a NATO exercise, the FCO now agreed with a proposition by Train (who in 1972 replaced Moynihan) that the UK project be transferred to the OECD.[38]

The diplomatic fiasco convinced Cottrell and FCO officials that engaging only with the non-environmental aspects in NATO's environmental programme had been wrong. They now considered a change of strategy. By now Labour Prime Minister Harold Wilson had been ousted from power following the general elections of June 1970 and his successor, the Conservative party leader Edward Heath, was as inclined as Nixon was to put the environment in the national political agenda.

The SS *Torrey Canyon* disaster was decisive to this policy transition,[39] and Heath agreed to establish a Department of the Environment. The Conservative MP Peter Walker was called to direct it.

Hamblin claims that in 1971 the 'whole NATO environmental enterprise left British environmental officials cold'.[40] But actually this was the time when they started to warm up a little. In an effort to avoid humiliation Walker's undersecretary reiterated in front of journalists assembled at NATO headquarters that US officials were too emotional in dealing with environmental problems.[41] But both Walker and Cottrell (who was about to replace Zuckerman as Heath's Chief Scientific Adviser) were by then pushing for promoting environmental initiatives. They had also shown considerable interest in the recent *Limits of Growth*, a forecast of the impact of pollution (among other environmental constraints) completed through computational analysis and modelling.[42] While rejecting open ententes at NATO, FCO officials now secretly agreed with their US colleagues to give in to their request that a high-profile delegate like Zuckerman take responsibility for CCMS business, something that brought the retiring science adviser back to Brussels.[43] They also agreed that the UK would contribute more to the committee's pilot projects.[44] Although one, on Advanced Health Care (which led to a study on automation in clinical laboratories), reiterated a UK commitment to CCMS projects not dealing with the physical environment, the British delegation also agreed to join a French-piloted Environmental and Regional Planning project, meant to provide environmental guidelines on land use. And by now, it had prepared a new submission for one on Advanced Waste Water Treatment.[45] Cottrell, who had acquired more knowledge on environmental issues, now sought to have a more prominent role in international environmental negotiations, representing the UK at the UN Conference on the Human Environment in Stockholm (June 1972). He also presided over an official committee on Future World Trends, and he proposed that London host an international conference on waste sea dumping (which resulted in a follow-up international convention).[46]

These démarches notwithstanding, the newly found environmentalism of Her Majesty's Government retained the distinctive lack of enthusiasm that had typified its origins, especially in so far as environmental ambitions hardly ever received priority in the presence of overarching business imperatives. It is especially the reactions in the UK to NATO plans for new international legislation abating oil spills at sea that vividly illustrates this ambivalence.

Make it sink! The oil spills ban and UK resistance

With the populace of the Cornish coast still struggling to get rid of polluting oil washing their beaches, the CCMS proposed to discuss a ban on oil discharges at sea. Now the Science and Technology Department liaised with other FCO branches such as that of Marine and Transport, as well as other UK agencies like the Board of Trade and the Department of Trade and Industry (DTI), to discuss the proposal. If on the one hand the civil servants involved agreed that the environment should find a space in international policy-making, they also resisted new legislation that would disadvantage UK-based oil and shipping concerns.

Fears of tougher oil shipping provisions were openly stated in a September 1970 memo on a Belgian proposal for a CCMS colloquium on oil spills.[47] What worried FCO officials was especially that a high-profile meeting could lead other delegations to advocate stronger measures to counter oil spills at the next NAC meeting, and that the council would, in turn, approve them.

The UK position was presented as an effort to avoid NATO interfering with the activities of other international organisations responsible for sea affairs, and especially the International Maritime Consultative Organisation (IMCO). But the UK stance wove in an ambition to dampen far-reaching plans to abate oil pollution. The IMCO was underfunded and proposals to raise its budget had been rejected several times. Furthermore, national delegates had been successful in jeopardising IMCO environmental propositions they did not agree with, given the lack of underlying Cold War allegiances in the organisation (the Soviet Union was a member). One DTI official even noted, with some satisfaction, that the Soviets had recently succeeded in resisting a US proposal on oil pollution so that the American application had to be withdrawn.[48]

When Belgian plans for an oil spills colloquium materialised, officials of the FCO Marine and Transport Department adopted a passive stance, deciding that no UK cabinet-rank minister should be present at the NATO meeting. Unsurprisingly the decision irritated US negotiators as it was rightly perceived as an attempt to avoid engaging with policy-level propositions. But on this specific item UK officials would not budge: 'we have expressed our opposition ... in forthright terms. We have also told the Americans that we would be bound to resist if they were to try and steer the Colloquium towards recommendations for "actions" by NATO.'[49]

Tensions between US and UK officials escalated again. Those of the FCO Marine and Transport Department sought to liaise with their Dutch colleagues to ramp up dissent within NATO.[50] Conversely, an attaché of the US embassy in London alleged that unless a high-ranking UK official was sent to the meeting and the attitude changed, the colloquium would end in a public row between delegations. The standoff led the UK Minister of Foreign Affairs, Alec Douglas Home, to intervene and, after an exchange with the DTI Secretary, Home instructed his deputy to attend the meeting.[51] The colloquium went exactly in the direction that the FCO personnel had feared. Recommendations for a ban on oil discharges were put forward. And in December 1970 the NAC endorsed them so that the plan for a NATO action gained momentum.[52]

While few in the UK government may have been against the provisions, they certainly feared the consequences they might have had for British business. Technological solutions were available but expensive, especially as *Torrey Canyon*-like accidents represented, metaphorically, just the tip of a polluting iceberg. If each year one million tons of oil poured in the sea as a result of accidental spills, three times as much leaked in routine operations such as the tankers' ballasting (increasing buoyancy by taking in and out sea water). Oil tankers in the 1970s had no segregated cargoes. Thus during these operations sea water mixed with oil, which was then released at sea. The UK solution to this problem was *load-on-top*, namely transferring the ballast mixture to a slop tank so that water and oil could separate before release. But many environmentalists, including Train, believed it not effective enough as some of the contaminating oil could still be released at sea after separation. Instead, what Train campaigned for was the overhaul of the existing oil-shipping industry and a transformation of existing tankers so as to segregate cargo and ballast water in different tanks. Officials of the UK shipping industry, whose interests were represented within the FCO Marine and Transport Department, had already concluded that the introduction of segregated tanks was 'unrealistic', as it would entail refurbishing the entire tanker fleet.[53]

When the NAC approved actions to abate oil pollution, UK officials became even more anxious about the traction of such a proposition. But (luckily for them), US diplomats were divided over the issue too, and in 1971 the influence of those opposing radical solutions had grown. When Train sought support from the US shipping industry's representatives, he was openly challenged.[54] Unsurprisingly, the NATO press release circulated after the NAC meeting of December 1970 left the question of implementations deliberately in the background so as to avoid upsetting the representatives of the shipping industry further.[55] The following April, when Train met with the president of the American Institute of Merchant Shipping (AIMS), he was told in clear terms that the entrepreneurs viewed the NATO resolution as 'beyond attainment'.[56] One of Train's collaborators now advised him to lobby with 'certain interested governments' for the swift implementation of the amended NATO resolution and was told that Britain was 'key in this context' as its officials were against the concept of segregated ballast.[57]

The UK position eventually prevailed as NATO delegations agreed to transfer the NAC recommendations to the IMCO and start a conversation with more countries on their enactment. The decision slowed down the implementation process significantly.[58] And if the 1973 IMCO draft enforced more stringent principles in dealing with oil spills, including segregation, then it did not press governments hard enough to comply. New provisions would initially apply only to new tankers over 70,000 tons deadweight, with smaller ones let free to continue polluting.[59] Furthermore the resolution was ratified only in the 1980s in line with the UK delegates' wish to postpone its implementation. In May 1971 one official of the FCO Marine and Transport Department could therefore satisfactorily note that the US–UK controversy about oil discharges was about to 'die a natural death'.[60] The decease demonstrated that business and industrial priorities were paramount in the UK government when considering environmental actions. And when other CCMS proposals did not die out naturally, UK officials sought to kill them through vigorous diplomatic work.

Don't let it fly! The CCMS project on air pollution and Concorde

In the early 1970s the upper atmosphere, as much as the seas, was a focus of environmental concerns especially because of the nitrate and sulphate oxides that jet planes released. A growing number of scientists viewed these pollutants as affecting both the formation of cloud cover and the ozone layer.[61] Yet when the Canadian delegation at NATO put forward a proposal to investigate how to combat pollution of the upper atmosphere at the CCMS meeting of April 1973, the proposition was met with silence by the British delegation. Its officers' unstated opposition derived from the forthcoming launch of the first supersonic passenger jet liner, the Concorde, as the new aircraft would be entering into service three years later and approval was sought from the US Federal Aviation

Administration to allow its landing in the USA.[62] The supersonic jet released a terrific amount of nitrous oxide, whose depleting impact on the ozone layer had been stated in a recent study by the Dutch atmospheric chemist Paul Jozef Crutzen.[63] This notwithstanding, FCO officials at the Science and Technology Department agreed to 'have a word' with the Canadians to dissuade them.[64] A few months later a two-pronged attack materialised. The British High Commission in Ottawa was asked to probe the Canadians' intentions so as to encourage them not to pursue their proposal at the CCMS without informing their British colleagues first. The French were also alerted to coordinate future initiatives at NATO headquarters, especially since the Concorde had by then become a staple of Anglo-French relations, intertwined with the early French opposition to, and then support for, Britain's entrance into the European common market.[65] Meanwhile DTI officials provided further scientific information on nitrous oxide emissions in the upper atmosphere in an effort to play down the polluting effects; evidence on the subject matter was scanty and inconclusive, they reiterated.[66]

Now even the UK's Meteorological Office got involved. When in April 1973 the NATO Advisory Group on Aeronautics Research and Development (AGARD) organised a conference on Atmospheric Pollution by Aircraft Engines in London, the office's research meteorologist Philip Goldsmith presented novel evidence ruling out a significant thinning of the ozone layer as a result of increased nitrous oxide emissions. Published in separate articles for *Nature* and *Science*, the study compared emissions from past nuclear tests and those from Concorde. It concluded that since 'no detectable changes' had occurred because of testing, there was no reason to believe that this would be the case for the Concorde. It would have taken an incredibly high number of Concorde flights each year (over 1,600) to match emissions due to nuclear tests.[67]

Harold S. Johnston, a US scientist who had originally warned about the negative impacts of supersonic flights, also gave a presentation at the same meeting. But he criticised these findings.[68] He was not the only one to be unconvinced. Scepticism existed in the UK diplomacy camp too, but was deliberately kept under wraps. One FCO memo claimed the study to have done little favour to the Concorde project due to the mounting scientific evidence about the thinning of the ozone layer as a result of atomic testing. Moreover, the Met Office team had provided different figures for the number of Concorde flights at the AGARD meeting and for those published in *Nature* so that it was 'particularly unfortunate to have been caught doing our sums wrong'.[69] As Jon Agar has shown, a few years later the Cabinet Office would also turn to the UK Meteorological Office for advice, and this time on the topic of climate change. And once again the agency provided the scientific evidence desperately sought to prove that no environmental actions were needed. So, as Concorde was about to take off, the office's response took the wind out of environmental initiatives, suggesting that the evidence on global warming due to CO_2 emissions was insufficient.[70]

Meanwhile Canadian officials had informed their UK colleagues at the High Commission in Ottawa that they were still mulling over the setting up of a CCMS pilot project, while DTI officials now relayed the Met Office results to show that no harm could come from a supersonic liner.[71] Eventually the Canadians decided to go ahead with their proposal for a CCMS project on pollution of the upper atmosphere as they had already announced it. But in the end, they agreed to appease the Britons and pulled their punches at the next CCMS meeting. One FCO official satisfactorily noted that: 'they spoke at not too great length and with not very much force; and we in turn contented ourselves with the modest response set out'.[72]

Epilogue: a contaminant?

In light of what this chapter has shown it is worth asking if current UK approaches to climate change diplomacy echo earlier government stances on environmental diplomacy. We now know that the British government's early overtures to international environmental provisions were not exactly displaying enthusiasm.[73] A distinctly hollow pragmatism failed to produce a UK agenda on international environmental protection even after the *Torrey Canyon* disaster vividly demonstrated the lack of adequate plans to abate sea pollution and some NATO delegates manifested the wish to play a role in reducing it. The decision to appoint, as NATO representatives on environmental negotiations, science advisers whose background was not in environmental studies (like Zuckerman and Cottrell) reiterated this cold approach.

This does not mean that UK representatives were entirely wrong in showing their reservations regarding an environmentally focused defence alliance. While CCMS actions were widely publicised and led to resounding propositions, they achieved much less than expected and not just in tackling oil spills. By 1974 Nixon's adviser Henry Kissinger informed Train that the CCMS programme was under review. The oil crisis and the president's impeachment accelerated a

US divestment from environmental affairs, which typified the administration at least up until the appointment of US president Jimmy Carter in 1977. Ironically, by then NATO had yet to deal with the pollution it produced in the form of military wastes. This meant that while advocating a number of new provisions that should have found application in NATO countries, the alliance would continue for several years to produce noxious land and sea pollutants in military exercises, and that nobody – within or outside the alliance – would seek to assess this environmental threat.[74]

But if the US and NATO positions were idiosyncratic (contradictory, even), the British one denoted an effort to win a diplomacy argument with US colleagues over attention to environmental actions at home and internationally. And while from 1971, partly because of the debacle such a strategy produced, the UK government showed some keenness for environmentalism, some of its departments' officials still considered environmental conservation to be second rank when compared with other government imperatives. In particular, the web of relations between Foreign Office departments materialised efforts to disengage from support to international environmental regulations that might have required reconsidering UK's business priorities. In particular, these government agencies supported only resolutions that would not stifle large technological projects (such as Concorde, for instance).

This is not to say that the UK's administration was entirely passive and over the years it found a much better way to harmonise diplomatic and environmental ambitions. At NATO, for instance, the British delegation championed more CCMS projects with a focus on water waste treatment (Drinking Water, 1977–82), hazardous substances (Disposal of Hazardous Wastes, 1974–81; Contaminated Land, 1981–4) and municipal sewage (Utilisation and Disposal of Municipal Sewage Sludge, 1979–85).[75]

But in the early days of environmental diplomacy the British stance at NATO betrayed a blasé attitude towards pollution, one imprudently abridged in the jargon utilised at the FCO. While scheming against CCMS propositions, seeking the support of UK scientists in denying environmental change and quarrelling with US colleagues, the British civil servants kept restating that, when dealing with environmental issues, the alliance should only have been a 'contaminant'. The word was used metaphorically to stress that the alliance should only invigorate the work of other international agencies and not take the lead. Yet, its routine adoption in diplomatic lingo admirably exemplifies the protracted ambivalence of Her Majesty's Government towards environmental protection.

Notes

1   Research for this chapter was generously funded by the European Research Council in the context of The Earth Under Surveillance – TEUS project (grant no. 241009). Archival collections analysed in the chapter are identified through the following acronyms: NATO (NATO Archive, Brussels, Belgium); ZUC (Solly Zuckerman Papers, University of East Anglia, Norwich, UK); TNA (The National Archives, Kew, London, UK); RABI (Isidor I. Rabi Papers) and TRA (Russell Train Papers), US Library of Congress, Washington, DC, USA; NARA (US National Archives, College Park, Maryland, USA).

2   'The Observer View on Triggering Article 50', *Observer* (26 March 2017). Criticism has surfaced in the Tories' camp too as a recent interview with the father of Boris Johnson, environmentalist Stanley Johnson, reveals: Karl Mathiesen, 'Brexit Could Put Britain's Environment at Risk, Says Stanley Johnson', *Guardian* (4 April 2016).

3   See, for instance, Bill Luckin, *Pollution and Control: A Social History of the Thames in the Nineteenth Century* (Bristol: Adam Hilger, 1986) and more recently, *Death and Survival in Urban Britain, 1800–1950* (London: Tauris, 2015).

4   John Sheail, 'Torrey Canyon: The Political Dimension', *Journal of Contemporary History* 42, no. 3 (2007): 485–504, 489. See also Alexander Hall, 'The Rise of Blame and Recreancy in the United Kingdom: A Cultural, Political and Scientific Autopsy of the North Sea Flood of 1953', *Environment and History* 17 (2011): 379–408; Jon Agar, '"Future Forecast – Changeable and Probably Getting Worse": The UK Government's Early Response to Climate Change', *Twentieth Century British History* 26, no. 4 (2015): 602–28. Stanley Johnson has written on environmental policy, nationally and internationally. See, for instance, S. Johnson, *The Politics of Environment: The British Experience* (London: Tom Stacey, 1973).

5   Stephen Bocking, *Nature's Experts: Science, Politics, and the Environment* (New Brunswick: Rutgers University Press, 2006), 38; and Andrew Jamison, *The Making of Green Knowledge: Environmental Politics and Cultural Transformation* (Cambridge: Cambridge University Press, 2001), 12. Specific to the UK case is: Meredith Veldman, *Fantasy, the Bomb, and the Greening of Britain: Romantic Protest, 1945–1980* (Cambridge: Cambridge University Press, 1994).

6   J. Brooks Flippen, 'Richard Nixon, Russell Train, and the Birth of Modern American Environmental Diplomacy', *Diplomatic History* 32 (2008): 613–38; Stephen J. Macekura, *Of Limits and Growth: The Rise of Global Sustainable Development in the Twentieth Century* (Cambridge: Cambridge University Press, 2015); Jacob Hamblin, 'Environmentalism for the Atlantic Alliance: NATO's Experiment with the "Challenges of Modern Society"', *Environmental History* 15 (2010): 54–75.

7   On NATO's origins and diplomatic shortcomings, see especially Marc Trachtenberg, *A Constructed Peace: The Making of the European Settlement, 1945–1963* (Princeton: Princeton University Press, 1999). See also Stephen E. Ambrose, *Rise to Globalism: American Foreign Policy since 1938* (New York: Penguin, 1971/1997), 95–105.

8   Jacob Darwin Hamblin, *Arming Mother Nature: The Birth of Catastrophic Environmentalism* (Oxford and New York: Oxford University Press, 2013), 129–49.

9   I discuss the circumstances of the history of this committee and NATO's science and environmental programmes in my forthcoming book *Greening the Alliance: Science, Environment, Diplomacy and the North Atlantic Treaty Organization* (Chicago: University of Chicago Press, Fall 2018).

10 Chief Scientific Adviser's Report, Confidential, 30 March 1967, SZ/CSA/163, ZUC.

11 Sheail, 'Torrey Canyon: The Political Dimension', 489. See also Adam Vaughan, 'Torrey Canyon Disaster – The UK's Worst-Ever Oil Spill 50 Years On', *Guardian* (18 March 2017); and for an interesting perspective 'from below', see Timothy Cooper and Anna Green, 'The Torrey Canyon Disaster, Everyday Life, and the "Greening" of Britain', *Environmental History* 22, 1 (2017): 101–26.

12 S. Zuckerman to Donald McLachlan, 5 April 1967, SZ/CSA/163, ZUC.

13 Philip Ziegler, 'Zuckerman, Solly, Baron Zuckerman (1904–1993), Scientist and Public Servant', *Oxford Dictionary of National Biography*. Last accessed 13 August 2017. www.oxforddnb.com/view/article/53466?docPos=2#. See also Solly Zuckerman, *Monkeys, Men, and Missiles: An Autobiography, 1946–88* (New York: Norton, 1989). On Zuckerman's wartime work, see William Thomas, *Rational Action: The Sciences of Policy in Britain and America, 1940–1960* (Cambridge, MA: MIT Press, 2015), 80–6.

14 'NATO Science', Limited Official Use, 6 January 1968, Box 41, Folder 5, RABI, 4.

15 'NATO Science', Limited Official Use, 6 January 1968, Box 41, Folder 5, RABI, 12.

16 A science adviser as renowned as Zuckerman, Rabi had had a similar career. Polish-born, he had pioneered radar studies in the United States before becoming an influential adviser in matters such as atomic weaponry and US–Western Europe scientific relations. For an overview, see John Krige, 'I. I. Rabi and the Birth of CERN', *Physics in Perspective* 7 (2005): 150–64.

17 Speeches commemorating the 10th anniversary of the foundation of the NATO Science Committee, 21 March 1968, AC/137-D/330, NATO, 22–5.

18 Train was also a founder of the World Wildlife Fund. Macekura, *Of Limits and Growth*, 105.

19 NATIS, Address by the President of the United States at the Commemorative NAC Session, 10 April 1969, NATO.

20 On the CCMS activities, see Hamblin, 'Environmentalism for the Atlantic Alliance'. See also Thorsten Schulz, 'Transatlantic Environmental Security in the 1970s? NATO's "Third Dimension" as an Early Environmental and "Human Security" Approach', *Historical Social Research* 35 (2010): 309–28; Patrick Kyba, 'CCMS: The Environmental Connection', *International Journal* 29 (1973): 256–67.

21 NATO Public Diplomacy Division, *Aspects of NATO – The Challenges of Modern Society* (Brussels: NATO, 1976), 5–7.

22 North Atlantic Council, Record of Meeting, 14 May 1969, C-R(69)22, NATO, 5–6; F. H. Capps, NATO: Support for CCMS Growing, Secret, 14 April 1970, in Record Group 59, Box 2894, NARA, ii.

23 Helen M. Rozwadowski, *The Sea Knows No Boundaries: A Century of Marine Science under ICES* (Copenhagen and Seattle: ICES/University of Washington Press, 2002). See also Samuel Robinson, 'Between the Devil and the Deep Blue Sea: Ocean Science and the British Cold War State' (PhD diss., University of Manchester, 2015).

24 Summary record of a meeting of the Council held at the NATO Headquarters, 25 March 1970, C-R(70)13, NATO, in Hamblin, 'Environmentalism for the Atlantic Alliance', 61.

25 Secretary General to Permanent Representatives, 29 May 1969, PO/69/265, NATO.

26 N. Elam, FCO to E.C. Appleyard, FCO, 22 August 1969, CAB 168/278, TNA.

27 Note of a Meeting at the Cabinet Office, 4 June 1969, CAB 168/278, TNA. A restricted document summarising the stance was prepared for the PM Harold Wilson.

28 J.N. Elam, handwritten notes on the paper 'CCMS Possible Topics for Discussion', 5 September 1969, CAB 168/278, TNA.

29 J.C. Thomas, FCO, to D.E. Richards, UKDELNATO, Confidential and Guard, 1 January 1970, FCO 55/408, TNA.

30 On Cottrell, see John Knott, 'Cottrell, Sir Alan Howard (1919–2012), Metallurgist'. Last accessed 10 September 2017. http://www.oxforddnb.com/view/10.1093/ref:odnb/9780198614128.001.0001/odnb-9780198614128-e-104660. See also R.E. Smallman and J.F. Knott, 'Sir Alan Cottrell, 17 July 1919–15 February 2012', *Biographical Memoir of the Fellows of the Royal Society* 59 (2013): 93–124.

31 See Thomas Turnbull, 'Simulating the Global Environment', Chapter 14 in this volume.

32 Lord Chalfont to Lord Kennet, 1 January 1970, FCO 55/408, TNA. FCO officials later commented that NATO 'had no business to discuss what makes people work'. M. Braithwaithe to J.C. Thomas, 10 February 1970, FCO 55/408, TNA.

33 C.J. Audland to A. Cottrell, 6 January 1970, FCO 55/408, TNA.

34 A. Cottrell to J.H. Galbraith, 20 January 1970, FCO 55/408, TNA.

35 P.B. Wheeler to J.C. Thomas, 2 February 1970, Restricted, FCO 55/408, TNA.

36 T.A.K. Elliott, British Embassy in Washington, DC, to Arculus, 13 April 1970, Personal and Confidential, FCO 55/455, TNA.

37 P.Q. by Gilbert Longden about 'Motivation in a Technological Society', 2 November 1970, FCO 55/455, TNA.

38   R. Train to A. Cottrell, 17 August 1972 and Train to Cottrell, 24 October 1972, Box 43, Folder 3, TRA. The study was eventually published as Norman A.B. Wilson, *On the Quality of Working Life: A Report Prepared for the Department of Employment* (London: HMSO, 1973).

39   Vaughan, 'Torrey Canyon Disaster – The UK's Worst-Ever Oil Spill 50 Years On'.

40   Hamblin, 'Environmentalism for the Atlantic Alliance', 60.

41   Hamblin, 'Environmentalism for the Atlantic Alliance', 60.

42   Agar, '"Future Forecast – Changeable and Probably Getting Worse"', 609–10. See Turnbull, 'Simulating the Global Environment'.

43   'We are supporting them up to the hilt in their efforts,' they stated in the briefing prepared for Zuckerman. Briefing for Sir Solly Zuckerman, 3 October 1970, CAB 168/284, TNA.

44   NATO CCMS, undated draft [late 1970], CAB 168/284, TNA.

45   NATO Scientific Affairs Division, *The Challenges of Modern Society* (Brussels: NATO, 1991), 8, 14.

46   See Agar, '"Future Forecast – Changeable and Probably Getting Worse"', 610.

47   F.B. Wheeler to Mr Campbell, 18 September 1970, FCO 55/443, TNA.

48   D.N. Byrne, DTI, to J. Ure, FCO, 21 April 1971, FCO 76/224, TNA.

49   H. Dudgeon, FCO, Colloquium on Oil Spills, Confidential, [undated], FCO 55/443, TNA.

50   D.A. Campbell, FCO, to H. Dudgeon, FCO, 5 October 1970, FCO 55/443, TNA.

51   A. Douglas Home to UKDELNATO, 30 October 1970, CAB 168/284, TNA.

52   The records of a Cabinet Office meeting stressed that: 'the implications of redeeming such a promise … are bound to be costly'. Report on Meeting on 19–20 October, Restricted, 9 November 1970, CAB 168/284, TNA.

53   '[W]e continued to regard the load-on-top system as a useful contribution to the solution of this problem. … We shall continue to advocate this system,' FCO to UK Embassy, Washington, DC, 11 November 1970, CAB 168/284, TNA. See also Douglas Cormack, *Response to Marine Oil Pollution: Review and Assessment* (Dordrecht: Kluwer, 1999), 30–1.

54   A. Denis Clift (White House), Memorandum of Meeting, 18 December 1970, Folder 3, Box 46, TRA.

55   NATIS, 'Resolution on the Pollution of the Sea by Oil Spills', 7 December 1970.

56   Blaney, Memorandum of Conversation, 19 April 1971, Folder 5, Box 46, TRA.

57   H. Blaney, Memorandum for Train, 4 May 1971, Folder 5, Box 46, TRA.

58   Margaret Rothwell, FCO to J. B. Ure, FCO, 15 April 1971, FCO 76/224, TNA.

59   R.E. Train, 'A New Approach to International Environmental Cooperation: The NATO CCMS', *Kansas Law Review* 22 (1973–4): 167–91, 175.

60   D.A. Campbell, FCO, 'NATO Council: CCMS, 14 May 1971', FCO 55/658, TNA.

61   On this see Spencer Weart, *The Discovery of Global Warming* (Cambridge, MA: Harvard University Press, 2008), 128–9.

62   G.D. Crane to J.A. Woolmer, 17 April 1973, FCO 55/1009, TNA. The FAA decision itself was the subject of media scrutiny. On this see Jon Anderson, 'Decision Analysis in Environmental Decision-making: Improving the Concorde Balance', *Columbia Journal of Environmental Law* 5 (1978–9): 156–88.

63   Paul J. Crutzen, 'The Influence of Nitrogen Oxides on the Atmosphere Ozone Content', *Quarterly Journal of the Royal Meteorological Society* 96 (1970): 320–5. Crutzen is also known for introducing the term 'Anthropocene'. See P. Crutzen, 'Geology of Mankind', *Nature* 415, no. 6867 (2002): 23.

64   J.L. Taylor to E. Burton, STD, 18 April 1973, FCO 55/1009, TNA.

65   On this see Peter Gillman, 'Supersonic Bust: The Story of the Concorde', *Atlantic* 239 (1977): 72–81. See also Jonathan Glancey, *Concorde: The Rise and Fall of the Supersonic Airliner* (London: Atlantic Books, 2015).

66   E. Burton, STD, to D.F. Downing, Ottawa, 11 June 1973, FCO 55/1009, TNA.

67   P. Goldsmith et al., 'Nitrogen Oxides, Nuclear Weapon Testing, Concorde and Stratospheric Ozone', *Nature* 244, no. 5418 (1973): 545–51. On the controversy, see also Brian Martin, *The Bias of Science* (Canberra: Society for the Social Responsibility in Science, 1979).

68   Harold S. Johnston, 'Supersonic Aircraft and the Ozone Layer', *Environmental Change* 2 (1974): 339–50. See also H.S. Johnston, 'Reduction of Stratospheric Ozone by Nitrogen Oxide Catalysts from Supersonic Transport Exhaust', *Science* 173 (1971): 517–22.

69   G.C. Lowe, British Embassy in the US to Burton, 21 June 1973, FCO 55/1009, TNA.

70   Agar, '"Future Forecast – Changeable and Probably Getting Worse"', 612.

71　'While there are conflicting views on this subject, the balance of evidence … comes down on the side of Concorde not causing harm.' P.J. Broomfield, DTI, to I. Mackley, FCO, 3 September 1973, FCO 55/1009, TNA.

72　J. Mellon, FCO, to Viscount Dunrossil, Ottawa, 9 November 1973, Restricted, FCO 55/1009, TNA.

73　This may have set an important precedent as later UK governments' approaches to environmentalism were not particularly encouraging either. In this respect see, for instance, the case in the 1980s discussed in Martin Mahony and Mike Hulme, 'Modelling and the Nation: Institutionalising Climate Prediction in the UK, 1988–92', *Minerva* 54, no. 4 (2016): 445–70.

74　Kyba, 'CCMS: The Environmental Connection', 256–67. NATO military wastes became the subject of CCMS projects only from the mid-1980s as discussed in Turchetti, *Greening the Alliance*.

75　Scientific Affairs Division, *The Challenges of Modern Society*, 13–14, 23.

14

Simulating the global environment: the British Government's response to *The Limits to Growth*

Thomas Turnbull

Published in 1972, *The Limits to Growth* argued that unless concerted international action was taken the Earth's population risked 'over-shoot'.[1] Collective demands on the environment would exceed the Earth's carrying capacity, resulting in demographic 'collapse' as a result of both 'pollution' and 'non-renewable resource depletion'.[2] The history of *Limits* has proved an important but well-worn story in the history of environmentalism. Encapsulated by American political scientist Lynton Caldwell, the report heralded a 'new environmental paradigm', in which the 'view of an earth unlimited in abundance' ended.[3] Instead of pursuing economic growth, the report bolstered a radical view, shared by some in positions of power, that post-war economic growth should end before disaster struck.[4] Much effort has subsequently been taken to describe the role *Limits* played in constructing the concept of a 'global' environmental space and divining a common future for 'mankind'.[5] Others have described how the technology upon which the book's thesis was based, computer simulation, granted long-held fears of environmental catastrophe a new-found credibility.[6] But the simplicity of the underlying model, 'World3', caused others to deride the exercise on methodological grounds and question the motives of those involved.[7] In recent years, however, some have defended the study, emphasising its role as a warning rather than a prediction, and asserting that the concept of limits to growth remains prescient,[8] particularly as it challenges the subsequent message of 'sustainable development', and its comforting mantra that economic growth can improve rather than denude the environment.[9] Moreover *Limits* can be seen as a precursor to contemporary earth systems science, and the vision of planetary management that has been reinvigorated by the proposed notion of the Anthropocene.[10] While these are important arguments, rather than looking at the planetary implications of this much-studied model, the chapter attempts to 'provincialise the Globe', in order to show how the idea of an enveloping global environmental system was something received, constructed and contested locally[11] – in this case, in a forward-looking, modern Britain.

While there have been attempts at tracing the public response to *Limits* in various countries, including Britain, the question addressed here is how the report, and the technology upon which it was based, was received and responded to by those in power. The aim of this chapter is to offer an account of the relation between forecasting simulations and the environment in this context.[12] *Limits* was remarkable in so far as it marked the hybridisation of two fields, environmentalism and 'future studies'.[13] Often pluralised, futures studies, or futurology, was a collection of practices, largely developed in the Cold War period, whose adherents claimed to be able to forecast the future in a scientific or otherwise systematic way, often beginning from the study of existing trajectories in demography, technology or other variables deemed suitably determinate.[14] Histories of these 'futures-of-the-past' have recently attracted a burst of scholarly attention, revealing a growing enthusiasm for such studies at the cusp of the 1970s across the world.[15] Emerging out of this wave, Paul Warde and Sverker Sörlin argue that *the* 'environment', as a definite article, was a concept that emerged alongside efforts to forecast the future of humankind, often speculatively, after the Second World War. Migrating from its original meaning in evolutionary biology, the forward-looking 'environment' was contrasted with the nostalgic term 'nature'.[16] *Limits*, in using a computer to forecast the dynamics of the global environment, adds weight to their argument, suggesting a co-constitutive relation between simulation technology and the future-oriented concept of the environment.

What was Britain's role in establishing this interaction between the future and the environment? Histories of British futurology are scant. One explanation is cultural. Britain had a long history of using science to guide policy, including forecasting, which emerged from wartime efforts to optimise military operations. In 1961, leading figures in British wartime Operations Research ('OR'), Patrick Blackett and Solly Zuckerman, became openly hostile to the work of RAND Corporation analyst Herman Kahn. Kahn had written a game-theoretical approach to nuclear strategy, published in 1960, that managed to be both dispassionate and sensationalist. Both Blackett and Zuckerman condemned Kahn's work in the press, and sought to distance their commitment to 'operational realities' from the abstracted and ill-considered 'futurology' being done by such American defence analysts.[17] As this chapter hopes to show, rather than revealing methodological archaism in British policy-making, this kind of boundary work indicated the existence of a distinct, somewhat oppositional, British school of futures studies. Attesting to this, Jon Agar has uncovered the forgotten work of the World Future Trends Committee (WFTC), a Cabinet research group established at the behest of Prime Minister Edward Heath. Agar suggests the interest around *Limits* kick-started this 'new interest' in futurology in government. Agar traces how this committee encouraged the government to undertake long-range climate forecasts, allowing the notion of climate change to enter Whitehall.[18] Contributing further to the history of British futures, this chapter re-constructs the earlier work of the WFTC, documenting how the committee reacted to *Limits* and the model on which it was based. In doing so, one intention is to add further weight to Warde and Sörlin's thesis about the constitutive relation between forecasting technologies and the environment.

This chapter will also argue the use of simulation in policy formation was well established in Britain before the arrival of *Limits*. By 1970, both the Labour government and Conservative opposition had shown enthusiasm for futurological research, whether or not they employed that term. The existence of the little-known 'Programmes Analysis Unit' (PAU) of the Labour government and the Conservative party's 'Conservative Systems Research Centre' (CSRC) suggests *Limits* was received into a political culture already familiar with computer-based forecasting. A fine-grained account of this reception is then told, based upon on two other sources. The first is an 'insiders' account of the work of the Club of Rome, the somewhat elusive organisation that commissioned the world model, found in the papers of British MP Jeremy Bray. There were other British members of the Club, including founding member Alexander King, Science Director at the Organisation for Economic Cooperation and Development (OECD), the physicist Dennis Gabor, biologist Conrad Waddington, and historian Asa Briggs. But only Bray was actively engaged in, and a vocal critic of, the technical development of World3, the version of the model upon which *Limits* was based.[19]

The second source of archival material is the Cabinet papers detailing the British government's attempts to imitate and then better the World3 model via the work of the WFTC. While other historians have acknowledged the Heath government's interest in *Limits*, no one has told an inside story of the British government's attempt to create its own computer model of the global environment.[20] In detailing this story, it is argued that simulation of a global 'system' was one of the principal ways in which the environment became understood as a political problem. Second, extending an argument first made by the historian Timothy Mitchell, it is suggested that in Britain during the 1970s the environment became 'an object of politics' that could rival the economy as a justification for political intervention.[21] A third and final argument is that government-led interrogation of the World3 model indicated that, with minor adjustment, the model could be used as a means of demonstrating the political conditions by which environmental limits could be transcended. In doing so, the environment was no longer understood in a biological sense, as the aggregate conditions that sustain organisms, but as a modifiable, human-directed system.[22]

British futures

The use of computers to ordain the future is not a practice one immediately associates with 1970s Britain, a period more popularly recognised for recession, strikes, blackouts, public disorder and the failure of the ideals of the 1960s.[23] Despite a wave of revisionism, today, the warning that a certain course of political action will 'take us back to the 70s' still chastens in some quarters.[24] One document that helped establish this pessimistic view was published in December 1974, just eight months after Heath had lost the general election during a strike by the National Union of Mineworkers, which resulted in the imposition of a three-day working week, as a result of fuel shortages. Hudson Europe, a branch of American think tank the Hudson Institute, had published a study, *The United Kingdom in 1980*, which forecast Britain's 'relative decline' in comparison with its neighbours, a gloomy portent given Heath's plan for Britain to enter the European Economic Community (EEC). The study claimed this decline was the result of 'a peculiarly British problem', as rather than looking to the future the political class were 'prone to archaism' and 'flights into pre-socialist, pre-capitalist fantasy'. Britain, it was argued, was stunted by its factional class system and aversion to technologically enhanced policy-making of the kind Hudson proffered. A 'shift in national style' was prescribed, involving the establishment of a future-oriented 'Planning Commission' consisting of 'Britain's best economists and administrators and engineers'.[25] But questions of the environment, raised by *Limits*, played only a minor role. Hudson's report considered the state of Britain's 'industrial experience, a greater threat to growth than exhausting our resource reserves', and that the 'real pollution is that which is caused by poverty and archaic industries'.[26]

Aside from its futurological perspective, the report was typically 'declinist', a genre of historical writing that denigrated post-war Britain and its commitment to scientific and technical progress as compared with other nations.[27] While declinism has proved influential, the historian David Edgerton has assembled a wealth of evidence to demonstrate that Britain's post-war investment in science and technology, by both industry and government, was prodigious, second only to that of the United States. As such, Edgerton argues that Britain was 'the scientific and technological powerhouse of Western Europe' by the 1960s, and declinist texts are 'anti-histories', inverse to evidence, that must be seen either as self-serving rhetoric for those seeking to provoke further investment in Britain's state-funded science and technology programmes, or as political polemics intended to denigrate the government of the time.[28] For example, Edgerton describes the American historian Martin Wiener's book *English Culture and the Decline of the Industrial Spirit*, published in 1981, as an 'ideologically significant' part of Thatcherite thought. Its account of the entropic effect of Britain's aristocratic culture on British industry emboldened an ascendant new right.[29] Less well known is that the thesis of Wiener's book had first been delivered at a conference organised by Hudson Europe, held at Hertfordshire's Ashridge Management School, during the report's research process.[30]

The Hudson Institute was largely the work of Herman Kahn, the infamous American policy analyst who had developed the pre-emptive strategies in case of thermonuclear war that Blackett and Zuckerman had been critical of. Based on the principles of game theory, the strategies had been developed on behalf of Californian think tank, and oftentimes client of the American military, the RAND Corporation. At RAND, Kahn had contributed to and employed some of futurology's most influential methods, from harnessing the unpredictability of random numbers using the 'Monte Carlo' method, to the more hypothetical and imaginative 'scenario planning', and the discursive the 'Delphi method'.[31] However, published in 1960, *On Thermonuclear War* not only irritated the elders of British OR, it also created media outcry in America, prompting Kahn to leave RAND to set up his own research organisation in New York.[32] Hudson's selling point, in light of the controversy his book had provoked, was to 'think the unthinkable' and offer uncomfortable truths to policymakers, which he did so with some success.[33] In 1973, with characteristic entrepreneurialism, Kahn oversaw the establishment of a European office, hoping to capitalise on the political uncertainties of the end of the Bretton Woods agreement and ongoing expansion of the EEC.[34] Britain seemed ripe for Hudson's style of heterodox, future-oriented, policy advice. Initially, they had some influence. In September 1973 Kahn, in Britain to promote Hudson's work, met with Victor Rothschild, head of the Central Policy Review Staff (CPRS). His gloomy prognoses of Britain's future found their way into a speech by Rothschild in September, to attendant press, at the Agricultural Research Council Laboratory in Wantage. His pessimism provoking a disciplinary rebuke from Heath.[35]

The report left the press irate. The *Sunday Times* claimed it was 'more a prolonged attempt to shake the national psyche than a piece of original research', the *Guardian* suggested it should be met with 'howls of derision', and the *Daily Mail* derided its presentation of 'the obvious and the speculative'.[36] However, on the left, Tom Nairn agreed with its portrayal of a self-serving ruling class retarding innovation.[37] But the generally poor reception the report received on publication seemed only to affirm Hudson's thesis. Lead author, the American political scientist Edmund Stillman, had warned that from the outset their style of research had been met with 'reactions ranging from dismissive scepticism to utter hostility'.[38] Was there a 'profound domestic intellectual resentment of economic and socio-political forecasting' as James Bellini, the sole British author, had warned, or did the report have an ulterior motive?[39] With parallels to Edgerton's anti-history thesis, writing in the journal *Futures*, Israeli futurologist Yehezkel Dror, a visiting scholar at the London School of Economics, revealed the report had emerged from a failed attempt to gain a research contract from Britain's government, as they had from the French.[40] Dror's review of the self-funded work argued the problem was not British archaism so much as the publication's 'shallow argumentation and spotty treatment' that seemingly borrowed from the French project for much of its content. Denigrating Britain's future, like its past, seemed to be driven by the pursuit of funding.[41]

For its detractors, the issue was not the report's futurological approach, so much as the feeling it did a disservice to futurology and underestimated Britain's forecasting capabilities. Another *Futures* reviewer regretted its prescriptive shortcomings, and elsewhere the fashionable British economist Ezra Mishan derided the report's 'naïve extrapolation'.[42] But to extend Edgerton's argument, just as declinist histories obscured Britain's scientific and technological capabilities, a similar case could be made for its futurological capacities in the 1970s. The journal *Futures*, in which Bellini and Dror argued about the quality of the Hudson report, had been established in Guildford in 1968.[43] In 1971, 'Political and Economic Planning' (PEP), a London-based think tank in all but name, carried out a survey of futures studies, identifying over six hundred organisations and individuals interested in all kinds of 'exploratory forecasting' globally, over a hundred of which were in Britain. These ranged from PEP itself, to Royal Dutch Shell, the University of Bradford, the Nature Conservancy and the Electricity Council's forecasting section.[44]

Forecasting was also being done by government. Though largely forgotten today, Harold Wilson's Labour government had established a future-oriented research organisation, the 'Programmes Analysis Unit' (PAU), in 1967. In opposition, Wilson's famed 'White Heat' speech had warned that Britain risked becoming a 'stagnant backwater' if government did not take full advantage of computation and automation in executing a programme of planned modernisation.[45] In power, his government sought to improve the productivity of Britain's industry through state intervention, overseen by an expansive Ministry of Technology ('Mintech'). Situated in Harwell, at the site of the Atomic Energy Agency (AEA), the PAU analysed how Mintech could maximise the returns on government investment in research by carrying out 'economic and market surveys as well as forecasts of future environment'.[46] One analyst, Derek Medford, described how futurology-like 'exploratory forecasts', 'temporal extrapolation' and the intriguing value-led 'normative forecasting' were carried out by 25 staff, often using the AEA's computers.[47] These methods were intended to produce disinterested evidence on the benefit of investing in new technologies, ranging from carbon fibres to cryogenics.[48] Echoing Blackett's earlier criticisms, Medford claimed the PAU's work was distinct from the 'pseudo-science' of Kahn and other 'hubristic' futurologists owing to its practical applications and lack of pretension, 'buttressed by Britain's invention of operations research'. Accordingly, Medford warned Britain's scientists to involve themselves in evaluating the future applications of their work, else 'they may awake too late and become the chattels of business school graduates'.[49]

In opposition, the Conservative party were also intrigued by the promises of forecasting, though of the kind Medford feared. In 1966, as opposition leader, Heath, already a self-acknowledged moderniser, issued a Conservative party pamphlet proposing a 'Central Cost Effectiveness Department' that would apply the latest 'American management science' to the problems of public spending, the most important of which was 'the Systems Analysis approach', which could be used to 'simulate activities with the help of a computer'.[50] Accordingly, in 1968 a Conservative Systems Research Centre (CSRC) was launched at the party's head office. The aim was to create a computer-based forecasting tool 'for scrutinising, comparing and discussing ranges of policy alternatives before rather than after the event'.[51] Overseen by MP Mervyn Pike, the organisation had four staff, including an analyst on loan from IBM. The group used the telephone network to connect to a time-sharing computer in London Victoria. Operated by Service in Informatics and Analysis Ltd, the '6600' was supposedly the world's most powerful computer at the time.[52] Led by the computer-savvy young Conservative Michael Spicer, the team developed a 'Policy and Information Control System', a linear programming matrix for simulating how proposed adjustments to tax and spending would affect different parts of the economy.[53] Techniques of this kind, described as American managerial science, were seen by many in the Conservative party as an antidote to the myopia and inefficiency of the public sector.[54]

Such aspersions came from the belief that British civil servants tended towards 'arts-based' training, producing generalists rather than technical experts. The 1965 All-Party Fulton Committee had investigated Civil Service reform, and concluded that 'new modes of specialised analysis' such as 'linear programming and cost-benefit analysis' should be used in the work of government.[55] This call for greater technical expertise led to an almost five fold increase in computing staff in the Civil Service. By 1972 there were 12,000 computer workers in central government, the majority of whom were involved in secretarial and clerical work.[56] But simulation remained a marginal activity.[57] Among the first and most sophisticated were models to forecast economic income and expenditure at the Treasury after the war, consisting of sets of equations calculated by hand. These became increasingly sophisticated in the following decades, and experimental computer models were intermittently trialled. In 1966 the London Business School produced the first fully computerised British macro-economic model, which the Treasury adopted in 1971.[58] But alongside the Treasury's work, with growing fears about human ecology, the government soon took an interest in modelling the environment as if it were a rival object of political knowledge.[59]

Simulating environments

By the end of the 1960s the environment was becoming an increasingly prominent concern in British public life. In 1963 the National Environmental Research Council (NERC) was established, in 1969 the Royal Commission on Environmental Pollution, and 1970 was declared a Year of Conservation by the European Council. But perhaps most significantly Prime Minister Edward Heath established a 'Department of the Environment' in 1970 when he formed his government, with responsibility for 'the whole range of functions which affect people's living environment'.[60] For those who had long studied the relation between the human and natural worlds, this sudden elevation of the environment in political discourse was disconcerting.[61] Timothy O'Riordan, a British geographer based in Canada, remarked that the 'academic world seems to be going topsy-turvy. While the political scientist talks about ecology ... the ecologist is discussing public policy.'[62] This semantic shift was manifest in the launch of the *Ecologist* magazine that year, a periodical that promoted ecologically sound rather than economically expansive ways of life.[63] One explanation was that the term 'ecology' had expanded in scope. It no longer meant just the study of biota and their environment, but also actions intended to preserve the environment against degradation. As historian Peter Taylor later suggested, in the Anglophone world more generally 'Ecology, and the prefix eco- became a call for social action as well as a science.'[64]

However, the World model's methodology drew on industrial management rather than ecology. As the many biographers of its creator, American electrical engineer Jay Forrester, have noted, his approach stemmed from wartime experience. He had worked on analogue servomechanisms, electrical control mechanisms that responded to informational feedback to fulfil a given task.[65] After the war, he worked at MIT's Digital Computer Laboratory, constructing an early form of digital computer, and later helped develop the US Air Force's 'Semi-automated Ground Environment' radar tracking system.[66] But in 1956 Forrester's career changed. He was offered a professorship at MIT's School of Management, one of the first centres for managerial education. There he attempted to apply the principles of control engineering to industry. Coded appropriately, industrial operations were represented by differential equations that linked temporally contingent reactions, via a mass of circular referents, or feedback loops, which could simulate change over time. Such models often depicted a 'rollercoaster' effect. An unexpected 10 per cent increase in demand, for example, could cause an increase in profit followed by a sharp decrease, like the stomach-churning ascent and descent of the funfair ride.[67]

Forrester increasingly believed his Systems Dynamics approach could be generalised. By simulating the behaviour of any given system, he believed you could learn how to manage it. In doing so, he increasingly stepped beyond observable systems, such as a single industry, towards 'systems' exceeding human perceptual abilities.[68] Accordingly, he began work on a global model, World2, following an undocumented prototype. His World models were based on the idea that the Earth's major dynamics could be symbolised numerically as a series of stocks and flows. In order to make this visually comprehensible, these global dynamics, such as 'natural resource use rate', could be presented graphically as nodes ('stocks') and their relation to other material and informational variables represented by vertices ('flows'). The relations between dynamics could be programmed into a computer as a sequence of differential equations, representing actual or hypothetical relations between nodes. Typed into over one hundred lines of code using DYNAMO programming language, Forrester claimed his model could be used to model all of Earth's most significant environmental dynamics.[69] Belying his engineering roots, Forrester's pragmatic approach seemed to suggest global dynamics could, in some sense, be programmed if the model's evidence was acted upon.[70]

In modelling the flows of resources, population, energy and information, Forrester's World models reiterated, seemingly unaware, the analytical approach systems ecologists had developed in the 1960s to model 'ecosystems' and his method was later employed by ecologists.[71] This easy migration of concepts was possible in a period in which both nature and society could be cast in the abstract language of 'systems', which as the American sociologist Talcott Parsons concisely defined it in 1968, meant 'a network of interconnection, the state or activity of one component influencing the state or activity of other components'.[72] Such systems could consist of heterogeneous components: mechanical, informational and biological. Society itself was a system, which, rendered into suitably abstract forms, could become amenable to mathematical or computational analysis, allowing the modeller to transform a complex reality into an instrumental tool to achieve a desired outcome.[73] For policy makers, this systemic rendering of society seemed to promise the ability to achieve pre-determined futures by altering behaviour in the present.[74]

Jeremy Bray

When it came to the technicalities of computer modelling, Jeremy Bray was well informed. Somewhat incongruously, given his physique and professorial manner, Bray had originally planned to become a coal miner, but soon realised he was entirely unsuited to the job. Harold Wilson's secretary, Marcia Williams, had once acknowledged Bray's brilliance but unkindly remarked that he looked like 'every mad professor of comic fiction'.[75] Instead of mining, he had achieved a double first in mathematics at Cambridge. Later, in the 1950s, he attended Claude Shannon's seminars at MIT, which had established the science of information theory as a distinct counterpart to cybernetics.[76] After some years at Imperial Chemical Industries (ICI), Bray joined the University of Cambridge's Department of Applied Economics, a group well recognised for their distinct approach to econometrics. Building on the university's strength in theoretical statistics, the group specialised in time series analysis, the statistical (probabilistic) analysis of dependent sequential values in temporal data. Another strength was detailed empirical studies based on the vast amounts of data on income, expenditure and consumer demand, first collected to support post-war recovery and rationing programmes.[77] The department also housed what Bray described as 'the most sophisticated model of its kind in the world', a computer-based model of Britain's economy for forecasting growth and the effects of policy interventions that he believed was superior to that of the Treasury.[78]

Alongside his econometric research, Bray had become a Labour parliamentary candidate in 1959, winning Middlesbrough West in 1962. Four years later he became junior minister at Mintech, where he oversaw the use of computer models to forecast future power demand.[79] He also became increasingly concerned with the problem of pollution, which he considered a manifestation of 'interacting, ecological, systemic problems'.[80] Drawing on these interests, in 1970 he had published *Decision in Government*, a highly technical proposal for a form of 'regenerative' socialism. Central to his plan was the idea of an 'adaptive economy', distinct from a liberalised economy as it was directed towards collective welfare, but similar, he hoped, in so far as 'control action' would be 'no more centralised than it need be'. He envisioned a decentralised network of computers able to relay information to central government, so that decision-making could be significantly decentralised, without losing the oversight of Westminster or the goal of redistributing wealth.[81] In essence, Bray believed he had developed a system of government that could harness the allocative capabilities of the market but in pursuit of a socialist agenda.[82] The book was densely written, full of detailed econometric arguments decipherable to few. But whatever its opacity, Wilson considered the book a criticism of his more conventional proposals for a planned economy and encouraged Bray's resignation.[83]

In January 1970, Aurelio Peccei, an Italian industrialist and the head of the Club of Rome, became aware of Bray's situation.[84] He invited the former MP to join. Bray, on his part, was impressed by the involvement of Turkish–American long-range planner Hasan Özbekhan, whom he had first met at a Long-Range Planning symposium in 1968. There, he had proposed a 'general theory of planning' in which forecasters abandoned the assumption that individuals acted in pursuit of the 'maximisation of self-interest', as most economic models suggested, lest the model become reality. Instead, Özbekhan claimed planning should be 'inspired by a higher principle'.[85] As a Methodist minister, Bray was moved by his suggestion that the principle should be 'love'.[86] Accordingly, Bray responded enthusiastically to Peccei's invitation, though he warned that the Club's stated proposal 'to contribute to the comprehension of the problems of modern society' using systems analysis ran counter to the traditions of British philosophy, politics and social science, which was 'highly empirical, and rather non-systematic'. Listing other nations' intellectual strengths, such as French 'logic and lucidity' and Japanese 'organisation', he claimed Britain's research tradition could offer only 'an empirical approach to specific problems', a warning that would prove prescient.[87]

The Club of Rome

In June 1970, around 40 members of the Club held a meeting at the Bellevue Palace in Bern, a luxurious hotel owned by the Swiss government. Described as an 'informal, multi-national, non-political group of scientists, economists, planners, educators and business leaders', they assembled alongside journalists in the hotel's impressive art nouveau meeting room with the ambitious agenda to address the 'Predicament of Mankind'.[88] This situation, Özbekhan had proposed, was the result of interconnected problems of overpopulation, malnutrition, poverty and pollution, which in combination, might result in a 'generalized meta-problem' with potentially deleterious consequences for humankind. He outlined a plan to address this 'problematique' using a complex of diagnostic computer models that would allow the extension of human perception beyond the 'fragmentation of reality into closed and well-bounded problems'.[89] By creating a model that could show the interactions between Earth's major problems, it was hoped that the most dangerous interactions between these various dynamics could become the focus of concerted global action.

One Club member, Caroll Wilson, a professor of management at MIT, had invited his colleague, Forrester. Following Özbekhan, the American engineer outlined his proposal to develop a single 'World' model to address this meta-problem. Explaining his existing work, he suggested he could extend his numerical model of the Earth's major dynamics to more accurately forecast the future conditions of human life on Earth. He seemed to offer a simple formalisation of the Club's concerns, which could give some indication as to when limits might be reached. Most accounts of the meeting suggest Forrester's proposal was positively received by Club members. However, Bray was at the meeting, and his handwritten notes revealed a deep scepticism towards Forrester's technique, his proposed use of data and the means by which the findings might be meaningfully implemented as policy. More fundamentally, his handwritten notes asked, 'at what level of aggregation is it possible to make meaningful forecast or arrive at action?'[90]

At Forrester's request, the Club's second meeting was held six weeks later at MIT's School of Management in Boston, Massachusetts. There, over 10 days of presentations and discussion, members were introduced to the principles of Forrester's 'System Dynamics' and its possible use in simulating global problems. The model that was presented to Club members, World2, contained only five variables and two empirical sets of data. But despite its shortcomings, every model run indicated a catastrophic collapse of the global system in the coming century. Both population and capital investment would increase, while resource availability remained fixed. Within these parameters, collapse was inevitable. Peccei was particularly impressed by the fact that each run of the model resulted in collapse, as he had predicted this in his own speculative book about the global future, *The Chasm Ahead*, published in 1968.[91]

Forrester had implored Bray to come to Boston, promising to deliver a 'more orderly presentation' of his model.[92] Bray had planned to attend, sending a dense book, *Time Series and Analysis Forecasting and Control*, published in 1970 by the British econometricians George Box and Gwilym Jenkins, which he suggested Forrester read.[93] But that June, Labour lost the general election, and Bray his seat.[94] Peccei offered to pay for Bray's travel to Boston, but the former MP regretfully cancelled.[95]

In his absence Bray became increasingly critical. His preferred candidate to lead the project, Özbekhan, felt his participation was 'no longer advisable' following the selection of Forrester's approach.[96] Forrester was not leading the project, in favour of his 29-year-old lead researcher Dennis Meadows. But, as the work developed, Bray, still a Club member, wrote to Peccei to reiterate the concerns about the World model he had first raised in Bern.[97] But in June 1971, the *Observer* published 'Shock Findings on the Environment Crisis', a report based on preliminary results leaked from MIT.[98] In response, Bray penned a letter to the paper describing the model as 'an exotic device for confirming the prejudices of the investigator'.[99] He then wrote to Peccei again to warn him about his letter, copying in Forrester. He repeated that Forrester had still failed to acknowledge his concerns, raised over a year earlier. Bray also argued the

Club needed greater balance and should include developing countries and trade unions. As it was, Bray warned that 'there is a danger that the Club of Rome will become known as a technocratic but unscientific right-wing organisation'.[100]

Bray's public letter prompted a response from Conrad Waddington, the Edinburgh University biologist famed for his work in theoretical biology. Waddington had also founded Britain's first Science Studies Unit in 1966, and he would establish another significant research group, drawing on his interest in futurology, Edinburgh University's 'School of the Man-Made Future' in 1972.[101] In his letter to Bray, Waddington agreed with his criticisms of the Club but pointed out that the mathematics used in World3 was almost identical to that used by biologists in his laboratory working on complex interacting systems of enzymes. He had asked them to check the model.[102] Systems science, in abstracting the objects of analysis, rendered all environments, from the global to the macromolecular, in such a way that they could be analysed in almost the same way. Reflecting in 1977, Waddington stated his belief that the World model had overdetermined the relations between dynamics. His experience was that 'we usually have almost no notion of how these strengths of interactions are going to change as the system itself changes'.[103]

In July Peccei responded to Bray's letters, acknowledging his concerns but also suggesting they were best channelled into criticising the enclosed draft of Meadows' report.[104] Meadows also contacted Bray, pointing out the refinements he had added to Forrester's work, and noting that he was 'particularly anxious' to receive Bray's well-informed comments. He also mentioned a forthcoming visit to Whitehall, and his hope they could meet.[105] Bray replied, reiterating his point about the model's lack of econometric sophistication, but agreed to meet, enclosing a copy of *Decision in Government*.[106] Meadows, somewhat provoked, responded by suggesting Bray had little sense of his refinements as he had not come to recent Club meetings. He also defended his model, suggesting its simplicity was intentional, mirroring the invariant structure of the modern world, and his belief that 'We have moved globally into a period where the gross behaviour modes are set by relatively few relationships and are insensitive to even medium changes in their precise co-efficient values.'[107]

In November, Forrester finally responded to Bray, in a less polite tone than his researcher; he accused Bray of wanting to 'wait for some utopian future when everything is known' and signed off asking him to 'demonstrate where [his] methods would lead to a better model'.[108] Spurning Forrester's pointed letter, Bray instead wrote to Meadows again with a detailed reiteration of his criticisms. Once more, he stated that the model was overly determinate and insufficiently probabilistic, and that too little had been done to test it with alternative parameters. In doing so, drawing on his econometric expertise, Bray articulated his fundamental disagreement with the very idea of limits to growth: 'bearing in mind that the percentage composition of gross domestic product changes, and that price relativities change, it is perfectly possible to maintain constant growth at constant prices indefinitely, without the volume of any physical resource consumed per head exceeding a fixed limit'.[109] Shifting relations in prices would encourage substitution, such as an increase in human labour or the use of a less scarce or recycled resource, altering the composition of the gross product of the economy, but not causing limits to growth.[110] The idea that price rises caused consumers to seek substitute goods or develop technological solutions was a highly orthodox position in neoclassical economics since the late nineteenth century.[111] Given that World3 ignored this very basic idea, Bray sounded a final warning, telling Meadows 'the methods you have used are not sound and your conclusions point in a potentially reactionary and potentially disastrous direction'.[112]

A British world model

In March 1972 *Limits* was launched at an event at the Smithsonian Institution in Washington. Peccei had hired a public relations company, Calvin Kytle Associates, to heighten the publication's impact. They had contacted the world's major newspapers and sent 12,000 copies of the report in six languages to world leaders.[113] The publication described Earth as a system of feedback loops, which gave the global system a clear trajectory: 'Whereas positive feedback loops generate runaway growth, negative feedback loops tend to regulate growth and to hold a system in some stable state.' However, a growing imbalance meant 'if present growth trends … continue unchanged, the limits to growth on this planet will be reached sometime in the next one hundred years'.[114] This message, given its concerted promotional effort, seemed to capture both public and policy makers' imaginations.[115]

In fact, Edward Heath had allowed his Chief Scientific Adviser Alan Cottrell to attend the Boston meeting of the Club. Cottrell had trained as a metallurgist, and had no specific interest in the environment aside from working at the AEA in the 1950s, though he had come into contact with environmentalist concerns following the *Torrey Canyon* oil spill in 1967.[116] Whereas Bray had been highly sceptical of the World model,

Cottrell had been enthused by its demonstration. Returning to Britain, he proposed that the British government develop a similar model, stating his belief that 'Forrester's approach is the most important development of its kind since Keynes' general theory'.[117]

Given the centrality of Keynesianism in post-war economic policy, this was a significant claim. Heath, as his early enthusiasm for management science had revealed, had some interest in forecasting and simulation, and gave his permission for a scoping study on the feasibility of a British world model. Cottrell held a meeting on the subject at the Cabinet Office in September 1971, in which he had told the assembled civil servants that developing a global model for British purposes would require £50,000 and four staff. In response, an unnamed civil servant argued that the Treasury had a more sophisticated econometric model that it used for forecasting. Despite this criticism, the general idea of a global environmental model was well received, and further work was proposed.[118] At a second meeting in November 1971, Forrester's lead researcher, Meadows, was flown in to explain the model at an event hosted by the American Embassy.[119]

The visit provoked considerable excitement, encouraging the retired Treasury economist James Meade, pupil of Keynes and 1977 Nobel laureate, to wryly request an audience with the 'computer forecasts of doom'.[120] British cybernetician Stafford Beer, British-Canadian OR pioneer Charles Goodeve and economist Wilfred Beckerman also attended.[121] Addressing the audience, Meadows confidently explained that 'numerical values ... were of little importance' when it came to modelling global dynamics. 'Much more critical was the structure.'[122] This was a familiar refrain in cybernetic science: structure rather than the values of variables determined the behaviour of a system.[123] The structure of the world, according to the model, indicated that continued economic growth, given environmental limits, would cause population growth to 'cease' or 'overshoot' and 'collapse back' in the near future. Meadows warned such feedback would mean large-scale famine, social disorder, and possibly the breakdown of government. He also cautioned the assembled civil servants that further results of the Club's study would be published in 1972 and, improperly handled, the findings might lead the press to 'alarmist conclusions'. Given the severity of this portent, critical voices were again raised by some attendees. A specific complaint was that the model failed to account for the role of the economy in mitigating against fixed limits.[124]

Five months later, as *Limits* was published in Washington, a sense of crisis pervaded Heath's government. In the preceding three months, the National Union of Mineworkers had gone on strike, unemployment had risen past one million, and a state of emergency had been declared owing to a lack of coal-power.[125] Against the backdrop of union-engineered scarcity, Environment Secretary Peter Walker wrote to the prime minister suggesting that 'a Research Group, within government should study the MIT method and produce its own forecast' as a matter of urgency owing to 'a mounting tide of public interest'.[126] Heath agreed to set up a committee to expand upon Cottrell's earlier investigation.[127] But as civil servants began contacting Bray about his involvement in the Club, to learn more about the veracity of the study, he distanced himself, telling Walter Marshall, head of research at the AEA, that the organisation was 'insubstantial' and 'a vehicle for Dr Peccei' and that he was 'rather ashamed of the whole business'.[128] By that time, correspondence between Bray and Meadows had descended into vitriol, with Meadows accusing Bray of treating *Limits* as a 'straw man' against which he was planning to launch his re-election bid.[129]

In June that year the United Nations convened what was understood to be the first global meeting on the environment, in Stockholm, and *Limits* dominated the agenda.[130] Keen to show off the British government's initiative, Walker announced the planned forecasting group. Speaking with both caution and foreboding, he told the assembled journalists that the British government was 'not persuaded by everything that has been said about these matters lately', but warned that 'if research teaches us that a tendency we now see could lead to devastation in 30 years, the fact that it can only be averted by unprecedented response is irrelevant'.[131] He seemed to imply the threat of unparalleled government intervention were the report's most dramatic conclusions proved credible.[132] Environmental futurology seemed to have found favour with the current government, while the former chief scientific adviser, Zuckerman, took his address at the conference as an opportunity to deride 'hysterical computerized gloom'.[133]

Unsurprisingly, given the timing, the press dubbed the group the 'limits-to-growth unit',[134] though the official title was the World Future Trends Committee (WFTC). Its staff consisted of natural scientists from various relevant departments, including Agriculture, the Treasury and Heath's 'think tank', the CPRS. At the first meeting, there was some debate as to where the group should be based, as some thought the DOE should only be concerned with Britain's environment rather than the global system envisioned by modellers.[135] The Foreign and Commonwealth Office had pushed the idea that WFTC should examine the European rather than world system, given Heath's recent attempts to persuade European states to grant British accession to the EEC, and on the presumption that European countries faced similar environmental problems.[136] This discussion seemed to open up the possible use of global models as tools of diplomacy, technological affirmations of shared goals.[137]

At a second meeting scale was raised again, though this time as a possible means of validating the model, by comparing observable disaggregated outcomes at a regional level with computer-based global forecasts. When Meadows had first presented the World model to the Civil Service he had explained that regional, national and city-level models would become necessary as there 'were no global decision makers'.[138] But there was a methodological problem with the idea of a British model. At that first meeting convened by Cottrell in September 1971, civil servants had complained that, because Forrester's model analysed the Earth as a 'closed system', it would be difficult to disaggregate the approach to analyse individual countries, as migration, trade and environmental dynamics did not recognise national boundaries. As such, it was decided the CPRS should explore what kind of 'self-contained' ecological problems could be fruitfully modelled.[139]

But instead of developing a model within Whitehall, one of the committee's first steps was to contact a number of universities, via Brian Flowers, Chairperson of the Science Research Council (SRC), to see if they might be commissioned to carry out a critical study of the World model.[140] This sub-contracting was in line with Heath's attempts to modernise government, as the conditions of state-funded research were undergoing reform. The British government was recast as a 'consumer' who would commission specific research from 'contractors', the idea being that scientists would become subject to a modicum of the competitive forces at play in a free market, with expected gains in efficiency and productivity.[141] In line with this change to the conditions of scientific funding, the government was keen to have MIT's model tested at a university, unaware that one British university was doing exactly that.

Cottrell received a response from Chris Freeman, director of Sussex University's Science Policy Research Unit (SPRU). Established in 1966, SPRU's aim was to take a sociologically informed approach to the study of scientific and industrial research. Broadly optimistic regarding the productive power of science, the group's ethos was also broadly in line with the radical politics of the university at the time, playing host to Marxists and radical scientists, promoting new modes of research in developing countries, interdisciplinarity and experimental approaches to teaching and learning.[142] The SPRU team had already acquired a copy of World3's code from MIT after a staff member had travelled there in January 1971, invited by Meadows, who had wanted to persuade the university to establish a forecasting group in England.[143] They had begun running the model on the PAU's IBM 360/165 computer at the AEA's Harwell laboratory.[144] Approving of the ongoing work, the SRC awarded an £11,000 contract to SPRU to formally review and evaluate current 'future' studies, including *Limits*.[145]

On closer inspection SPRU researchers were concerned that Meadows' conclusions were over determined, and the Club's adoption of them risked an overly technocratic response. To counter this, a research group called STAFF ('Social and Technological Alternatives for the Future') was established in 1971 with government funding, to allow SPRU researchers to interrogate the model as both 'technologists and social scientists'. In contrast to American futurological efforts, SPRU staff felt their work addressed the 'social' determinants of the future, in contrast to 'technological assessment', the extrapolation of technological trends, which evinced a crude conception of the social. STAFF had access to World2 and World3 and ran the simulation several hundred times in a process of 'sensitivity analysis' to identify the assumptions that underlay the two models. An important finding was that if parameters were shifted, so that the model assumed a 1–2 per cent rate of natural resource discovery and a similar rate of increase in the technical ability to reduce pollution, neither scarcity nor pollution appeared deleterious.[146] In the MIT models these parameters were fixed and limits pre-ordained. Given the gravity of their conclusion, the output of both Forrester's and Meadows' models seemed highly sensitive to plausible developments.

Alongside the SPRU grant, Scottish economist James Mirrlees, at the University of Oxford, was given £2,000 to write a general critique of 'system dynamics'.[147] Like Bray, Mirrlees was a highly accomplished econometrician, recognised for his influential work on optimum taxation rates.[148] His study compared system dynamics models with those used in econometrics. He came to three conclusions: modellers mistook complexity for realism; unlike many econometric models, system dynamics models lacked means of statistical validation; but most damning, he noted the conclusions reached in *Limits* were obvious, and had not needed a computer to demonstrate: if an economy grew and the resource base was fixed, of course a limit would be reached. As such, he concluded that 'this kind of scientific activity is, at best, worthless', and that the work was akin to that of 'oracles, clairvoyants and other charlatans throughout the ages'.[149]

Bray and Mirrlees were part of a national econometric tradition that considered itself more statistically rigorous and empirically grounded than the American or Dutch schools. In this vein, leaning on developments in control theory, time series analysis was a particular British strength that the World modellers had ignored.[150] George Box and Gwilym Jenkins, the British econometricians Bray had encouraged both MIT researchers to read, had developed a method for making forecasts from temporal data in a probabilistic way, rather than using co-efficient sums as World3 had done. This allowed forecasts to be made in a more mathematically defensible way from extant data.[151] Bray and Mirrlees also argued that the World model lacked the means of verification by which econometric models attained validity. Testing was required, whereby multiple runs could be used to compare simulated values with actual values. Once calibrated, a forecaster could then claim a forecasted outcome was *probable* rather than asserting it as an inevitable result of determinate relations.[152] Worse still, as Bray had complained, whereas economists formulate questions, develop models and then produce a narrative, the Club of Rome seemed to have begun with a narrative – that of a coming crisis – and then used a model to bolster its authority.[153]

Rather than reject global models entirely, from 1974 onward, the Department of the Environment maintained a Systems Analysis Research Unit (SARU). Led by the physicist Peter Roberts, SARU monitored other global models, tested their feasibility and explored disaggregation of specific resources or geopolitical regions. In January, SARU's observations were outlined to the WFTC. Their central claim was that if world models accounted for 'economic feedback mechanisms', the propensity for consumers to reduce the rate or find a substitute for their demand for resources, in response to price rises, for example. The models 'result in a much more robust structure'.[154] In the course of their work, SARU and the WFTC had been well aware of Bray's similar criticisms of World3, as he had communicated these to Cottrell and others in Whitehall, while Forrester and Meadows had attempted to pre-empt him by similarly copying Whitehall into their correspondence.[155] These criticisms found their way into the British government's final response to *Limits*, published in 1976. With regard to the environment, *Future World Trends* claimed there were 'no hard and fast physical limits to resources; the limits are economic and technological and can vary widely'.[156]

SARU's own model, 'SARUM 76', was launched at the Royal Society in 1977.[157] The model was disaggregated into over 15 regions, each containing 13 economic sectors. Despite its greater sophistication, the model was free of the hubris and ambition heaped on World3. The aim was to learn about the relationships between dynamics, rather than to produce forecasts; SPRU researchers were employed to critique the work; and little effort was made to publicly promote its findings.[158] However, what truly distinguished the model was the inclusion of feedback loops representing the effect of 'technology plus the market', two dynamics that, as Bray, SPRU and others working on behalf of the British government had argued, could iteratively transcend environmental limits if left to respond to a freely operating price mechanism.[159] As Donella Meadows, co-author and partner of Dennis, noted, the theoretical assumptions underlying SARUM, and its forecast of a more abundant future, were those of neoclassical economics. The model assumed individuals maximised the value they could derive from scarce resources, which would mean substitution and increased technological efficiency would take place as resources became scarce.[160] In effect, the environment had been reduced to a subset of economic theory. But the idea that the freely operating markets could transcend resource limits was a distant prospect for Heath's government which, despite coming to power with a plan for liberalisation, had been forced to impose wage and price controls in November 1972 in an attempt to fight inflation and manage the demands of the Trades Unions Congress and the nationalised industries.[161]

Conclusion

In his opening chapter, Agar proposes eight ways in which technology and the environment intersected in modern Britain.[162] This chapter documented the arrival of a representation of the global environment, represented by just 150 lines of code.[163] This was not an 'artificial world' of the kind mentioned in Agar's type (4), as World3 was not intended to impose a border between nature and society so much as draw attention to the wrongheadedness of those who believed they were not subject to natural limits. This simulated environment perhaps sits best somewhere between Agar's type (6) and (7) interactions. Global models represent the dynamics of the environment just as prose or photography can do, but their mode of representation is analogical, consisting of code and numbers. As a 'mediating' technology (6), computer simulations of dynamic systems were not designed to represent the environment so much as allow its analysis. But owing to its abstractive form of mediation, simulations are able to simulate a wide range of historical processes from the denaturing of enzymes to the growth of global industry. But interaction (7) implies that technology can provide a means of registering and organising environmental knowledge. Here simulation fits well, as by rendering various dynamics into mathematical relations, and feeding these relations with numerical data, the intention was to both register and organise pertinent elements of global environmental change in such a way that the global dynamics of environmental change might be understood, critically, over time.

Bray published his final thoughts on *Limits* in a Fabian Society pamphlet. Drawing on the information-theoretical vision outlined in *Decision in Government*, he suggested the 'process of decision in human society', involving individuals, parliamentary debates, boardroom discussions and myriad other interactions, was itself a 'vast "computer model"' of which the World model was just one part. Computer models were 'intelligence amplifiers' but could not hope to rival the greater reasoning power of the collective 'human decision system'.[164] In effect, this explanation nicely characterises what happened when World3 arrived in Britain. The model, and the objectives of the Club, were received with initial enthusiasm, which was soon tempered by British expertise in forecasting and modelling, which stemmed from fields ranging from econometrics to biology and OR, and it involved a wide range of private and public institutions, from the PAU to Cambridge University. And aside from being able to dissect the model on a technical and theoretical basis, Britain's futurological capacities suggested this was a nation that was self-consciously modern, in so far as policy based upon forecasting, done correctly, was seen as a means of breaking from past mistakes.

A concern with 'the environment' was also seen as something discontinuous with past modes of government. The problem, from the British government's perspective, was that World3's conception of the environment was too limited. Its creators understood it as simply the intersection between ecology and technology, and that this could be captured in three figures of computer code.[165] But Britain's experts argued World3 was too crude. It failed to account for the effects of market forces and technological change. In effect, the MIT team had failed to account for the dynamism of human society that neoclassical economic theory assumed. As a result, the British attempt to remodel global dynamics countered concern for the environment with a re-assertion of liberal economic theory, which ran against the interventionist post-war consensus. The result was a contrasting model that mathematically formalised a future means for achieving environmental abundance through the free operation of the market. This became one of the central tenets of 'modern' environmentalism, the idea that environmental change could, in some sense, be programmed by collective human action, manifest, in this case, in the aggregated agency of a liberalised market.

Notes

1 Archival collections consulted in the chapter are identified through the following acronyms: Conservative Party Archive, Bodleian Library, Oxford (CPA); The National Archives, Kew, London (TNA); the papers of Jeremy Bray MP, Churchill Archives Centre, Cambridge University (BRAY).

2 Dennis Meadows, Donella Meadows, Jorgen Randers and William Behrens William, *The Limits to Growth: A Report for the Club of Rome's Project on the Predicament of Mankind* (New York: Potomac Books, 1972).

3 Lawrence Caldwell, *International Environmental Policy: From the Twentieth to the Twenty First Century* (Durham, NC and London: Duke University Press, 1996), 48.

4 Matthias Schmelzer, '"Born in the Corridors of the OECD": The Forgotten Origins of the Club of Rome, Transnational Networks, and the 1970s in Global History', *Journal of Global History* 12, no. 1 (2017): 26–48; Francis Sandbach, 'The Rise and Fall of the Limits to Growth Debate', *Social Studies of Science* 8, no. 4 (1978): 495–520.

5 Fernando Elichirigoity, *Planet Management: Limits to Growth, Computer Simulation, and the Emergence of Global Spaces* (Evanston: Northwestern University Press, 1999); Peter Taylor and Frederick Buttel, 'How Do We Know We Have Global Environmental Problems? Science and the Globalisation of Environmental Discourse', *Geoforum* 23, no. 3 (1992): 405–16; Sheila Jasanoff, 'Image and Imagination: The Formation of Global Environmental Consciousness', in *Changing the Atmosphere: Expert Knowledge and Environmental Governance*, ed. Paul Edwards and Clark Miller (Cambridge, MA: MIT Press, 2001); Jenny Andersson and Sibylle Duhautois, 'Futures of Mankind: The Emergence of the Global Future', in *The Politics of Globality since 1945*, ed. Rens van Munster and Casper Sylvest (London and New York: Routledge, 2016), 106–25.

6 Elodie Vielle Blanchard, 'Modelling the Future: An Overview of the "Limits to Growth" Debate', *Centaurus* 52, no. 2 (2010): 91–116.

7 Paul Edwards, 'The World in a Machine: Origins and Impacts of Early Computerized Global Systems Models', in *Systems, Experts, and Computers: The Systems Approach in Management and Engineering, World War II and After*, ed. Agatha Hughes and Thomas Hughes (Cambridge, MA: MIT Press, 2000), 221–54.

8 William Thomas and Lambert Williams, 'The Epistemologies of Non-Forecasting Simulations, Part 1: Industrial Dynamics and Management Pedagogy at MIT', *Science in Context* 22, no. 2 (2009): 245–70; Ugo Bardi, *The Limits to Growth Revisited* (New York: Springer, 2011).

9 Elodie Vielle Blanchard, 'Technoscientific Cornucopian Futures versus Doomsday Futures: The World Models and the Limits to Growth', in *The Struggle for the Long-Term in Transnational Science and Politics: Forging the Future*, ed. Jenny Andersson and Egle Rindzevičiūtė (New York and London: Routledge, 2015), 92–114.

10 Clive Hamilton and Jacques Grinevald, 'Was the Anthropocene Anticipated?', *The Anthropocene Review* 2, no. 1 (2015): 59–72; Will Steffen, Jacques Grinevald, Paul Crutzen and John McNeil, 'The Anthropocene: Conceptual and Historical Perspectives', *Philosophical Transactions of the Royal Society A* 369, no. 1938 (2011): 842–67.

11 Bruno Latour, '*Onus Orbis Terrarum*: About a Possible Shift in the Definition of Sovereignty', *Millennium: Journal of International Studies* 44, no. 3 (2016): 305–20.

12 Elke Seefried, 'Towards the Limits to Growth? The Book and its Reception in West Germany and Britain, 1972–1973', *German Historical Institute London Bulletin* 33, no. 1 (2011): 3–37.

13 Peter Moll, *From Scarcity to Sustainability. Future Studies and the Environment: The Role of the Club of Rome* (Frankfurt: Peter Lang: 1991).

14 Edward Cornish, *The Study of the Future: An Introduction to the Art and Science of Understanding and Shaping Tomorrow's World* (Washington, DC: World Future Society, 1977).

15 Jenny Andersson, 'The Great Future Debate and the Struggle for the World', *American Historical Review* 117, no. 5 (2012): 1411–30; Jenny Andersson and Egle Rindzevičiūtė (eds), *The Struggle for the Long-Term in Transnational Science and Politics: Forging the Future* (London and New York: Routledge, 2015).

16 Paul Warde and Sverker Sörlin, 'Expertise for the Future: The Emergence of Environmental Prediction c.1920–1970', in Andersson and Rindzevičiūtė, *The Struggle for the Long-Term*, 38–62.

17 William Thomas, *Rational Action: The Sciences of Policy in Britain and America, 1940–1960* (Cambridge, MA: MIT Press, 2015), 295.

18 Jon Agar, '"Future Forecast – Changeable and Probably Getting Worse": The UK Government's Early Response to Anthropogenic Climate Change', *Twentieth Century British History* 26, no. 4 (2015): 602–28.

19 All had an interest in forecasting, which supports the preceding argument. See Dennis Gabor, *Inventing the Future* (London: Secker & Warburg, 1963); Conrad Waddington (ed.), *Towards a Theoretical Biology: Prolegomena* (Edinburgh: Edinburgh University Press, 1968); Asa Briggs, 'The Historian and the Future', *Futures* 10, no. 6 (1978): 445–51.

20 Agar, 'Future Forecast'; Seefried, 'Towards the Limits to Growth?'

21 Timothy Mitchell, *Carbon Democracy: Political Power in the Age of Oil* (New York and London: Verso, 2011), 176.

22 Paul Warde, 'The Environment', in *Local Places, Global Processes: Histories of Environmental Change in Britain and Beyond*, ed. Peter Coates, David Moon and Paul Warde (Oxford: Windgather, 2016), 32–46.

23 Rosaleen Hughes, '"Governing in Hard Times": The Heath Government and Civil Emergencies – The 1972 and the 1974 Miners' Strikes' (PhD diss., Queen Mary University of London, 2015).

24 Lawrence Black, 'Review Essay: An Enlightening Decade? New Histories of 1970s Britain', *International Labor and Working-Class History* 82 (2012): 174–86; for example, David Blunkett, 'Corbyn Will Take Us Back to the 70s', *Sun* (12 September 2015).

25 Edmund Stillman, James Bellini, William Pfaff, Laurence Schloesing and Michael Barth, *The United Kingdom in 1980: The Hudson Report* (London: Association Business Programmes, 1974), 113–33.

26 Stillman et al., *The United Kingdom in 1980*, 56–69.

27 David Edgerton, *Science, Technology and the British Industrial 'Decline' 1870–1970* (Cambridge: Cambridge University Press, 1996).

28 David Edgerton, 'The White Heat Revisited: British Government and Technology in the 1960s', *Twentieth Century British History* 7, no. 1 (1996): 53–82; David Edgerton, 'C.P. Snow as Anti-Historian of British Science: Revisiting the Technocratic Moment', *History of Science* 43, no. 2 (2005): 187–208.

29 David Edgerton, *Warfare State: Britain, 1920–1970* (Cambridge: Cambridge University Press, 2006), 301.

30 Martin Wiener, *English Culture and the Decline of the Industrial Spirit, 1850–1980* (Cambridge: Cambridge University Press, 1981). The preface to the first edition notes, 'Edmund Stillman, of the Hudson Institute Europe, offered support and an occasion to present some of these ideas for the first time', p. xi., see also p. xv. On Ashridge, see Clarisse Berthezène, *Training Minds for the War of Ideas: Ashridge College, The Conservative Party and the Cultural Politics of Britain, 1929–54* (Manchester: Manchester University Press, 2015).

31 Paul Dragos Aligica, 'The Challenge of the Future and the Institutionalisation of Interdisciplinarity: Notes on Herman Kahn's Legacy', *Futures* 36, no. 1 (2004): 67–83; Sharon Ghamari-Tabrizi, *The Worlds of Herman Kahn: The Intuitive Science of Thermonuclear War* (Cambridge, MA: Harvard University Press, 2005).

32 Thomas, *Rational Action*, 285; Ghamari-Tabrizi, *The Worlds of Herman Kahn*, 17–23.

33 Peter Galison, 'The Future of Scenarios: State Science Fiction', in *The Subject of Rosi Braidotti: Politics and Concepts*, ed. Bolette Blaagard and Iris van der Tuin (London and New York: Bloomsbury, 2014), 39.

34 James Bellini, 'Hudson Institute's Year of Europe', *Futures* 6, no 1 (1974): 84.

35 Jon Davis, *Prime Ministers and Whitehall* (London: Hambledon Continuum, 2007).

36 Ivan Klimeš, 'Journalism in Forecasting and Vice Versa', *Futures* 7, no. 1 (1975): 2.

37 Tom Nairn, 'The Future of Britain's Crisis', *New Left Review* 113 (1979): 43.

38 Stillman et al., *The United Kingdom in 1980*, 109.

39 James Bellini, 'Hudson Institute to Study Future of the UK', *Futures* 5, no. 4 (1973): 421–9.

40 Edmund Stillman, James Bellini, William Pfaff, Laurence Schloesing, Jonathan Story, Jean-Jacques De Peretti and Herman Kahn, *L'Envol de la France: Portrait de la France dans les années 80* (Paris: Hachette, 1973).

41 Yehezkel Dror, 'The Misfired Report', *Futures* 7, no. 1 (1975): 64–8.

42 Tom Price, 'The British Economy: A Call for National Planning', *Futures* 7, no. 3 (1975): 258–60; E. Mishan, 'Review: The United Kingdom in 1980 – The Hudson Report', *The Political Quarterly* 46, no. 2 (1975): 206–9.

43 Editorial, 'Futures, Confidence from Chaos', *Futures* 1, no. 1 (1968): 2.

44 John Pinder, Max Nicholson and Kenneth Lindsay, *Fifty Years of Political and Economic Planning: Looking Forward, 1931–1981* (London: Heinemann, 1981); Charles de Houghton, William Page and Guy Streatfield, *'... and Now the Future': A PEP Survey of Futures Studies* (London: Policy Studies Institute, 1971).

45 Harold Wilson, *Labour's Plan for Science: Reprint of Speech by the Rt Hon. Harold Wilson, MP, Leader of the Labour Party at the Annual Conference Scarborough, Tuesday, October 1* (London: The Labour Party, 1963), 2.

46 Statement by the Minister of Technology, 15 January 1969, cited in Derek Medford, *Environmental Harassment or Technology Assessment?* (London: Elsevier, 1973).

47 Derek Medford, 'Some Remarks on the Application of Technical Forecasting', in *Technological Forecasting. Proceedings of University of Strathclyde Conference on Technological Forecasting*, ed. R.V. Arnfield (Edinburgh: Edinburgh University Press, 1969).

48 Derek Medford, *Environmental Harassment or Technology Assessment?* (Amsterdam and London: Elsevier Science), ch. 4.

49 Derek Medford, 'The New Thaumaturgy of Governmental Research and Development', *Futures* 1, no. 6 (1968): 510–26. The claim Britain invented OR is contentious; see Thomas, *Rational Action*, for an account of transatlantic developments.

50 The Conservative Goal: A Call to Action, by Rt Hon. Edward Heath MP, 1966, PUB 171/ 31, CPA.

51 Heath to Pike, 28 March 1969, CCO 20/26/7, CPA.

52 CSRC The Control Date 210 Video Display Unit, undated, CCO 20/26/7, CPA.

53 Michael Spicer, *The Spicer Diaries* (London: Biteback, 2012), 30–7; Michael Spicer, 'Towards a Policy Information and Control System', *Public Administration* 48, no. 4 (1970): 443–8.

54 Though linear programming was not something that originated in the private sector, it was a desire to reorganise government in line with private sector managerial techniques that led Heath to establish the CPRS. See Hugh Heclo and Aaron Wildavsky, *The Private Government of Public Money* (London: Macmillan, 1974), 271–3.

55 *Cmnd. 3638. Fulton Committee: The Civil Service, Vol. 1, Report of the Committee, 1966–1968* (London: HMSO, 1968).

56 Jon Agar, *The Government Machine: A Revolutionary History of the Computer* (Cambridge, MA: MIT Press, 2003); Marie Hicks, *Programmed Inequality: How Britain Discarded Women Technologists and Lost its Edge in Computing* (Cambridge, MA: MIT Press, 2017), 173–4.

57 The Meteorological Office began making use of global climate simulations as a means of discerning the Earth's atmospheric properties in the 1960s: Martin Mahony and Mike Hulme, 'Modelling and the Nation: Institutionalising Climate Prediction in the UK, 1988–92', *Minerva* 54, no. 4 (2016): 445–70.

58 Giles Keating, *The Production and Use of Economic Forecasts* (London: Routledge, 1985).

59 A point made in relation to forecasting during the energy crisis in the United States. Mitchell, *Carbon Democracy*, 189–93.

60 John McCormick, *The British Government and the Environment* (London: Earthscan, 1991), 16.

61 David Stoddart, 'Our Environment', *Area* 2, no. 1 (1970): 1–4; Frederick Hare, 'Environment: Resuscitation of an Idea', *Area* 1, no. 4 (1970): 52–5.

62 Timothy O'Riordan 'New Conservation and Geography', *Area* 2, no. 5 (1970): 33–6.

63 Edward Goldsmith, 'Living with Nature', *The Ecologist* 1, no. 1 (1970): 2.

64 Peter Taylor, *Unruly Complexity: Ecology, Interpretation, Engagement* (Chicago: University of Chicago Press, 2005), xv.

65 Andrew Pickering, 'Cyborg History and the World War II Regime', *Perspectives on Science* 3, no. 1 (1995): 1–48. For biographies of Forrester, see Bloomfield (note 68), Edwards (note 7), Kwa (note 66), inter alia.

66 Chunglin Kwa, 'Modelling Technologies of Control', *Science as Culture* 4, no. 3 (1994): 363–91.

67 William Thomas and Lambert Williams, 'The Epistemologies of Non-Forecasting Simulations, Part 1: Industrial Dynamics and Management Pedagogy at MIT', *Science in Context* 22, no. 2 (2009): 245–70.

68 Brian Bloomfield, *Modelling the World: The Social Constructions of Systems Analysts* (Oxford: Blackwell, 1986); Edwards, 'The World in a Machine'.

69 Jay Forrester, *World Dynamics* (Cambridge, MA: Wright Allen Press, 1971).

70 David Mindell, *Between Human and Machine: Feedback, Control, and Computing before Cybernetics* (Baltimore: Johns Hopkins University Press, 2002); Jennifer Gabrys, *Program*

Earth: Environmental Sensing Technology and the Making of a Computational Planet (Minneapolis and London: University of Minnesota Press, 2016).

71 Chunglin Kwa notes mathematician George Innis used Forrester's Industrial Dynamics in his EML 'Grasslands' model in 1973: Chunglin Kwa, 'Modeling the Grasslands', *Historical Studies in the Physical and Biological Studies* 24, no. 1 (1993): 125–55; Taylor notes in *Unruly Complexity*, 111, that William Seifert used Forrester's approach to model the Sahel.

72 Talcott Parsons (1968) cited in Arthur Mason, 'Images of the Energy Future', *Environmental Research Letters* 1, no. 1 (2006): 20–5.

73 Egle Rindzevičiūtė, *The Power of Systems: How Policy Sciences Opened Up the Cold War World* (Ithaca: Cornell University Press, 2016), 79.

74 Alexander Schmidt-Gernig, 'The Cybernetic Society: Western Future Studies of the 1960s and 1970s and their Predictions for the Year 2000', in *What the Future Holds: Insights from Social Science*, ed. Richard Cooper and Richard Layard (Cambridge, MA: MIT Press, 2003), 233–60.

75 Andrew Roth, 'Obituary: Jeremy Bray', *Guardian* (5 June 2002).

76 Jeremy Bray, *Standing on the Shoulders of Giants: Science, Politics and Trust, A Parliamentary Life* (Cambridge: Elizabeth Bray, 2004). On the differences between information theory and cybernetics, see Ronald Kline, *The Cybernetics Moment: Or Why We Call Our Age the Information Age* (Baltimore: Johns Hopkins University Press, 2015), 10–11.

77 Judy Klein, *Statistical Visions in Time: A History of Time Series Analysis, 1662–1938* (Cambridge: Cambridge University Press, 1997); Roy Epstein, *A History of Econometrics* (Amsterdam: Elsevier, 1987), 141–7.

78 Jeremy Bray, 'Economic Modelling with Computers', *New Scientist* 27, no. 453 (1965): 196.

79 Nuno Madureira, 'The Confident Forecaster: Lessons from the Upscaling of the Electricity Industry in England and Wales', *Business History* 59, no. 3 (2017): 408–30.

80 Jeremy Bray, 'Tools for the Seventies', *The Political Quarterly* 41, no. 2 (1970): 151–5.

81 Jeremy Bray, *Decision in Government* (London: Victor Gollancz, 1970).

82 There were parallels to Stafford Beer's proposed cybernetic socialism in Allende's Chile: Eden Medina, *Cybernetic Revolutionaries: Technology and Politics in Allende's Chile* (Cambridge, MA: MIT Press, 2011).

83 Richard Stone, 'Review: Decision in Government', *The Economic Journal* 81, no. 323 (1971): 668–70.

84 Peccei congratulated Bray for 'relinquishing' his duties to write his book. 'Club of Rome' Peccei to Bray, 1970, Acc. 610 36/406, BRAY; for an account of Peccei's work, see Robert Golub and Joe Townsend, 'Malthus, Multinationals, and the Club of Rome', *Social Studies of Science* 7, no. 2 (1977): 201–22; Blanchard, 'Technoscientific Cornucopian Futures', 92–114.

85 Hasan Özbekhan, 'Toward a General Theory of Planning', in *Perspectives of Planning*, ed. Erich Jantsch (Bellagio, Italy: OECD, 1968), 132.

86 Jeremy Bray, 'Bellagio Symposium Papers Published', *Futures* 2, no. 1 (1970): 76–88.

87 Club of Rome, 1970: Letter to Peccei from Bray, 6 March 1970, File 406, Acc. 610, BRAY.

88 Moll, *From Scarcity to Sustainability*; Elichirigoity, *Planet Management*.

89 Hasan Özbekhan, *The Predicament of Mankind: A Quest for Structured Responses to World Wide Complexities and Uncertainties* (Geneva: Club of Rome, 1970).

90 Notes on Bern Meeting, undated, Acc. 610 36/406, BRAY.

91 Moll, *From Scarcity to Sustainability*; Blanchard, 'Technoscientific Cornucopian Futures'.

92 Telegram from Forrester, 13 July 1970, Acc. 610 36/406, BRAY.

93 Letter to Forrester, 15 July 1970, Acc. 610 36/406, BRAY.

94 Tam Dayell, 'Bray, Jeremy William, 1930–2002, Mathematician and Politician', in *Oxford Dictionary of National Biography* (Oxford: Oxford University Press, 2006).

95 Handwritten note, undated, Acc. 610 36/406, BRAY.

96 Thiemann to Bray, notes from MIT, 21 August 1970, Acc. 610 36/406, BRAY.

97 Letter to Peccei, 18 February 1971, Acc. 610 34/379–80, BRAY.

98 Gerald Leach, 'Shock Findings on the Environment Crisis', *Observer* (June 1971).

99 Jeremy Bray, 'Row over Global Crisis Report', *Observer* (July 1971).

100 Letter to Peccei, 30 June 1971, Acc. 610 34/379–80, BRAY.

101 John Henry, 'Historical and Other Studies of Science, Technology and Medicine in the University of Edinburgh', *Notes and Records of the Royal Society* 62, no. 2 (2008): 223–35; Conrad Waddington, 'School of the Man-Made Future at Scottish University', *Futures* 4, no. 4 (1972): 378.

102 Waddington to Bray, 5 July 1971, Acc. 610 34/379–80, BRAY.
103 Conrad Waddington, *Tools for Thought: How to Understand and Apply the Latest Scientific Techniques of Problem Solving* (New York: Basic Books, 1977).
104 Peccei to Bray, July 1971, Acc. 610 34/379–80, BRAY.
105 Meadows to Bray, 23 July 1971, Acc. 610 34/379–80, BRAY.
106 Bray to Meadows, 3 August 1971, Acc. 610 34/379–80, BRAY.
107 Meadows to Bray, 24 September 1971, Acc. 610 34/379–80, BRAY.
108 Letter from Forrester to Bray, Forrester cc. Cottrell et al., 8 November 1971, Acc. 610 34/379–80, BRAY.
109 Letter from Bray to Meadows, 10 November 1971, Acc 610 34/379-80 BRAY.
110 See Mat Paskins, 'The Woods for the State', Chapter 12 this volume, on the role the substitutability of wood played in an earlier period of fears of resource shortages.
111 Blaug notes that the 'The dominant role of the concept of substitution at the margin' was the major development of the nineteenth-century 'Marginal Revolution' in economic thought: Mark Blaug, *Economic Theory in Retrospect* (Cambridge: Cambridge University Press, 1962), 311.
112 Letter from Bray to Meadows, 10 November 1971, Acc. 610 34/379–80, BRAY.
113 Robert Gillette, 'The Limits to Growth: Hard Sell for a Computer View of Doomsday', *Science* 175, no. 4026 (1972): 1088–92.
114 Meadows et al., *The Limits to Growth*, 23.
115 Francis Sandbach, 'The Rise and Fall of the Limits to Growth Debate', *Social Studies of Science* 8 (1978): 495–520.
116 Lindsay Greer and Frans Spaepen, 'Sir Alan Cottrell (1919–2012)', *Proceedings of the National Academy of Sciences* 109, no. 42 (2012): 16753. See also Simone Turchetti, 'The UK Government's Environmentalism: Britain, NATO and the Origins of Environmental Diplomacy', Chapter 13 this volume.
117 Cottrell to Heath, Zuckerman, 28 July 1971, CAB 168/291, TNA; see also Seefried, 'Towards the Limits to Growth'.
118 Minutes, 'Future World Trends', 9 September 1971, CAB 134/3584, TNA.
119 Minutes, 'Future World Trends', 24 November 1971, CAB 168/291, TNA.
120 Note, Meade to Cottrell, 30 September 1971, CAB 168/291, TNA.
121 Patrick Ryan, 'Meadows under Scrutiny', *New Scientist* (20 April 1972): 166.
122 'Future World Trends', 24 November 1971, CAB 168/291, TNA.
123 Norbert Wiener, *The Human Use of Human Beings: Cybernetics and Society* (Garden City: Doubleday, 1954), 57: 'the structure of the machine or of the organism is an index of the performance that may be expected from it'.
124 Minutes, 'Future World Trends', 24 November 1971, CAB 168/291, TNA.
125 Stuart Ball, 'A Chronology of the Heath Government', in *The Heath Government, 1970–1974: A Reappraisal*, ed. Stuart Ball and Anthony Seldon (New York and London: Longman, 1996), 396.
126 Walker to Heath et al., 13 February 1972, CAB 164/1083, TNA.
127 Heath to Walker et al., 28 February 1972, CAB 164/1083, TNA.
128 Club of Rome, 1972: Letter from Bray to Marshall, 25 July 1972, File 359, Acc. 610, BRAY.
129 Note, Meade to Cottrell, 30 September 1971, CAB 168/291, TNA.
130 An earlier event of this kind was the UN conference on the conservation and use of resources, in the early years of the Cold War: Shoko Mizuno, 'Global Governance of Natural Resources and the British Empire: A Study on the United Nations Scientific Conference on the Conservation and Utilisation of Resources, 1949', in *Environmental History in the Making, Volume II: Acting*, ed. Cristina Joanaz de Melo, Estelita Vaz and Ligia M. Costa Pinto (Cham: Springer, 2016), 291–308.
131 Jon Tinker, 'Britain to Start a Limits to Growth Unit', *New Scientist*, 15 June 1972, 615.
132 Hamblin terms this 'catastrophic environmentalism', the idea that ecological imbalance might legitimise unprecedented government intervention, just as the threat of thermonuclear war had earlier done: Jacob Darwin Hamblin, *Arming Mother Nature: The Birth of Catastrophic Environmentalism* (Oxford: Oxford University Press, 2013).
133 Rindzevičiūtė, *Power of Systems*, 135.
134 Jon Tinker, 'Britain to Start a Limits to Growth Unit'.
135 Minutes of Cabinet Meeting, 'Future World Trends', 14 April 1972, Folio 46, CAB 164/1083, TNA.

136 CAB 164/1083, letter to Cottrell from Denis Greenhill (FCO), Future World Trends, 2 March 1972.

137 A point also made in reference to East–West collaboration at the International Institute for Applied Systems Analysis (IIASA) modelling work: Rindzevičiūtė, *Power of Systems*.

138 Minutes of Cabinet Meeting Official Committee on Future World Trends, 24 October 1972, CAB 134/3584, TNA.

139 Note of a meeting held in the Cabinet Office, 9 September 1971, CAB 168/291, TNA.

140 Letter from Sir Brian Flowers to Asa Briggs, University of Sussex, 15 March 1972, CAB 164/1083, TNA.

141 Jon Agar, 'Thatcher, Scientist', *Notes and Records of the Royal Society* 65, no. 3 (2011): 215–32.

142 Simone Turchetti, 'Looking for the Bad Teachers: The Radical Science Movement and its Transnational History', in *Science Studies during the Cold War and Beyond*, ed. Elena Aranova and Simone Turchetti (New York: Palgrave Macmillan; 2016), 77–102; Bloomfield, *Modelling the World*, 86.

143 World Dynamics, Note from Sam Cole, SPRU, 29 February 1973, CAB 164/1083, TNA.

144 Hugh Cole, Chris Freeman, M. Marie Jahoda and K.R.L. Pavit, 'Preface' in *Thinking about the Future: A Critique of The Limits to Growth*, ed. Hugh Cole, Chris Freeman, M. Marie Jahoda and K.R.L. Pavit (London: Chatto & Windus, 1973).

145 'Application of Dynamic Analysis and Forecasting to World Problems', A Summary Report to the UK Research Councils, with funding recommendations, Science Policy Research Unit (SPRU), January 1974, CAB 134/3859, TNA.

146 Clive Sinclair, 'Technological Change = Social Change? The Work of the Science Policy Research Unit at Sussex University in the Field of Forecasting', *Industrial Marketing Management* 2, no. 4 (1973): 375–90.

147 James Mirrlees, 'Simulation Modelling for World Problems', Final Report on work done for the SSRC and SRC on the application of dynamic analysis and forecasting to world problems, January 1974, CAB 134/3859, TNA.

148 James Mirrlees, 'Nobel Lecture: Information and Incentives: The Economics of Carrots and Sticks', *The Economic Journal* 107, no. 444 (1997): 1311–29.

149 James Mirrlees, 'Simulation Modelling for World Problems', Final Report on work done for the SSRC and SRC on the application of dynamic analysis and forecasting to world problems, January 1974, CAB 134/3859, TNA.

150 Epstein notes, 'the specification of many of the English models was most often the rigorous and successful development of communication and optimal control theory during the 1950s by mathematical engineers and statisticians, not economists': Epstein, *A History of Econometrics*, 151.

151 Clive Gilbert, 'The British Approach to Time Series Econometrics', *Oxford Economic Papers* 41, no. 1 (1989): 108–28.

152 Jeremy Bray, 'The Logic of Scientific Method in Economics', *Journal of Economic Studies* 4, no. 1 (1977): 1–28.

153 Mary Morgan, *The World in the Model: How Economists Work and Think* (Cambridge: Cambridge University Press, 2012), 31.

154 In concluding, SARU suggested that limits were not fixed, and collapse was unlikely, but they presciently added that their analysis might not have accounted for all possible risks, such as 'severe climatological upset': Minutes 'Modelling', Cabinet Official Committee on Future World Trends, 10 January 1974, CAB 134/3859, TNA.

155 For example, Letter from Forrester to Bray, Forrester cc. Cottrell, 8 November 1971, Acc. 610 34/379–80, BRAY.

156 Future World Trends, a discussion paper on world trends in population, resources, pollution, etc., and their implications, HMSO, 1976, CAB 134/4103, TNA, 10

157 Michael Kenward, 'This Week: Model Behaviour by the British Government', *New Scientist*, 6 October 1977, 7.

158 Donella Meadows, John Richardson and Gerhart Bruckmann, *Groping in the Dark: The First Decade of Global Modelling. Proceedings of the Sixth International Institute for Applied Systems Analysis Symposium on Global Modelling; Laxenburg* (Chichester: Wiley, 1982), 66.

159 Alan Mowle, 'SARUM 76 Global Modelling Project by SARU', *The Journal of the Operational Research Society* 29, no. 8 (1978): 828.

160   SARU Staff, *SARUM Handbook* (London: Department of the Environment, 1978), cited in Meadows et al., *Groping in the Dark*, 8.

161   Robert Taylor, 'The Heath Government and Industrial Relations: Myth and Reality', in *The Heath Government, 1970–1974: A Reappraisal*, ed. Stuart Ball and Anthony Seldon (New York and London: Longman, 1996), 161–90.

162   Jon Agar, 'Historiography and Intersections', Chapter 1 this volume.

163   Vaclav Smil, 'Perils of Long-Range Energy Forecasting: Reflections on Looking Far Ahead', *Technological Forecasting and Social Change* 65, no. 3 (2000): 251–64.

164   Jeremy Bray, *The Politics of the Environment* (London: Fabian Society, 1972), 22–3.

165   Dennis Meadows, William Behrens, Donella Meadows, Roger Naill, Jorgen Randers and Erich Zahn, *Dynamics of Growth in a Finite World* (Cambridge, MA: Wright Allen Press, 1974), 4–16.

Bibliography

Adams, Annemarie, Kevin Schwartzman and David Theodore. 'Collapse and Expand: Architecture and Tuberculosis Therapy in Montreal, 1909, 1933, 1954'. *Technology and Culture* 49, 2008, 908–42.

Adams, W.M. *Future Nature: A Vision for Conservation*, revised edition. London: Earthscan, 1996/2003.

Agar, Jon. 'Making a Meal of the Big Dish: The Construction of the Jodrell Bank Mark 1 Radio Telescope as a Stable Edifice, 1946–57'. *The British Journal for the History of Science* 27, no. 1, 1994, 3–21.

Agar, Jon. 'Bodies, Machines, Noise'. In *Bodies/Machines*, edited by Iwan Rhys Morus. Oxford: Berg, 2002, 197–220.

Agar, Jon. *The Government Machine: A Revolutionary History of the Computer*. Cambridge, MA: MIT Press, 2003.

Agar, Jon. 'Thatcher, Scientist'. *Notes and Records of the Royal Society* 65, no. 3, 2011, 215–32.

Agar, Jon. '"Future Forecast – Changeable and Probably Getting Worse": The UK Government's Early Response to Anthropogenic Climate Change'. *Twentieth-Century British History* 26, no. 4, 2015, 602–28.

Ahlmann, Hans W. 'Researches on Snow and Ice, 1918–40'. *The Geographical Journal* 107, no. 1/2, 1946, 11–25.

Aikin, Arthur. 'Plantations'. *Transactions of the Society for the Encouragement of Arts, Manufactures and Commerce for the Session 1831–1832* 49, 1832, 4.

Alberti, Samuel J.M.M. 'Amateurs and Professionals in One County: Biology and Natural History in Late Victorian Yorkshire'. *Journal of the History of Biology* 34, 2001, 115–47.

Alberti, Samuel J.M.M. 'Placing Nature: Natural History Collections and their Owners in Nineteenth-Century Provincial England'. *British Journal for the History of Science* 35, 2002, 291–311.

Alberti, Samuel J.M.M. 'Objects and the Museum'. *Isis* 96, 2005, 559–71.

Albion, Robert. *Forests and Seapower: The Royal Dockyards during the Revolutionary and Napoleonic Wars*. Cambridge, MA: Harvard University Press, 1926.

Aligica, Paul Dragos. 'The Challenge of the Future and the Institutionalisation of Interdisciplinarity: Notes on Herman Kahn's Legacy'. *Futures* 36, no. 1, 2004, 67–83.

Allan, D.G.C. *RSA: A Chronological History of the Royal Society for the Encouragement of Arts, Manufactures, and Commerce: Founded 1754, Royal Charter 1847, Royal 'Prefix' 1908.* Private printing, 1998.

Allen, David E. *The Naturalist in Britain: A Social History.* London: Allen Lane, 1976.

Allen, David E. 'On Parallel Lines: Natural History and Biology from the Late Victorian Period'. *Archives of Natural History* 25, 1998, 361–71.

Ambrose, Stephen E. *Rise to Globalism: American Foreign Policy since 1938.* New York: Penguin, 1971/1997.

Ammon, Francesca Russello. *Bulldozer: Demolition and Clearance of the Postwar Landscape.* New Haven: Yale University Press, 2016.

Anderson, Jon. 'Decision Analysis in Environmental Decision-making: Improving the Concorde Balance'. *Columbia Journal of Environmental Law* 5, 1978–9, 156–88.

Andersson, Jenny. 'The Great Future Debate and the Struggle for the World'. *American Historical Review* 117, no. 5, 2012, 1411–30.

Andersson, Jenny and Sibylle Duhautois. 'Futures of Mankind: The Emergence of the Global Future'. In *The Politics of Globality since 1945*, edited by Rens van Munster and Casper Sylvest. London and New York: Routledge, 2016, 106–25.

Andersson, Jenny and Egle Rindzevičiūtė, eds. *The Struggle for the Long-Term in Transnational Science and Politics: Forging the Future.* London and New York: Routledge, 2015.

Anon. 'Scheme for Improvement of Bogs and Other Waste Lands in Ireland' (written by W.V.). *Transactions of the Royal Dublin Society*, 2, 1801, 178–82.

Anon. 'Commissioners Appointed to Enquire into the Nature and Extent of the Several Bogs in Ireland'. *Journals of the House of Commons* 64, Sess. 1809.

Anon. 'On Economy in Planting'. *Journal of Agriculture* 2, 1831, 415.

Anon. 'Review of British Trees and Shrubs'. *Edinburgh Review* 69, 1839, 210.

Anon. 'Etude clinique de l'emploi et des effets du bain d'air comprimé dans la traitement de diverses maladies selon les procédés de M. Emile Tabarié; par M.E. BERTIN. Paris, 1855'. *The British Journal of Homoeopathy* 14, 1856, 127.

Anon. 'A Day at Ben Dhrypping'. *The Leisure Hour* 9, 2 February 1860, 73.

Anon. 'Compressed Air Baths'. *The Manufacturer and Builder* 7, 1875, 91.

Anon. *The Port of London and the Thames Barrage: A Series of Expert Studies and Reports.* London: S. Sonnenschein & Co., 1907.

Anon. 'The Outlook'. *British Journal of Tuberculosis* 18, 1924, 89.

Anon. *Somerset Ways.* Second edition. London: The Great Western Railway Company, 1928.

Anon. 'Radiation and Biology'. *Nature* 180, no. 4587, 1957, 629.

Anon. *Cereals in the West and South: Papers Given at Fellows Conferences at Harper Adams Agricultural College and the Hampshire Agricultural College, Sparsholt, January 1978.* Cambridge: National Institute of Agricultural Botany, 1978.

Anon. *The Fresno Scraper, Invented in 1883: A National Historic Mechanical Engineering Landmark*. Fresno, CA: American Society of Mechanical Engineers, 1991.

Archer, Frank H. 'Earth-Moving Machinery in Motorway Construction'. *Proceedings of the Institution of Mechanical Engineers, Conference Proceedings*, 179, part 3F, 1964–5, 1–8.

Arnold, Lorna. *Windscale 1957: Anatomy of a Nuclear Accident*. Second edition. Basingstoke: Macmillan, 1995.

Atkinson, Harriet. *The Festival of Britain*. London: IB Tauris, 2012.

Ausubel, Jesse H., Robert A. Frosch and Robert Herman. 'Technology and Environment: An Overview'. In *Technology and Environment*, edited by Jesse H. Ausubel and Hedy E. Sladovich. Washington, DC: National Academy Press, 1989, 1–20.

Baigent, Elizabeth. '"God's Earth Will Be Sacred": Religion, Theology, and the Open Space Movement in Victorian England'. *Rural History* 22, no. 1, 2011, 31–58.

Bailey, John and G. Culley. *A General View of the Agriculture of the County of Northumberland*. Newcastle: Sol Hodgson, 1797.

Baird, James. *Ben Rhydding: Its Amenities, Hygiene, and Therapeutics*. London: A.G. Dennant, 1871.

Ball, Stuart. 'A Chronology of the Heath Government'. In *The Heath Government, 1970–1974: A Reappraisal*, edited by Stuart Ball and Anthony Seldon. New York and London: Longman, 1996, 396.

Balmer, Andrew S., Katie Bulpin and Susan Molyneux-Hodgson. *Synthetic Biology: A Sociology of Changing Practices*. Basingstoke: Palgrave Macmillan, 2016.

Barber, T.W., Clayton Beadle, William Joseph Dibdin, Thomas Hennell and D. Urquhart. *The Port of London and the Thames Barrage. A Series of Expert Studies and Reports*. London: S. Sonnenschein & Co., 1907.

Bardi, Ugo. *The Limits to Growth Revisited*. New York: Springer, 2011.

Bartrip, Peter. *The Way from Dusty Death: Turner and Newall and the Regulation of Occupational Health in the British Asbestos Industry, 1890s–1970*. London: Athlone, 2001.

Bates, Captain O. 'Ploughing the Seabed'. *Post Office Telecommunications Journal* 22, no. 4, 1970, 2–4.

Baxter, Peter J. 'The East Coast Big Flood, 31 January–1 February 1953: A Summary of the Human Disaster'. *Philosophical Transactions of the Royal Society of London A: Mathematical, Physical and Engineering Sciences* 363, no. 1831, 2005, 1293–312.

Beinart, William and Lotte Hughes. *Environment and Empire*. Oxford: Oxford University Press, 2007.

Bell, George Douglas Hutton. *Cultivated Plants of the Farm*. Cambridge: Cambridge University Press, 1948.

Bell, George Douglas Hutton. 'Plant Breeding for Crop Improvement in Britain: Methods, Achievements and Objectives'. *Proceedings of the Royal Society of London. Series B, Biological Sciences* 171, 1968, 148.

Bellini, James. 'Hudson Institute to Study Future of the UK'. *Futures* 5, no. 4, 1973, 421–9.

Bellini, James. 'Hudson Institute's Year of Europe'. *Futures* 6, no. 1, 1974, 84.

Benediktsson, Karl. '"Scenophobia", Geography and the Aesthetic Politics of Landscape'. *Geografiska Annaler: Series B, Human Geography* 89, no. 3, 2007, 203–17.

Bennett, Brett M. *Plantations and Protected Areas: A Global History.* Cambridge, MA: MIT Press, 2015.

Berghoff, Hartmut and Adam Rome, eds. *Green Capitalism?: Business and the Environment in the Twentieth Century.* Philadelphia: University of Pennsylvania Press, 2017.

Bernhardt, Christoph, ed. *Environmental Problems of European Cities of the 19th and 20th Centuries.* Münster: Waxmann, 2000.

Berry, Dominic J. 'Genetics, Statistics, and Regulation at the National Institute of Agricultural Botany 1919–1969'. PhD diss., University of Leeds, 2014.

Berry, Dominic J. 'The Plant Breeding Industry after Pure Line Theory: Lessons from the National Institute of Agricultural Botany'. *Studies in History and Philosophy of Science Part C: Studies in History and Philosophy of Biological and Biomedical Sciences* 46, 2014, 25–37.

Berry, Dominic J. 'Agricultural Modernity as a Product of the Great War: The Founding of the Official Seed Testing Station for England and Wales, 1917–1921'. *War & Society* 34, no. 2, 2015, 121–39.

Bert, Paul. 'L'anesthésie par le protoxyde d'azote: travaux récents'. *Revues scientifiques publiées par le journal 'La République Française'* 1, 1880, 318.

Berthezène, Clarisse. *Training Minds for the War of Ideas: Ashridge College, The Conservative Party and the Cultural Politics of Britain, 1929–54.* Manchester: Manchester University Press, 2015.

Bijsterveld, Karin. *Mechanical Sound: Technology, Culture, and Public Problems in the Twentieth Century.* Cambridge, MA: MIT Press, 2008.

Black, Lawrence. 'Review Essay: An Enlightening Decade? New Histories of 1970s Britain'. *International Labor and Working-Class History* 82, 2012, 174–86.

Blanchard, Elodie Vielle. 'Modelling the Future: An Overview of the "Limits to Growth" Debate'. *Centaurus* 52, no. 2, 2010, 91–116.

Blanchard, Elodie Vielle. 'Technoscientific Cornucopian Futures versus Doomsday Futures: The World Models and the Limits to Growth'. In *The Struggle for the Long Term in Transnational Science and Politics: Forging the Future,* edited by Jenny Andersson and Egle Rindzevičiūtė. New York and London: Routledge, 2015, 92–114.

Blaug, Mark. *Economic Theory in Retrospect.* Cambridge: Cambridge University Press, 1962.

Blaxter, Kenneth and Noel Robertson. *From Dearth to Plenty: The Modern Revolution in Food Production*. Cambridge: Cambridge University Press, 1995.

Bleasdale, J.K.A. 'Britain's Green Revolution'. In *Advances in Agriculture: Proceedings of Section M Agriculture, 139th Annual Meeting of the British Association for the Advancement of Science*, edited by W.A. Hayes. Birmingham: University of Aston, 1978, 1–2.

Bloomfield, Brian. *Modelling the World: The Social Constructions of Systems Analysts*. Oxford: Blackwell, 1986.

Bocking, Stephen. *Ecologists and Environmental Politics: A History of Contemporary Ecology*. New Haven: Yale University Press, 1997.

Bocking, Stephen. *Nature's Experts: Science, Politics, and the Environment*. New Brunswick: Rutgers University Press, 2006.

Bocking, Stephen. 'Nature on the Home Front: British Ecologists' Advocacy for Science and Conservation'. *Environment and History* 18, 2012, 261–81.

Boli, John. 'Contemporary Developments in World Culture'. *International Journal of Comparative Sociology*, 42, no. 5/6, 2005, 383–404.

Bonacina, L.C.W. and E.C. Hawkes. *Climatic Change and the Retreat of Glaciers*. Royal Meteorological Society, 1947.

Bondansky, Daniel. 'The History of the Global Climate Change Regime'. In *International Relations and Global Climate Change*, edited by Urs Luterbacher and Detlef F. Sprinz. Cambridge, MA: MIT Press, 2001.

Bone, V.W. 'Modern Developments in Tractor-Drawn Excavating Equipment'. *Proceedings of the Institution of Mechanical Engineers*, 137, 1937, 345–53.

Bone, V.W., W. Savage and R.M. Wynne-Edwards. 'Public Works Contractors' Plant'. *Proceedings of the Institution of Mechanical Engineers*, 157, part 1, 1947, 313–17.

Bowden, K.F. 'Storm Surges in the North Sea'. *Weather* 8, no. 3, March 1953, 82.

Bowen, H.J.M., et al. 'The Induction of Sports in Chrysanthemums by Gamma Radiation'. *Radiation Botany* 2, 1962, 303.

Bowler, Catherine and Peter Brimblecombe. 'Control of Air Pollution in Manchester prior to the Public Health Act, 1875'. *Environment and History* 6, 2000, 71–98.

Bowles, Oliver and K.G. Warner. 'Asbestos'. In *Minerals Yearbook*, edited by Herbert Hughes. Washington, DC: United States Government Printing Office, 1939, 1309.

Bowman, Waldo G., Harold W. Richardson, Nathan A. Bowers, Edward J. Cleary and Archie N. Carter. *Bulldozers Come First: The Story of US War Construction in Foreign Lands*. New York: McGraw-Hill, 1944.

Brace, Catherine. 'Finding England Everywhere: Regional Identity and the Construction of National Identity, 1890–1940'. *Ecumene* 6, no. 1, 1999, 90–109.

Brace, Catherine. 'A Pleasure Ground for Noisy Herds? Incompatible Encounters with the Cotswolds and England, 1900–1950'. *Rural History* 11, no. 1, 2000, 75–94.

Bradley, James, Marguerite Dupree and Alastair Durie. 'Taking the Water-Cure: The Hydropathic Movement in Scotland, 1840–1940'. *Business and Economic History* 26, 1997, 426.

Bradley, Simon. *The Railways: Nation, Network and People*. London: Profile, 2015.

Brassley, Paul. 'Output and Technical Change in Twentieth-Century British Agriculture'. *The Agricultural History Review* 48, 2000, 60–84.

Brassley, Paul, David Harvey, Matt Lobley and Michael Winter. 'Accounting for Agriculture: The Origin of the Farm Management Survey'. *Agricultural History Review* 61, 2013, 135–53.

Bray, Jeremy. 'Economic Modelling with Computers'. *New Scientist* 27, no. 453, 1965, 196.

Bray, Jeremy. 'Bellagio Symposium Papers Published'. *Futures* 2, no. 1, 1970, 76–88.

Bray, Jeremy. *Decision in Government*. London: Victor Gollancz, 1970.

Bray, Jeremy. 'Tools for the Seventies'. *The Political Quarterly* 41, no. 2, 1970, 151–5.

Bray, Jeremy. *The Politics of the Environment*. London: Fabian Society, 1972.

Bray, Jeremy. 'The Logic of Scientific Method in Economics'. *Journal of Economic Studies* 4, no. 1, 1977, 1–28.

Bray, Jeremy. *Standing on the Shoulders of Giants: Science, Politics and Trust, A Parliamentary Life*. Cambridge: Elizabeth Bray, 2004.

Briggs, Asa. 'The Historian and the Future'. *Futures* 10, no. 6, 1978, 445–51.

Bright, Charles. *Submarine Telegraphs: Their History, Construction and Working*. Cambridge: Cambridge University Press, 2014 [1898].

Brimblecombe, Peter. *The Big Smoke: A History of Air Pollution in London since Medieval Times*. London: Methuen, 1988.

Brimblecombe, Peter and Catherine Bowler. 'Air Pollution in York 1850–1900'. In *The Silent Countdown: Essays in European Environmental History*, edited by Christian Pfister and Peter Brimblecombe. Berlin: Springer Verlag, 1990, 182–95.

Brogan, D.W. 'The Bulldozer'. In *American Themes*. London: Hamish Hamilton, 1948, 202.

Browne, Janet. 'Spas and Sensibilities: Darwin at Malvern'. *Medical History* 10 Supplement, 1990, S102–13, S103.

Brunt, Liam. 'Rehabilitating Arthur Young'. *Economic History Review* 56, 2003, 265–99.

Bryan, Harold. 'Wart Disease Infection Tests'. *Journal of Agricultural Science* 18, no. 3, 1928, 507–14.

Buchanan, R. Angus. 'Reflections on the Decline of the History of Technology in Britain'. *History of Technology* 22, 2001, 211–22.

Bud, Robert. 'Penicillin and the New Elizabethans'. *British Journal for the History of Science* 31, no. 3, 1998, 305–33.

Bunge, J.H.O. *Tideless Thames in Future London*. London: Thames Barrage Assn., 1944.

Cadman, W.A. 'Forestry and Silvicultural Developments in North Wales'. *Forestry*, 26, no. 2, 1953, 65–80.

Caldwell, Lawrence. *International Environmental Policy: From the Twentieth to the Twenty First Century*. Durham, NC: Duke University Press, 1996.

Cannadine, David. 'Engineering History, or the History of Engineering? Re-Writing the Technological Past'. *Transactions of the Newcomen Society* 74, 2004, 163–80.

Cantor, Geoffrey and Sally Shuttleworth. *Science Serialized: Representations of the Sciences in Nineteenth-Century Periodicals*. Cambridge, MA: MIT Press, 2004.

Carlson, Robert H. *Biology is Technology: The Promise, Peril and New Business of Engineering Life*. Cambridge, MA: Harvard University Press, 2011.

Carson, Rachel. *Silent Spring*. London: Hamish Hamilton, 1963.

Charnley, Berris. 'Agricultural Science, Plant Breeding and the Emergence of a Mendelian System in Britain, 1880–1930'. PhD diss., University of Leeds, 2011.

Charnley, Berris. 'Experiments in Empire-Building: Mendelian Genetics as a National, Imperial, and Global Agricultural Enterprise'. *Studies in History and Philosophy of Science Part A* 44, no. 2, 2013, 292–300.

Charnley, Berris. 'Geneticists on the Farm: Agriculture and the All-English Loaf'. In *Scientific Governance in Britain, 1914–79*, edited by Don Leggett and Charlotte Sleigh. Manchester: Manchester University Press, 2016, 181–98.

Charnley, Berris and Gregory Radick. 'Intellectual Property, Plant Breeding and the Making of Mendelian Genetics'. *Studies in History and Philosophy of Science Part A* 44, no. 2, 2013, 222–33.

Choffin, Morgan. 'The Cunningham Sanatorium'. Cleveland Historical, 2017. Last accessed 7 August 2017. https://clevelandhistorical.org/items/show/378.

Cirkel, Fritz. *Asbestos: Its Occurrence, Exploitation and Uses*. Ottawa: Mines Branch, Department of the Interior, 1905.

Clapp, B.W. *An Environmental History of Britain since the Industrial Revolution*. London: Taylor & Francis, 1994.

Clark, John. 'Pesticides, Pollution and the UK's Silent Spring, 1963–64: Poison in the Garden of England'. *Notes and Records of the Royal Society*, published online 15 February 2017. Last accessed 15 August 2017. http://rsnr.royalsocietypublishing.org/content/71/3/297

Clucas, T.M. 'The Contribution of Plant Breeders to Vegetable Production in the 70s'. *Journal of the National Institute of Agricultural Botany* 12, Supplement, 1970, 48.

Coates, Peter. *Nature: Western Attitudes since Ancient Times*. Cambridge: Polity, 1998.

Coates, Peter. 'Can Nature Improve Technology?' In *The Illusory Boundary: Environment and Technology in History*, edited by Martin Reuss and Stephen H. Sutcliffe. Charlottesville: University of Virginia Press, 2010.

Cobbett, William. *The Woodlands, or, a Treatise on the Preparing of Ground for Planting, on the Planting, on the Cultivating, on the Pruning, and on the Cutting Down of Forest Trees and Underwoods.* London: William Cobbett, 1825.

Cobbett, William. *Rural Rides.* Vol. 1. London: William Cobbett, 1830.

Cole, Hugh, Chris Freeman, M. Marie Jahoda and K.R.L. Pavitt. 'Preface'. In *Thinking about the Future: A Critique of the Limits to Growth*, edited by Hugh Cole, Chris Freeman, M. Marie Jahoda and K.R.L. Pavitt. London: Chatto & Windus, 1973.

Cole, Tim. '"Beauty and the Motorway – The Problem for All": Motoring through the Quantocks Area of Natural Beauty'. in *Local Places, Global Processes. Histories of Environmental Change in Britain and Beyond*, edited by Peter Coates, David Moon and Paul Warde. Oxford: Windgather Press, 2016, 171–83.

Collins, Edward J.T. and Joan Thirsk, eds. *The Agrarian History of England and Wales, Vol. 7, 1850–1914.* Cambridge: Cambridge University Press, 2000.

Collison Black, R.D. *Economic Thought and the Irish Question 1817–1870.* Cambridge: Cambridge University Press, 1960, 182.

Conekin, Becky E. *'The Autobiography of a Nation': The Festival of Britain.* Manchester: Manchester University Press, 2003.

Conford, Philip. *The Development of the Organic Network 1945–95.* Edinburgh: Floris Books, 2011.

Connolly, Cyril. *The Unquiet Grave: A Word Cycle by Palinurus.* London: Hamish Hamilton, 1944.

Cooke, George, ed. *Agricultural Research, 1931–81.* London: Agricultural Research Council, 1981.

Cooke, W.E. 'Fibrosis of the Lungs Due to the Inhalation of Asbestos Dust'. *British Medical Journal*, 1924, 140–2, 147.

Cooper, Tim. 'British Environmental History', 2008. Last accessed 25 September 2017. www.history.ac.uk/makinghistory/resources/articles/environmental_history.html

Cooper, Timothy and Anna Green. 'The *Torrey Canyon* Disaster, Everyday Life, and the "Greening" of Britain'. *Environmental History* 22, 1, 2017, 101–26.

Cormack, Douglas. *Response to Marine Oil Pollution – Review and Assessment.* Dordrecht: Kluwer, 1999.

Cornish, Edward. *The Study of the Future: An Introduction to the Art and Science of Understanding and Shaping Tomorrow's World.* Washington DC: World Future Society, 1977.

Cosier, J.E.H. 'Getting to Grips with Undersea Cables'. *Post Office Telecommunications Journal* 30, no. 1, Spring 1978, 7–8.

Crawford, A.B., et al. 'The Research Background of the Telstar Experiment'. *The Bell System Technical Journal* 42, no. 4, July 1963, 747–51.

Creager, Angela. *Life Atomic: A History of Radioisotopes in Science and Medicine*. Chicago and London: University of Chicago Press, 2013.

Cresswell, Beatrice F. *The Quantock Hills, their Combes and Villages: The Homeland Handbooks Vol. 35. Fifth Edition*. London: The Homeland Association, 1922.

Cronon, William. *Nature's Metropolis: Chicago and the Great West*. New York: Norton, 1991.

Cronon, William. 'The Trouble with Wilderness; or, Getting Back to the Wrong Nature'. *Environmental History*, 1996, 1, 7–28.

Crosby, Alfred W. *The Columbian Exchange: Biological and Cultural Consequences of 1492*. Westport: Greenwood Press, 1972.

Crosby, Alfred W. *Ecological Imperialism: The Biological Expansion of Europe, 900–1900*. Cambridge: Cambridge University Press, 1986.

Crutzen, Paul J. 'The Influence of Nitrogen Oxides on the Atmosphere Ozone Content'. *Quarterly Journal of the Royal Meteorological Society* 96, 1970, 320–5.

Crutzen, Paul J. 'Geology of Mankind'. *Nature* 415, no. 6867, 2002, 23.

Curry, Helen Anne. 'Industrial Evolution: Mechanical and Biological Innovation at the General Electric Research Laboratory'. *Technology and Culture* 54, 2013, 746–81.

Curry, Helen Anne. 'Atoms in Agriculture: A Study of Scientific Innovation between Technological Systems'. *Historical Studies in the Natural Sciences* 46, 2016, 119–53.

Curry, Helen Anne. *Evolution Made to Order: Plant Breeding and Technological Innovation in Twentieth-Century America*. Chicago and London: University of Chicago Press, 2016.

Curzon, John J. 'Changing Mining Methods at the Holden Mine'. *Transactions of the American Institute of Mining and Metallurgical Engineers*, 163, 1946, 75–7.

Dalyell, Tam. 'Bray, Jeremy William, 1930–2002, Mathematician and Politician'. In *Oxford Dictionary of National Biography*, Oxford: Oxford University Press, 2006.

Darbishire, David H. 'Address'. *The Journal of the National Institute of Agricultural Botany* 12, 1972, 519–23.

Darlington, Cyril Dean. 'The Problem of Chromosome Breakage'. In *Symposium on Chromosome Breakage*, John Innes Horticultural Institution. London and Edinburgh: Oliver & Boyd, 1953.

Darlington, Cyril Dean. 'Genetics and Plant Breeding, 1910–80'. *Philosophical Transactions of the Royal Society of London. Series B, Biological Sciences* 292, 1981, 401–5.

David, Paul. 'Clio and the Economics of QWERTY'. *American Economic Review* 75, no. 2, 1985, 332–7.

Davies, C.R. 'Effects of Gamma Irradiation on Growth and Yield of Agricultural Crops – I. Spring Sown Wheat'. *Radiation Botany* 8, 1968, 29.

Davies, C.R. 'Effects of Gamma Irradiation on Growth and Yield of Agricultural Crops – III. Root Crops, Legumes and Grasses'. *Radiation Botany* 13, 1973, 134.

Davies, C.R. and D.B. Mackay. 'Effects of Gamma Irradiation on Growth and Yield of Agricultural Crops – VI. Effects on Yields of the Second Generation in Cereals and Potato'. *Radiation Botany* 13, 1973, 137–44.

Davies, Gail. 'Science, Observation and Entertainment: Competing Visions of Postwar British Natural History Television, 1946–1967'. *Ecumene* 7, 2000, 432–59.

Davis, Jon. *Prime Ministers and Whitehall*. London: Hambledon Continuum, 2007.

Dawkins, Richard. *The Selfish Gene*. Oxford: Oxford University Press, 1976.

Dawson, James. *Canal Extensions in Ireland, Recommended to the Imperial Legislature, as the Best Means of Promoting the Agriculture – Draining the Bogs – and Employing the Poor*. Dublin: Carrick, 1819, Letters 1 and 3.

de Chadarevian, Soraya and Nick Hopwood, eds. *Models: The Third Dimension of Science*. Stanford: Stanford University Press, 2004.

de Houghton, Charles, William Page and Guy Streatfield. '*… And Now the Future': A PEP Survey of Futures Studies*. London: Policy Studies Institute, 1971.

de Vries, Hugo. *Species and Varieties: Their Origin by Mutation*. Chicago: The Open Court, 1905.

Di Palma, Vittoria. *Wasteland: A History*. New Haven: Yale University Press, 2014.

Dow, Andrew. *The Railway: British Track since 1804*. Barnsley: Pen & Sword, 2015.

Dror, Yehezkel. 'The Misfired Report'. *Futures* 7, no. 1, 1975, 64–8.

Dudley, Marianna. 'A Fairy (Shrimp) Tale of Military Environmentalism: The "Greening" of Salisbury Plain'. In *Militarized Landscapes: From Gettysburg to Salisbury Plain*, edited by Chris Pearson, Peter Coates and Tim Cole. London: Bloomsbury, 2010, 135–50.

Dudley, Marianna. *An Environmental History of the UK Defence Estate, 1945 to the Present*. London: Continuum, 2012.

Dudley, R. 'Report of the Second Crop Conference'. *The Journal of the National Institute of Agricultural Botany* 7, 1954, 198.

Dujardin-Beaumetz, Prof. 'Hygienic Therapeutics: A Lecture on Aerotherapy, trans. E.P. Hurd'. *The Therapeutic Gazette* 12, 15 May 1888, 292.

Duncan, C.C. 'Communications and Defense'. *Bell Telephone Magazine* 37, no. 1, Spring 1958, 15–24.

Durie, Alastair J. 'The Business of Hydropathy in the North of England, c.1850–1930'. *Northern History* 39, 2002, 46–7.

Duvick, Donald N. 'Biotechnology in the 1930s: The Development of Hybrid Maize'. *Nature Reviews Genetics* 2, 2001, 73.

Dyck, Ian. *William Cobbett and Rural Popular Culture*. Cambridge: Cambridge University Press, 1992.

Easterbrook, Laurence, ed. *The Future of Agriculture*. London: Todd, 1943.

Edgerton, David. *Science, Technology and the British Industrial 'Decline' 1870–1970*. Cambridge: Cambridge University Press, 1996.

Edgerton, David. 'The White Heat Revisited: British Government and Technology in the 1960s'. *Twentieth Century British History* 7, no. 1, 1996, 53–82.

Edgerton, David. 'Reflections on the History of Technology in Britain'. *History of Technology*, 2001, 22.

Edgerton, David. 'C.P. Snow as Anti-Historian of British Science: Revisiting the Technocratic Moment'. *History of Science* 43, no. 2, 2005, 187–208.

Edgerton, David. *The Shock of the Old: Technology and Global History since 1900*. London: Profile, 2006.

Edgerton, David. *Warfare State: Britain, 1920–1970*. Cambridge: Cambridge University Press, 2006.

Edgerton, David. 'Innovation, Technology, or History: What is the Historiography of Technology about?' *Technology and Culture* 51, no. 3, 2010, 680–97.

Edgerton, David. *England and the Aeroplane: An Essay on a Militant and Technological Nation*. London: Penguin, 2013.

Editorial. 'Futures, Confidence from Chaos'. *Futures* 1, no. 1, 1968, 2.

Edwards, Paul N. 'The World in a Machine: Origins and Impacts of Early Computerized Global Systems Models'. In *Systems, Experts, and Computers: The Systems Approach in Management and Engineering, World War II and After*, edited by Agatha Hughes and Thomas Hughes. Cambridge, MA: MIT Press, 2000, 221–54.

Edwards, Paul N. 'Infrastructure and Modernity: Force, Time, and Social Organisation in the History of Sociotechnical Systems'. In *Modernity and Technology*, edited by Thomas J. Misa, Philip Brey and Andrew Feenberg. Cambridge, MA and London: MIT Press, 2003, 189.

Elichirigoity, Fernando. *Planet Management: Limits to Growth, Computer Simulation, and the Emergence of Global Spaces*. Evanston, IL: Northwestern University Press, 1999.

Elina, Olga, Susanne Heim and Nils Roll-Hansen. 'Plant Breeding on the Front: Imperialism, War, and Exploitation'. *Osiris* 20, 2005, 178–9.

Ellis, F.B. 'The Contamination of Grassland with Radioactive Strontium – II. Effect of Lime and Cultivation on the Levels of Strontium-90 in Herbage'. *Radiation Botany* 8, 1968, 269–84.

Ellis, Markman. *The Politics of Sensibility. Race, Gender and Commerce in the Sentimental Novel*. Cambridge: Cambridge University Press, 1996.

Elmendorf, C.H. and B.C. Heezen. 'Oceanographic Information for Engineering Submarine Cable Systems'. *The Bell System Technical Journal* 36, no. 5, September 1957, 1035–94.

Endersby, Jim. *A Guinea Pig's History of Biology: The Plants and Animals Who Taught Us the Facts of Life*. London: Heinemann, 2007.

Envirotech. 'Are Animals Technology?' – archive of messages posted on Envirotech, Envirotech@lists.Stanford.EDU, 2001. Last accessed 15 September 2017. http://envirotechweb.org/wp-content/uploads/2007/05/animaltech.pdf

Epstein, Roy. *A History of Econometrics*. Amsterdam: Elsevier, 1987.

Etheridge, J.H. 'Compressed Air: Synopsis of a Lecture Delivered Before the Rush Medical College Students, Dec. 1872'. *Chicago Medical Journal* 30, 1873.

Evans, David. *A History of Nature Conservation in Britain*. Second edition. London: Routledge, 1997.

Ewen, Shane. 'Sheffield's Great Flood of 1864: Engineering Failure and the Municipalisation of Water'. *Environment and History* 20, 2014, 177–207.

Fairbrother, Nan. *New Lives, New Landscapes*. London: The Architectural Press, 1970.

Fiege, Mark. *Irrigated Eden: The Making of an Agricultural Landscape in the American West*. Seattle: University of Washington Press, 1999.

Finn, Michael A. 'The West Riding Lunatic Asylum and the Making of the Modern Brain Sciences in the Nineteenth Century'. PhD diss., University of Leeds, 2012.

Fitch, James Marston. 'New Uses for Artistic Patrimony'. *Journal of the Society of Architectural Historians*, 30, no. 1, March 1971, 3–16.

Fitzgerald, Deborah. *The Business of Breeding: Hybrid Corn in Illinois, 1890–1940*. Ithaca: Cornell University Press, 1989.

Fitzgerald, Deborah. 'Farmers Deskilled: Hybrid Corn and Farmer's Work'. *Technology and Culture* 34, 1993, 328–9.

Fleming, L.D. *The Air Cure; or Atmospheric Therapeutics*. Rochester: Wm. S. Falls, 1867.

Flippen, J. Brooks. 'Richard Nixon, Russell Train, and the Birth of Modern American Environmental Diplomacy'. *Diplomatic History* 32, 2008, 613–38.

Forrester, Jay. *World Dynamics*. Cambridge, MA: Wright Allen Press, 1971.

Fradkin, Philip L. *The Seven States of California: A Natural and Human History*. Berkeley: University of California Press, 1997.

Fraser, Robert. *Sketches and Essays on the Present State of Ireland, Particularly Regarding the Bogs, Waste Lands & Fisheries; with Observations on the Best Means of Employing the Population in Useful and Productive Labour*. Dublin: Carrick, 1822.

Frick, Carlos. 'The Movement for Smoke Abatement in 19th-Century Britain'. *Technology and Culture* 21, 1980, 29–50.

Fyfe, Aileen. 'Natural History and the Victorian Tourist: From Landscapes to Rock-Pools'. In *Geographies of Nineteenth-Century Science*, edited by David N. Livingstone and Charles W.J. Withers. Chicago: University of Chicago Press, 2011.

Gabor, Dennis. *Inventing the Future*. London: Secker & Warburg, 1963.

Gabrys, Jennifer. *Program Earth: Environmental Sensing Technology and the Making of a Computational Planet*. Minneapolis and London: University of Minnesota Press, 2016.

Galison, Peter. 'The Future of Scenarios: State Science Fiction'. In *The Subject of Rosi Braidotti: Politics and Concepts*, edited by Bolette Blaagard and Iris van der Tuin. London and New York: Bloomsbury, 2014, 38–46.

Garlick, Steve. 'Given Time: Biology, Nature and Photographic Vision'. *History of the Human Sciences* 22, 2009, 81–101.

Ghamari-Tabrizi, Sharon. *The Worlds of Herman Kahn: The Intuitive Science of Thermonuclear War*. Cambridge, MA: Harvard University Press, 2005.

Gifford, Margaret Jeune, ed. *Pages from the Diary of an Oxford Lady 1843–1862*. Oxford: The Shakespeare Head Press/B. Blackwell, 1932.

Gijswijt-Hofstra, Marijke. 'Conversions to Homoeopathy in the Nineteenth Century: The Rationality of Medical Deviance'. In *Illness and Healing Alternatives in Western Europe*, edited by Marijke Gijswijt-Hofstra, Hilary Marland and Hans De Waardt. London: Routledge, 1997, 161–82.

Gilbert, Clive. 'The British Approach to Time Series Econometrics'. *Oxford Economic Papers* 41, no. 1, 1989, 108–28.

Gilbert, Stuart and Ray Horner. *The Thames Barrier*. London: Thomas Telford, 1984.

Gillette, Robert. 'The Limits to Growth: Hard Sell for a Computer View of Doomsday'. *Science* 175, no. 4026, 1972, 1088–92.

Gillman, Peter. 'Supersonic Bust: The Story of the Concorde'. *Atlantic* 239, 1977, 72–81.

Ginsberg, Alexandra Daisy, et al. *Synthetic Aesthetics: Investigating Synthetic Biology's Designs on Nature*. Cambridge, MA: MIT Press, 2014.

Glancey, Jonathan. *Concorde: The Rise and Fall of the Supersonic Airliner*. London: Atlantic Books, 2015.

Gleichmann, T.F., A.H. Lince, M.C. Wooley and F.J. Braga. 'Repeater Design for the North Atlantic Link'. *The Bell System Technical Journal* 36, no. 1, January 1957, 91.

Glynne, Mary D. 'Infection Experiments with Wart Diseases of Potatoes. *Synchytrium Endobioticum* (Schilb) Perc'. *Annals of Applied Biology* 12, no. 1, 1924, 34–60.

Goddard, Nicholas. 'A Mine of Wealth: The Victorians and the Agricultural Value of Sewage'. *Journal of Historical Geography* 22, 1996, 274–90.

Goldsmith, Edward. 'Living with Nature'. *The Ecologist* 1, no. 1, 1970, 2.

Goldsmith, P., A.F. Tuck, J.S. Foot, E.L. Simmons and R.L. Newson. 'Nitrogen Oxides, Nuclear Weapon Testing, Concorde and Stratospheric Ozone'. *Nature* 244, no. 5418, 1973, 545–51.

Golub, Robert and Joe Townsend. 'Malthus, Multinationals, and the Club of Rome'. *Social Studies of Science* 7, no. 2, 1977, 201–22.

Gooday, Graeme. '"Nature" in the Laboratory: Domestication and Discipline with the Microscope in Victorian Life Science'. *British Journal for the History of Science* 24, 1991, 307–41.

Gooday, Graeme. 'The Flourishing History of Technology in the United Kingdom: A Critique of Antiquarian Complaints of "Neglect"'. *History of Technology* 22, 2001, 189–202.

Gorman, Alice. 'Cultural Landscape of Space'. In *Handbook of Space Engineering, Archaeology, and Heritage*, edited by Ann Garrison Darrin and Beth Laura O'Leary. Boca Raton: CRC Press, 2009, 335–46.

Gorman, Hugh S. and Betsy Mendelsohn. 'Where Does Nature End and Culture Begin? Converging Themes in the History of Technology and Environmental History'. In *The Illusory Boundary: Environment and Technology in History*, edited by Martin Reuss and Stephen Cutcliffe. Charlottesville: University of Virginia Press, 2010, 265–90.

Graduate of the Edinburgh University [James Baird]. 'A Graduate of the Edinburgh University'. *Ben Rhydding. The Principles of Hydropathy and the Compressed-Air Bath*. London: Hamilton, Adams and Co., 1858.

Gough, George C. 'Wart Disease of Potatoes'. *Journal of the Royal Horticultural Society* 45, 1919, 301–12.

Gouyon, Jean-Baptiste. 'The BBC Natural History Unit: Instituting Natural History Film-Making in Britain'. *History of Science* 49, 2011, 425–51.

Gouyon, Jean-Baptiste. 'From Kearton to Attenborough: Fashioning the Telenaturalist's Identity'. *History of Science* 49, 2011, 25–60.

Graham, Loren R. *What Have We Learned about Science and Technology from the Russian Experience?* Stanford: Stanford University Press, 1998.

Green, Nicholas. *The Spectacle of Nature: Landscape and Bourgeois Culture in Nineteenth-Century France*. Manchester: Manchester University Press, 1990.

Greene, Anne N. 'War Horses: Equine Technology in the Civil War'. In *Industrializing Organisms: Introducing Evolutionary History*, edited by Susan Schrepfer and Philip Scranton. London: Routledge, 2004, 143–65.

Greene, Anne N. *Horses at Work: Harnessing Power in Industrial America*. Cambridge, MA: Harvard University Press, 2008.

Greer, Lindsay and Frans Spaepen. 'Sir Alan Cottrell (1919–2012)'. *Proceedings of the National Academy of Sciences* 109, no. 42, 2012, 16753.

Griffith, Richard. *Practical Domestic Politics: Being a Comparative and Prospective Sketch of the Agriculture and Population of Great Britain and Ireland, Including Some Suggestions on the Practicability and Expediency of Draining and Cultivating the Bogs of Ireland*. London: Sherwood, Neely & Jones, 1819.

Grigson, Geoffrey, ed. *About Britain No. 5. Chilterns to Black Country. A New Guide Book with a Portrait by W.G. Hoskins*. London: Collins, 1951.

Grigson, Geoffrey, ed. *About Britain No. 8. East Midland and the Peak. A New Guide Book with a Portrait by W.G. Hoskins*. London: Collins, 1951.

Grindrod, Ralph Barnes. *The Compressed Air Bath: A Therapeutic Agent in Various Affections of the Respiratory Organs, and other Diseases*. London: Simpkin, Marshall and Co., 1860.

Grindrod, Ralph Barnes. *Malvern: Past and Present – Its History, Legends, Topography, Climate, etc.* Malvern: H.W. Lamb, 1865.

Grindrod, Ralph Barnes. *Malvern: Its Claims as a Health Resort with Notes on Climate, in Its Relation to Health And Disease, Also an Exposition of the Physiological and Therapeutic Influence of Compressed Air*. London: Robert Hardwicke, 1871.

Grove, Richard. *Green Imperialism: Colonial Expansion, Tropical Island Edens and the Origins of Environmentalism, 1600–1860*. Cambridge: Cambridge University Press, 1994.

Gruffudd, Pyrs, David T. Herbert and Angela Piccinni. 'In Search of Wales: Travel Writing and Narratives of Difference, 1918–50'. *Journal of Historical Geography* 26, no. 4, 2000, 589–604.

Haddock, Keith. *The Earthmover Encyclopedia*. St Paul, MN: Motorbooks, 2002.

Hahn, Barbara. *Making Tobacco Bright: Creating an American Commodity, 1617–1937*. Baltimore: Johns Hopkins University Press, 2011.

Haine, K.E. 'Improved Varieties of Brassica Crops in NIAB Trials'. *The Journal of the National Institute of Agricultural Botany* 12, Supplement, 1970, 1–4.

Hall, Alexander. 'The Rise of Blame and Recreancy in the United Kingdom: A Cultural, Political and Scientific Autopsy of the North Sea Flood of 1953'. *Environment and History* 17, 2011, 379–408.

Hall, Alexander. 'Risk, Blame, and Expertise: The Meteorological Office and Extreme Weather in Post-War Britain'. PhD diss., University of Manchester, 2012.

Hall, Rupert. 'Where is the History of Technology?' *History of Technology* 22, 2001, 203–10.

Hamblin, Jacob Darwin. *Oceanographers and the Cold War: Disciples of Marine Science*. Seattle: University of Washington Press, 2005.

Hamblin, Jacob Darwin. *Poison in the Well: Radioactive Waste in the Oceans at the Dawn of the Nuclear Age*. New Brunswick and London: Rutgers University Press, 2008.

Hamblin, Jacob Darwin. 'Environmentalism for the Atlantic Alliance: NATO's Experiment with the "Challenges of Modern Society"'. *Environmental History* 15, 2010, 54–75.

Hamblin, Jacob Darwin. *Arming Mother Nature: The Birth of Catastrophic Environmentalism*. Oxford and New York: Oxford University Press, 2013.

Hamblin, Jacob Darwin. 'Quickening Nature's Pulse: Atomic Agriculture at the International Atomic Energy Agency'. *Dynamis* 35, 2015, 396–7.

Hamilton, Clive and Jacques Grinevald. 'Was the Anthropocene Anticipated?' *The Anthropocene Review* 2, no. 1, 2015, 59–72.

Hamlin, Christopher. *A Science of Impurity: Water Analysis in Nineteenth-Century Britain*. Bristol: Adam Hilger, 1990.

Hamlin, Christopher. 'Edwin Chadwick and the Engineers, 1842–1854: Systems and Anti-Systems in the Pipe-and-Brick Sewers War'. *Technology and Culture*, 1992, 33, 680–709.

Hamlin, Christopher. *Public Health and Social Justice in the Age of Chadwick: Britain, 1800–1854*. Cambridge: Cambridge University Press, 1998.

Hamlin, Christopher. *Cholera: The Biography*. Oxford: Oxford University Press, 2009.

Hamlin, Christopher. 'Surgeon Reginald Orton and the Pathology of Deadly Air: The Contest for Context in Environmental Health'. In *Toxic Airs. Body, Place, Planet in Historical Perspective*, edited by James Rodger Fleming and Ann Johnson. Pittsburgh: University of Pittsburgh Press, 2014, 23–49.

Hare, Frederick. 'Environment: Resuscitation of an Idea'. *Area* 1, no. 4, 1970, 52–5.

Harman, Oren Solomon. *The Man Who Invented the Chromosome: A Life of Cyril Darlington*. Cambridge: Harvard University Press, 2004.

Harris, Miriam Coles. *A Corner of Spain*. Boston: Houghton, Mifflin & Company, 1898.

Harrison, Ruth. *Animal Machines*. London: Vincent Stuart, 1964.

Harwood, Jonathan. 'Did Mendelism Transform Plant Breeding? Genetic Theory and Breeding Practice, 1900–1945'. In *New Perspectives on the History of Life Sciences and Agriculture*, edited by Denise Phillips and Sharon Kingsland. Heidelberg, New York, Dordrecht and London: Springer, 2015, 345–70.

Hassan, John. 'Were Health Resorts Bad for Your Health? Coastal Pollution Control Policy in England, 1945–76'. *Environment and History* 5, 1999, 53–73.

Haughton, Edward. 'On the Use of the Compressed Air Bath'. *Dublin Hospital Gazette* 5, 15 February 1858.

Hauser, Kitty. *Shadow Sites: Photography, Archaeology, and the British Landscape 1927–1955*. Oxford: Oxford University Press, 2007.

Hauser, Kitty. *Bloody Old Britain: O.G.S. Crawford and the Archaeology of Modern Life*. London: Granta Books, 2008.

Hawes, Richard. 'The Control of Alkali Pollution in St Helens, 1862–1890'. *Environment and History* 1, 1995, 159–71.

Hawes, Richard. 'The Municipal Regulation of Smoke Pollution in Liverpool, 1853–1866'. *Environment and History* 4, 1998, 75–90.

Haycraft, William, R. *Yellow Steel: The Story of the Earthmoving Equipment Industry*. Champaign: University of Illinois Press, 2000.

Hayden, Thomas. 'On the Respiration of Compressed Air'. *Proceedings of the Royal Irish Academy: Science* 1, 1870–4, 199–208.

Heclo, Hugh and Aaron Wildavsky. *The Private Government of Public Money*. London: Macmillan, 1974.

Heidegger, Martin. 'The Question Concerning Technology' [1949], published in German 1954, reprinted in *Philosophy of Technology: The Technological Condition. An Anthology*, edited by Robert C. Scharff and Val Dusek. Oxford: Blackwell, 2003, 252–64.

Henry, John. 'Historical and Other Studies of Science, Technology and Medicine in the University of Edinburgh'. *Notes and Records of the Royal Society* 62, no. 2, 2008, 223–35.

Herring, Horace. 'The Conservation Society: Harbinger of the 1970s Environment Movement in the UK'. *Environment and History* 7, 2001, 381–401.

Hicks, Marie. *Programmed Inequality: How Britain Discarded Women Technologists and Lost its Edge in Computing*. Cambridge, MA: MIT Press, 2017.

Higgs, John. 'The Museum of English Rural Life'. *Nature* 175, no. 4468, 18 June 1955, 1061.

Higuchi, Toshihiro. 'Atmospheric Nuclear Weapons Testing and the Debate on Risk Knowledge in Cold War America, 1945–1963'. In *Environmental Histories of the Cold War*, edited by J.R. McNeill and Corinna R. Unger. Cambridge: Cambridge University Press, 2010, 319.

Hipkins, S. and S.F. Watts. 'Estimates of Air Pollution in York: 1381–1891'. *Environment and History* 2, 1996, 337–45.

Hiscox, Gardiner Dexter. *Mechanical Movements, Powers, Devices, and Appliances*. New York: N.W. Henley, 1904.

Hissey, James J. *Untravelled England*. London: Macmillan, 1906.

Hissey, James J. *An English Holiday with Car and Camera*. London: Macmillan, 1908.

Hissey, James J. *A Leisurely Tour in England*. London: Macmillan, 1913.

Hissey, James J. *The Road and the Inn*. London: Macmillan, 1917.

HMSO. Cmnd. 3638. Fulton Committee: The Civil Service, Vol. 1, Report of the Committee, 1966–1968. London: HMSO, 1968.

Hodgson, Guy. *War Torn: Manchester, Its Newspapers and the Luftwaffe's Christmas Blitz of 1940*. Chester: University of Chester Press, 2015.

Hoffman, William and Leo Furcht. *The Biologist's Imagination: Innovation in the Biosciences*. Oxford: Oxford University Press, 2014.

Högselius, Per, Arne Kaijser and Erik van der Vleuten. *Europe's Infrastructure Transition: Economy, War, Nature*. New York: Palgrave Macmillan, 2016.

Holderness, Brian. *British Agriculture since 1945*. Manchester: Manchester University Press, 1985.

Hollister-Short, Graham, ed. 'The Current State of History of Technology in Britain'. *History of Technology* 22, 2001, 173–80.

Horne, Frank R. 'Some Aspects of Crop and Seed Improvement since 1945'. *The Journal of the National Institute of Agricultural Botany* 12, 1971, 400.

Hoskins, William G. *The Face of Britain. Midland England: A Survey of the Country between the Chilterns and the Trent*. London: B.T. Batsford, 1949.

Hoskins, William G. 'Chilterns to Black Country: A Portrait'. In *About Britain No. 5. Chilterns to Black Country: A New Guide Book with a Portrait by W.G. Hoskins*, edited by Geoffrey Grigson. London: Collins, 1951, 27.

Hoskins, William G. 'East Midlands and the Peak: A Portrait by W.G. Hoskins'. In *About Britain No. 8. East Midland and the Peak: A New Guide Book with a Portrait by W.G. Hoskins*, edited by Geoffrey Grigson. London: Collins, 1951, 22.

Hoskins, William G. *The Making of the English Landscape*. London: Hodder & Stoughton, 1955.

House of Commons. *The First Report of the Commissioners Appointed to Enquire Into Nature and Extent of the Several Bogs in Ireland; and the Practicability of Draining and Cultivating Them*. London: The House of Commons. 1810.

House of Commons. *The Second Report of the Commissioners Appointed to Enquire Into Nature and Extent of the Several Bogs in Ireland; and the Practicability of Draining and Cultivating Them*. London: The House of Commons, 1811.

House of Commons. *The Fourth Report of the Commissioners Appointed to Enquire Into Nature and Extent of the Several Bogs in Ireland; and the Practicability of Draining and Cultivating Them*. London: The House of Commons, 1814.

House of Commons. 'Land Cultivation'. *Hansard*, 383, col. 732, 29 September 1942.

Hovent, Dr 'Pneumo-Therapeutics'. *The Medical News* 59, 18 July 1891, 64.

Howell, Joel D. *Technology in the Hospital: Transforming Patient Care in the Early Twentieth Century*. Baltimore: Johns Hopkins University Press, 1995.

Howkins, Alun. 'The Discovery of Rural England'. In *Englishness: Politics and Culture*, edited by Robert Colls and Philip Dodd. London: Croom Helm, 1986, 62–88.

Hubbard, Jennifer M. *A Science on the Scales: The Rise of Canadian Atlantic Fisheries Biology, 1898–1939*. Toronto: University of Toronto Press, 2006.

Hubbard, Jennifer. 'Mediating the North Atlantic Environment: Fisheries Biologists, Technology, and Marine Spaces'. *Environmental History* 18, 2013, 88–100.

Hudson, Derek and Kenneth Luckhurst. *The Royal Society of Arts, 1754–1954*. London: John Murray, 1954.

Hughes, Rosaleen. '"Governing in Hard Times": The Heath Government and Civil Emergencies – The 1972 and the 1974 Miners' Strikes'. PhD diss., Queen Mary University of London, 2015.

Hughes, Thomas P. *Networks of Power: Electrification in Western Society*. Baltimore: Johns Hopkins University Press: 1983.

Hughes, Thomas P. 'The Evolution of Large Technological Systems'. In *The Social Construction of Large Technological Systems*, edited by Wiebe Bijker Hughes and Trevor Pinch. Cambridge, MA: MIT Press, 1987, 51–82.

Hughes, Thomas P. *Human-Built World: How to Think about Technology and Culture*. Chicago: University of Chicago Press, 2004.

Huish, Robert. *Memoirs of the Late William Cobbett Esq. MP for Oldham*. London: John Saunders, 1836.

Hume, Joseph. *Brief Notices Respecting the Origin and Purpose of Some of the Departments of the Royal Dublin Society, Referring Also to a Scheme for Improving Bogs and Other Waste Lands in Ireland*. Pamphlets, Hume Tracts, 1831.

Inkpen, Rob. 'Atmospheric Pollution and Stone Degradation in Nineteenth-Century Exeter'. *Environment and History* 5, 1999, 209–20.

Insley, Jane. 'Little Landscapes: Agriculture, Dioramas and the Science Museum'. *International Commission for the History of Technology* 12, 2006, 5–14.

Insley, Jane. 'Little Landscapes: Dioramas in Museum Displays'. *Endeavour* 32, 2008, 27–31.

Jack, J.S., C.W.H. Leech and H.A. Lewis. 'Route Selection and Cable Laying for the Transatlantic Cable System'. *The Bell System Technical Journal* 36, no. 1, January 1957, 293–326.

Jackson, John Brinckerhoff. *Discovering the Vernacular Landscape*. New Haven and London: Yale University Press, 1984.

Jacobs, Nancy. *Environment, Power, and Injustice: A South African History*. Cambridge: Cambridge University Press, 2003.

Jamison, Andrew. *The Making of Green Knowledge. Environmental Politics and Cultural Transformation*. Cambridge: Cambridge University Press, 2001.

Jardine, Nicholas, James A. Secord and E.C. Spary, eds. *Cultures of Natural History*. Cambridge: Cambridge University Press, 1996.

Jasanoff, Sheila. 'Image and Imagination: The Formation of Global Environmental Consciousness'. In *Changing the Atmosphere: Expert Knowledge and Environmental Governance*, edited by Paul Edwards and Clark Miller, Cambridge, MA: MIT Press, 2001.

Jasanoff, Sheila. *Designs on Nature: Science and Democracy in Europe and the United States*. Princeton: Princeton University Press, 2007.

Jasanoff, Sheila. 'Future Imperfect: Science, Technology, and the Imaginations of Modernity'. In *Dreamscapes of Modernity: Sociotechnical Imaginaries and the Fabrication of Power*, edited by Sheila Jasanoff and Sang-Hyun Kim. Chicago: University of Chicago Press, 2015, 1–33.

Johnson, Paige. 'Safeguarding the Atom: The Nuclear Enthusiasm of Muriel Howorth'. *British Journal for the History of Science* 45, 2012, 553.

Johnson, Stanley. *The Politics of Environment: The British Experience*. London: Tom Stacey, 1973.

Johnston, Harold S. 'Reduction of Stratospheric Ozone by Nitrogen Oxide Catalysts from Supersonic Transport Exhaust'. *Science* 173, 1971, 517–22.

Johnston, Harold S. 'Supersonic Aircraft and the Ozone Layer'. *Environmental Change* 2, 1974, 339–50.

Jones, Amanda T. *A Psychic Autobiography*. New York: Greaves, 1910.

Jonsson, Fredrik. *Enlightenment's Frontier: The Scottish Highlands and the Origins of Environmentalism*. New Haven: Yale University Press, 2013.

Jørgensen, Dolly. 'Not By Human Hands: Five Technological Tenets for Environmental History in the Anthropocene'. *Environment and History* 20, 2014, 4.

Jørgensen, Dolly, Finn Arne Jørgensen and Sara B. Pritchard, eds. *New Natures: Joining Environmental History with Science and Technology Studies*. Pittsburgh: University of Pittsburgh Press, 2013.

Kane, Robert. *The Industrial Resources of Ireland*. Dublin: Carrick, 1844.

Kasson, John F. *Civilizing the Machine: Technology and Republican Values in America, 1776–1900*. Harmondsworth: Penguin, 1977.

Kavanagh, Gaynor. 'Mangles, Muck and Myths: Rural History Museums in Britain'. *Rural History* 2, 1991, 187–203.

Keal, E.E. 'Asbestosis and Abdominal Neoplasms'. *Lancet* 276, no. 7162, December 1960, 1211–6.

Keating, Giles. *The Production and Use of Economic Forecasts*. London: Routledge, 1985.

Keller, Evelyn Fox. 'What Does Synthetic Biology Have to Do with Biology?'. *BioSocieties* 4, 2009.

Kelly, Matthew. *Quartz and Feldspar: Dartmoor – A British Landscape in Modern Times*. London: Jonathan Cape, 2015.

Kenward, Michael. 'This Week: Model Behaviour by the British Government'. *New Scientist*, 6 October 1977, 7.

Ketabgian, Tamara Siroone. *The Lives of Machines: The Industrial Imaginary in Victorian Literature and Culture*. Ann Arbor: University of Michigan Press, 2011.

King, William. 'Of the Bogs and Loughs of Ireland by Mr William King, Fellow of the Dublin Society, as It was Presented to That Society'. *Philosophical Transactions* 15, 1685, 948–60.

Kingsbury, Noel. *Hybrid: The History and Science of Plant Breeding*. London and Chicago: University of Chicago Press, 2009.

Kipling, Rudyard. 'The Deep Sea Cables'. In *The Seven Seas*. London: Methuen & Co., 1896.

Kirsch, Scott and Don Mitchell. 'Earth-Moving as the "Measure of Man": Edward Teller, Geographical Engineering, and the Matter of Progress'. *Social Text* 16, no. 1, Spring 1998, 100–34.

Klein, Cornelis. 'Rocks, Minerals, and a Dusty World'. In *Reviews in Mineralogy: Health Effects of Mineral Dusts*, edited by Brooke T. Mossman and George D. Guthrie Jr. Chelsea: Mineralogical Society of America, 1993, 7–59.

Klein, Judy. *Statistical Visions in Time: A History of Time Series Analysis, 1662–1938*. Cambridge: Cambridge University Press, 1997.

Klimeš, Ivan. 'Journalism in Forecasting and Vice Versa'. *Futures* 7, no. 1, 1975, 2.

Kline, Ronald. *The Cybernetics Moment: Or Why We Call Our Age the Information Age*. Baltimore: Johns Hopkins University Press, 2015.

Kloppenburg, Jack Ralph. *First the Seed: The Political Economy of Plant Biotechnology, 1492–2000*. Cambridge: Cambridge University Press 1988.

Knight, Roger. 'New England Forest and British Seapower: Albion Revisited'. *American Neptune* 46, 1986, 221–9.

Knight, Roger and Martin Wilcox. *Sustaining the Fleet, 1793–1815: War, the British Navy and the Contractor State*. Martlesham: Boydell & Brewer, 2010.

Knights, Ben. 'In Search of England: Travelogue and Nation between the Wars'. In *Landscape and Englishness*, edited by Robert Burden and Stephan Kohl. Amsterdam: Rodopi, 2006, 165–84.

Kohl, Stephen. 'Rural England: An Invention of the Motor Industries'. In *Landscape and Englishness*, edited by Robert Burden and Stephan Kohl. Amsterdam: Rodopi, 2006, 185–206.

Krige, John. 'I. I. Rabi and the Birth of CERN'. *Physics in Perspective* 7, 2005, 150–64.

Krige, John. *American Hegemony and the Postwar Reconstruction of Science in Europe*. Cambridge, MA; London: MIT Press, 2008.

Kwa, Chunglin. 'Modeling the Grasslands'. *Historical Studies in the Physical and Biological Studies* 24, no. 1, 1993, 125–55.

Kwa, Chunglin. 'Modelling Technologies of Control'. *Science as Culture* 4, no. 3, 1994, 363–91.

Kyba, Patrick. 'CCMS: The Environmental Connection'. *International Journal* 29, 1973, 256–67.

Lamb, H.H. *Climate History and the Future*. Princeton: Princeton University Press, 1977.

Land, H.A. 'Crawler-Mounted Tractors and Attachments for Scraping and Dozing'. *Proceedings of the Institution of Mechanical Engineers*, 179, part 3F, 1964–5, 89–105.

Larsen, Jonas. 'Tourism Mobilities and the Travel Glance: Experiences of Being on the Move'. *Scandinavian Journal of Hospitality and Tourism* 1, no. 2, 2011, 80–9.

Latour, Bruno. *Politics of Nature*. Cambridge, MA: Harvard University Press, 2009.

Latour, Bruno. 'Onus Orbis Terrarum: About a Possible Shift in the Definition of Sovereignty'. *Millennium: Journal of International Studies* 44, no. 3, 2016, 305–20.

Launius, Roger D. 'Writing the History of Space's Extreme Environment'. *Environmental History* 15, no. 3, 2010, 526–32.

Lawrence, Anna and Norman Dandy. 'Private Landowners' Approaches to Planting and Managing Forests in the UK: What's the Evidence?'. *Land Use Policy* 36, 2014, 351–60.

Lawrence, C.W. 'Genetic Control of Radiation-Induced Chromosome Exchange in Rye'. *Radiation Botany* 3, 1963, 89–94.

Lee, Charles A. *The Physiological and Therapeutical Effects of Compressed Air Baths*. Buffalo: Jones S. Leavitt, 1868.

Leffingwell, Randy. *Caterpillar: Farm Tractors and Bulldozers*. Osceola: Motorbooks International, 1994.

Leggett, Don and Charlotte Sleigh, eds. *Scientific Governance in Britain, 1914–79*. Manchester: Manchester University Press, 2016.

Lemov, Rebecca. *The World as Laboratory: Experiments with Mice, Mazes, and Men*. New York: Hill & Wang, 2005.

Loudon, J.C. *Arboretum et fruticetum Britannicum, or, The Trees and Shrubs of Britain*. Vol. 2. London: Longman, 1838.

Louter, David. *Windshield Wilderness: Cars, Roads, and Nature in Washington's National Parks*. Seattle: University of Washington Press, 2006.

Luckin, Bill. *Pollution and Control: A Social History of the Thames in the Nineteenth Century*. Bristol: Adam Hilger, 1986.

Luckin, Bill. 'At the Margin: Continuing Crisis in British Environmental History?'. *Endeavour* 23, 2004, 97–100.

Luckin, Bill. *Death and Survival in Urban Britain, 1800–1950*. London: Tauris, 2015.

Mabey, Richard. *Weeds: The Story of Outlaw Plants*. London: Profile Books, 163–4.

McCormick, John. *The British Government and the Environment*. London: Earthscan, 1991.

McCulloch, Jock and Geoffrey Tweedale. *Defending the Indefensible: The Global Asbestos Industry and Its Fight for Survival*. Oxford: Oxford University Press, 2008.

Macekura, Stephen J. *Of Limits and Growth. The Rise of Global Sustainable Development in the Twentieth Century*. Cambridge: Cambridge University Press, 2015.

McEvoy, Arthur. 'Working Environments: An Ecological Approach to Industrial Health and Safety'. *Technology and Culture* 36, 1995 (supplement), S145–73.

MacKenzie, John M. *The Empire of Nature: Hunting, Conservation, and British Imperialism*. Manchester: Manchester University Press, 1988.

Mackintosh, Will B. 'Mechanical Aesthetics: Picturesque Tourism and the Mechanical Revolution in Pennsylvania'. *Pennsylvania History: A Journal of Mid-Atlantic Studies* 81, 2014, 88–105.

MacLeod, Christine and Gregory Radick. 'Claiming Ownership in the Technosciences: Patents, Priority and Productivity'. *Studies in History and Philosophy of Science Part A* 44, no. 2, 2013, 188–201.

McManus, John. 'Ballast and the Tay Eider Duck Populations'. *Environment and History* 5 1999, 237–44.

MacNeice, Louis. 'New Jerusalem' [1962]. In *Collected Poems*. London: Faber, 1979, 529.

McNeill, John Robert. *Something New under the Sun: An Environmental History of the Twentieth-Century World*. London: Allen Lane, 2000.

McNeill, John Robert and Corinna R. Unger. 'Introduction: The Big Picture'. In *Environmental Histories of the Cold War*, edited by J.R. McNeill and Corinna R. Unger. Cambridge: Cambridge University Press, 2010.

Madureira, Nuno. 'The Confident Forecaster: Lessons from the Upscaling of the Electricity Industry in England and Wales'. *Business History* 59, no. 3, 2017, 408–30.

Mahony, Martin and Mike Hulme. 'Modelling and the Nation: Institutionalising Climate Prediction in the UK, 1988–92'. *Minerva* 54, no. 4, 2016, 445–70.

Maines, Rachel. *Asbestos and Fire: Technological Tradeoffs and the Body at Risk*. New Brunswick: Rutgers University Press, 2005.

Mandler, Peter. 'Against Englishness: English Culture and the Limits to Rural Nostalgia, 1850–1940'. *Transactions of the Royal Historical Society* 7, 1997, 155–75.

Marshall, D.R. 'The Advantages and Hazards of Genetic Homogeneity'. *Annals of the New York Academy of Sciences* 287, 1977, 17–18.

Marshall, William. *A Review of the Reports to the Board of Agriculture from the Eastern Department of England*. York: Thomas Wilson & Sons, 1811.

Marshall, William. *The Review and Abstract of the County Reports to the Board of Agriculture*. Vol. 1. York: Thomas Wilson & Sons, 1818.

Marsham, Robert. 'A Supplement to the Measures of Trees, Printed in the Philosophical Transactions for 1759'. *Philosophical Transactions of the Royal Society* 87, 1759, 128–32.

Marsham, Robert. 'Observations on the Growth of Trees'. *Philosophical Transactions of the Royal Society* 51, 1759–60, 7–12.

Martin, Brian. *The Bias of Science*. Canberra: Society for the Social Responsibility in Science, 1979.

Marx, Leo. *The Machine in the Garden: Technology and the Pastoral Ideal in America*. New York: Oxford University Press, 1964.

Mason, Arthur. 'Images of the Energy Future'. *Environmental Research Letters* 1, no. 1, 2006, 20–5.

Massard-Guilbaud, Geneviève and Christoph Bernhardt, eds. *Le démon moderne: La pollution dans les sociétés urbaines et industrielles d'Europe*. Clermont-Ferrand: Presses Universitaires Blaise-Pascal, 2002.

Matless, David. 'One Man's England: W.G. Hoskins and the English Culture of Landscape'. *Rural History* 4, no. 2, 1993, 187–207.

Matless, David. 'Moral Geographies of English Landscape'. *Landscape Research* 22, no. 2, 1997, 141–55.

Matless, David. *Landscape and Englishness*. London: Reaktion, 1998.

Matless, David. 'Versions of Animal-Human. Broadland, c.1945–1970'. In *Animal Spaces, Beastly Places*, edited by Chris Philo and Chris Wilbert. London: Routledge, 2000, 117–41.

Matless, David. *In the Nature of Landscape: Cultural Geography on the Norfolk Broads*. Chichester: Wiley-Blackwell, 2014.

Matthew, Patrick. *On Naval Timber and Arboriculture*. London: Longman, 1831.

Matthews, John A. and K.R. Briffa. 'The "Little Ice Age": Re-Evaluation of an Evolving Concept'. *Geografiska Annaler: Series A, Physical Geography* 87, no. 1, 2005, 17.

Mauch, Christof and Thomas Zeller, eds. *The World beyond the Windshield: Roads and Landscapes in the United States and Europe*. Athens, OH: Ohio University Press, 2008.

Meacham, Standish. *Regaining Paradise: Englishness and the Early Garden City Movement*. New Haven: Yale University Press, 1999.

Meadows, Dennis, Donella Meadows, Jorgen Randers and William Behrens. *The Limits to Growth: A Report for the Club of Rome's Project on the Predicament of Mankind*. New York: Potomac Books, 1972.

Meadows, Dennis, William Behrens, Donella Meadows, Roger Naill, Jorgen Randers and Erich Zahn. *Dynamics of Growth in a Finite World*. Cambridge, MA: Wright Allen Press, 1974.

Meadows, Donella, John Richardson and Gerhart Bruckmann. *Groping in the Dark: The First Decade of Global Modelling*. Chichester: Wiley, 1982.

Medford, Derek. 'The New Thaumaturgy of Governmental Research and Development'. *Futures* 1, no. 6, 1968, 510–26.

Medford, Derek. 'Some Remarks on the Application of Technical Forecasting'. In *Technological Forecasting. Proceedings of University of Strathclyde Conference on Technological Forecasting*, edited by R.V. Arnfield. Edinburgh: Edinburgh University Press, 1969, 268–89.

Medford, Derek. *Environmental Harassment or Technology Assessment?* Amsterdam and New York: Elsevier Science, 1973.

Medina, Eden. *Cybernetic Revolutionaries: Technology and Politics in Allende's Chile*. Cambridge, MA: MIT Press, 2011.

Melville, Henry Dundas. *A Letter from Lord Viscount Melville to the Right Hon. Spencer Perceval, on the Subject of Naval Timber*. London: S. Bagster, 1810.

Merriman, Peter. '"Operation Motorway": Landscapes of Construction on England's M1 Motorway'. *Journal of Historical Geography* 31, 2005, 113–33.

Merriman, Peter. 'A New Look at the English Landscape: Landscape Architecture, Movement and the Aesthetics of Motorways in Early Postwar Britain'. *Cultural Geographies* 13, 2006, 78–105.

Merriman, Peter, George Revill, Tim Cresswell, Hayden Lorimer, David Matless, Gillian Rose and John Wylie. 'Landscape, Mobility, Practice'. *Social and Cultural Geography* 9, no. 2, 2008, 191–212.

Miller, Joseph A. 'From Bulls to Bulldozers: A Memoir on the Development of Machines in the Western Woods from Letters of Ted P. Flynn'. *Forest History*, 7, no. 3, Autumn 1963, 14–7.

Mindell, David. *Between Human and Machine: Feedback, Control, and Computing before Cybernetics*. Baltimore: Johns Hopkins University Press, 2002.

Mirrlees, James. 'Nobel Lecture: Information and Incentives: The Economics of Carrots and Sticks'. *The Economic Journal* 107, no. 444, 1997, 1311–29.

Mishan, E. 'Review: The United Kingdom in 1980: The Hudson Report'. *The Political Quarterly* 46, no. 2, 1975, 206–9.

Mitchell, Charles Thomas. 'Some Economical Aspects of Modern Earthmoving Equipment'. *Institution of Civil Engineers Engineering Division Papers*, 4, no. 2, 1946, 3.

Mitchell, Timothy. *Rule of Experts: Egypt, Techno-Politics, Modernity*. Berkeley and Los Angeles: University of California Press, 2002.

Mitchell, Timothy. *Carbon Democracy: Political Power in the Age of Oil*. New York and London: Verso, 2011.

Mitchison, Rosalind. 'The Old Board of Agriculture, 1793–1822'. *The English Historical Review* 74, 1959, 41–69.

Mitman, Gregg. *Reel Nature: America's Romance with Wildlife on Film*. Cambridge, MA: Harvard University Press, 1999.

Mitman, Gregg, Michelle Murphy and Christopher Sellers, eds. Landscapes of Exposure: Knowledge and Illness in Modern Environments. *Osiris* 19, 2004.

Mizuno, Shoko. 'Global Governance of Natural Resources and the British Empire: A Study on the United Nations Scientific Conference on the Conservation and Utilisation of Resources, 1949'. In *Environmental History in the Making, Volume II: Acting*, edited by Cristina Joanaz de Melo, Estelita Vaz and Ligia M. Costa Pinto. New York: Springer, 2016, 291–308.

Moll, Peter. *From Scarcity to Sustainability. Future Studies and the Environment: The Role of the Club of Rome*. Frankfurt: Peter Lang: 1991.

Monteath, Robert. *A New and Easy System of Draining and Reclaiming the Bogs and Marshes of Ireland: With Plans for Improving Waste Lands in General*. Dublin: Carrick, 1829.

Morgan, Mary. *The World in the Model: How Economists Work and Think*. Cambridge: Cambridge University Press, 2012.

Morgan, Mary. 'Nature's Experiments and Natural Experiments in the Social Sciences'. *Philosophy of the Social Sciences* 43, 2013, 3.

Morris, G.P. 'Descriptions of Wheat and Barley Varieties'. *Journal of the National Institute of Agricultural Botany* 7, 1953, 460.

Morris, Peter, ed. *Science for the Nation*. Basingstoke: Palgrave Macmillan, 2010.

Morrison, Sara. 'Forests of Masts and Seas of Trees: The English Royal Forests and the Restoration Navy'. In *English Atlantics Revisited: Essays Honouring Ian K. Steele*, edited by Nancy L. Rhoden. Kingston, ON: McGill-Queen's University Press, 2014, 151.

Mosley, Stephen. *The Chimney of the World: A History of Smoke Pollution in Victorian and Edwardian Manchester*. Cambridge: White Horse Press, 2001.

Mosley, Stephen. '"A Network of Trust": Measuring and Monitoring Air Pollution in British Cities, 1912–1960'. *Environment and History* 15, 2009, 273–302.

Mowle, Alan. 'SARUM 76 Global Modelling Project by SARU'. *Journal of the Operational Research Society* 29, no. 8, 1978, 828.

Muller, Hermann Joseph. 'Variation Due to Change in the Individual Gene'. *The American Naturalist* 56, 1922, 15.

Mumford, Lewis. *The Highway and the City*. New York: New American Library, 1963.

Murray, Alexander. 'Report of Alexander Murray, Esq., Assistant Provincial Geologist, Addressed to W.E. Logan, Esq., Provincial Geologist,' *Canadian Geological Survey*. Montreal: Lovell & Gibson, 1849, 388.

Nahum, Andrew. 'Exhibiting Science: Changing Conceptions of Science Museum Display'. In *Science for the Nation*, edited by Peter Morris. Basingstoke: Palgrave Macmillan, 2010, 176–209.

Nairn, Tom. 'The Future of Britain's Crisis'. *New Left Review* 113, 1979, 43.

NATO Public Diplomacy Division. *Aspects of NATO – The Challenges of Modern Society*. Brussels: NATO, 1976.

NATO Scientific Affairs Division. *The Challenges of Modern Society*. Brussels: NATO, 1991.

Nead, Lynda. *The Haunted Gallery: Painting, Photography, Film c.1900*. New Haven: Yale University Press, 2007.

Nead, Lynda. 'The Age of the "Hurrygraph": Motion, Space and the Visual Image c.1900'. In *The Edwardian Sense: Art, Design, and Performance in Britain 1901–10*, edited by Morna O'Neill and Michael Hatt. New Haven and London: The Paul Mellon Centre for Studies in British Art and Yale University Press, 2010, 99–113.

Nevell, Michael. *Cottonopolis: An Archaeology of the Cotton Industry of North West England*. Manchester: Manchester University Press, 2017.

Newell, Edmund and Simon Watts. 'The Environmental Impact of Industrialisation in South Wales in the Nineteenth Century: "Copper Smoke" and the Llanelli Copper Company. *Environment and History* 2, 1996, 309–36.

Newenham, Thomas. *A View of the Natural, Political, and Commercial Circumstances of Ireland*. London: T. Cadell & W. Davies, 1809.

Newson, Robert, Peter Wade-Martins and Adrian Little. *Farming in Miniature: A Review of British-Made Toy Farm Vehicles up to 1980: Volume 1 Airfix to Denzil Skinner*. Sheffield: Old Pond, 2012.

Newson, Robert, Peter Wade-Martins and Adrian Little. *Farming in Miniature: A Review of British-Made Toy Farm Vehicles up to 1980: Volume 2 Dinky to Wend-al*. Sheffield: Old Pond, 2014.

Nielsen, Kristian. '"What Things Mean in Our Daily Lives": A History of Museum Curating and Visiting in the Science Museum's Children's Gallery from *c*.1929 to 1969'. *British Journal for the History of Science* 47, 2014, 505–38.

Nuffield Council on Bioethics. *Genome Editing: An Ethical Review*. London: Nuffield Council on Bioethics, 2016.

Nye, David E. *American Technological Sublime*. Cambridge, MA: MIT Press, 1994.

Nye, David E. 'Technologies of Landscape'. In *Technologies of Landscape: From Reaping to Recycling*, edited by David E. Nye. Amherst: University of Massachusetts Press, 1999, 3–20.

Ó Gráda, Cormac. *Ireland: A New Economic History 1780–1939*. Oxford: Oxford University Press, 1995.

O'Dea, William. 'The Science Museum's Agricultural Gallery'. *The Museums Journal* 51, 1952, 300.

O'Hanley, David S. 'The Origin of the Chrysotile Asbestos Veins in Southwestern Quebec'. *Canadian Journal of Earth Sciences* 24, no. 1–3, 1987, 1–9.

O'Riordan, Timothy. 'New Conservation and Geography'. *Area* 2, no. 5, 1970, 33–6.

Olby, Robert C. 'Social Imperialism and State Support for Agricultural Research in Edwardian Britain'. *Annals of Science* 48, no. 6, 1991, 509–26.

Olby, Robert C. 'Horticulture: The Font for the Baptism of Genetics'. *Nature Reviews Genetics* 1, 2000, 65–70.

Oosthoek, Jan. *Conquering the Highlands: A History of the Afforestation of the Scottish Uplands*. Canberra: ANU Press, 2013.

Orland, Barbara. 'Turbo-Cows: Producing a Competitive Animal in the Nineteenth and Early Twentieth Centuries'. In *Industrializing Organisms: Introducing Evolutionary History*, edited by Susan R. Schrepfer and Phillip Scranton. New York: Routledge, 2004.

Orlemann, Eric C. *Super-Duty Earthmovers*. Osceola: Motorbooks International, 1999.

Orwin, Charles. *Problems of the Countryside*. Cambridge: Cambridge University Press, 1945.

Orwin, Christabel S. and Edith H. Whetham. *History of British Agriculture 1846–1914*. Newton Abbot: David & Charles, 1971.

Özbekhan, Hasan. 'Toward a General Theory of Planning'. In *Perspectives of Planning*, edited by Erich Jantsch. Bellagio, Italy: OECD, 1968, 45–175.

Özbekhan, Hasan. *The Predicament of Mankind: A Quest for Structured Responses to World Wide Complexities and Uncertainties*. Club of Rome, 1970.

Palladino, Paolo. 'Science, Technology and the Economy: Plant Breeding in Great Britain, 1920–1970'. *The Economic History Review* 49, 1996, 119.

Palladino, Paolo. *Plants, Patients and the Historian: (Re)membering in the Age of Genetic Engineering*. Manchester: Manchester University Press, 2002.

Passmore, John. *Man's Responsibility for Nature: Ecological Problems and Western Traditions*. London: Duckworth, 1974.

Pauly, Philip J. *Controlling Life: Jacques Loeb and the Engineering Ideal in Biology*. Oxford: Oxford University Press, 1987.

Phillips, Horace. 'Power-Farming and Game Birds'. *Sport & Country*, 11 January 1950, 2.

Phillips, John L. *The Bends: Compressed Air in the History of Science, Diving and Engineering*. New Haven: Yale University Press, 1998.

Phythian, Graham. *Blitz Britain: Manchester and Salford*. Stroud: The History Press, 2015.

Pickering, Andrew. 'Cyborg History and the World War II Regime'. *Perspectives on Science* 3, no. 1, 1995, 1–48.

Pickstone, John V. 'Museological Science? The Place of the Analytical/Comparative in 19th-Century Science, Technology and Medicine'. *History of Science*, 1994, 32, 111–38.

Pinder, John, Max Nicholson and Kenneth Lindsay. *Fifty Years of Political and Economic Planning: Looking Forward, 1931–1981*. London: Heinemann, 1981.

Plumb, Henry. 'Address'. *Journal of the National Institute of Agricultural Botany* 14, 1977, 363.

Pomeranz, Kenneth. *The Great Divergence: China, Europe, and the Making of the Modern World Economy*. Princeton: Princeton University Press, 2000.

Potter, Margaret and Alexander Potter. *A History of the Countryside*. Harmondsworth: Puffin, 1944.

Powell, Christopher. *The British Building Industry since 1800: An Economic History*. London: Routledge, 2013.

Poyser, Thomas and Dr Milliet. 'On the Treatment of Chronic and other Diseases by Baths of Compressed Air'. *Association Medical Journal* 1, 9 September 1853, 798.

Pratt, Mary Louise. *Imperial Eyes: Travel Writing and Transculturation*. New York: Routledge, 1992.

Price, Tom. 'The British Economy: A Call for National Planning'. *Futures* 7, no. 3, 1975, 258–60.

Priestley, John B. 'The Beauty of Britain'. In *The Beauty of Britain: A Pictorial Survey*. London: B.T. Batsford, 1937, 1–10.

Prioleau, John. *Car and Country: Week-End Signposts to the Open Road*. London: J.M. Dent and Sons, 1929.

Pritchard, Sara B. *Confluence: The Nature of Technology and the Remaking of the Rhône*. Cambridge, MA: Harvard, 2011.

Purdon, James. 'Landscapes of Power'. Apollo Magazine, 21 December 2012. Last accessed 14 September 2017. https://web.archive.org/web/20130121141715/www.apollo-magazine.com/features/7920953/landscapes-of-power.thtml

Purdon, James. 'Electric Cinema, Pylon Poetry'. *Amodern*, October 2013.

Pyne, S. 'Extreme Environments'. *Environmental History* 15, no. 3, 2010, 509–13.

Rable, George C. *But There Was No Peace: The Role of Violence in the Politics of Reconstruction*. Athens, GA: University of Georgia Press, 1984/2007.

Rackham, Oliver. *The History of the Countryside*. London: J.M. Dent, 1986.

Radkau, Joachim. *Wood: A History*. Cambridge: Polity, 2011.

Rand, Lisa Ruth. 'Orbital Decay: Space Junk and the Environmental History of Earth's Planetary Borderlands'. PhD diss., University of Pennsylvania, 2016.

Rasmussen, Nicolas. *Gene Jockeys: Life Sciences and the Rise of Biotech Enterprise*. Baltimore: Johns Hopkins University Press, 2014.

Redclift, Michael. *Frontiers: Histories of Civil Society and Nature*. Cambridge, MA: MIT Press, 2006.

Reed, Peter. *Acid Rain and the Rise of the Environmental Chemist in Nineteenth-Century Britain: The Life and Work of Robert Angus Smith*. London: Ashgate, 2014.

Rendel, Palmer & Tritton. *Yare Basin Flood Control Study*. London: Rendel, Palmer & Tritton, 1978.

Reuss, Martin. 'Afterword'. In *The Illusory Boundary: Environment and Technology in History*, edited by Martin Reuss and Stephen H. Sutcliffe. Charlottesville: University of Virginia Press, 2010, 298.

Reuss, Martin and Stephen Cutcliffe, eds. *The Illusory Boundary: Environment and Technology in History*. Charlottesville: University of Virginia Press, 2010

Reynolds, J.D. 'Improved Varieties of Peas and Carrots in NIAB Trials'. *The Journal of the National Institute of Agricultural Botany* 12, Supplement, 1970, 16.

Rich, Jennifer. 'Sound, Mobility and Landscapes of Exhibition: Radio-Guided Tours at the Science Museum, London, 1960–1964'. *Journal of Historical Geography* 52, 2016, 61–73.

Richards, Stewart. '"Masters of Arts and Bachelors of Barley": The Struggle for Agricultural Education in Mid-Nineteenth-Century Britain'. *History of Education* 12, 1983, 3.

Richards, Stewart. 'The South-Eastern Agricultural College and Public Support for Technical Education, 1894–1914'. *Agricultural History Review* 36, 1988, 2.

Richmond, Marsha. 'Women in the Early History of Genetics: William Bateson and the Newnham College Mendelians, 1900–1910'. *Isis* 92, no. 1, 2001, 55–90.

Rindzevičiūtė, Egle. *The Power of Systems: How Policy Sciences Opened up the Cold War World*. Ithaca: Cornell University Press, 2016.

Robinson, Samuel. 'Between the Devil and the Deep Blue Sea: Ocean Science and the British Cold War State'. PhD diss., University of Manchester, 2015.

Robson, E.G. 'The Development and Testing of Earth-Moving Machinery'. *Proceedings of the Institution of Mechanical Engineers*, 179, part 3F, 1964–5, 55–64.

Rom, William N. and Markowitz, Steven B. *Environmental and Occupational Medicine*. Lippincott: Williams & Wilkins, 2007.

Rooney, David. '"A Worthy and Suitable House": The Science Museum Buildings and the Temporality of Space'. In *Science for the Nation*, edited by Peter Morris. Basingstoke: Palgrave Macmillan, 2010, 157–75.

Rose, Gene. *The San Joaquin: A River Betrayed*. Clovis: Quill Driver Books/Word Dancer Press, 1992/2000.

Rosen, Christine Meisner. 'Industrial Ecology and the Greening of Business History'. *Business and Economic History* 26, no. 1, 1997, 123–37.

Rosen, Christine Meisner. 'The Business–Environment Connection'. *Environmental History* 10, no. 1, 2005, 77–9.

Rosen, Christine Meisner and Christopher C. Sellers. 'The Nature of the Firm: Towards an Ecocultural History of Business'. *The Business History Review* 73, no. 4, 1999, 577–600.

Rossiter, J.R. 'The North Sea Storm Surge of 31 January and 1 February 1953'. *Philosophical Transactions of the Royal Society A* 246, no. 915, 1954, 371–400.

Rotherham, Ian D. 'Peat Cutters and their Landscapes: Fundamental Change in a Fragile Environment'. *Landscape Archaeology and Ecology*, 1999, 4, 28–51.

Rotherham, Ian D. *The Lost Fens: England's Greatest Ecological Disaster*. Stroud: The History Press, 2013.

Rothery, David A. 'Obduction'. In *The Oxford Companion to the Earth*, edited by Paul Hancock and Brian J. Skinner. Oxford: Oxford University Press, 2000.

Routley, Richard. 'Metaphysical Fallout from the Nuclear Predicament'. *Philosophy & Social Criticism* 10, 1984, 19–34.

Rozwadowski, Helen M. 'Technology and Ocean-scape: Defining the Deep Sea in Mid-Nineteenth Century'. *History and Technology* 17, no. 3, 2001, 217–47.

Rozwadowski, Helen M. *The Sea Knows No Boundaries: A Century of Marine Science under ICES*. Copenhagen and Seattle: ICES/University of Washington Press, 2002.

Rozwadowski, Helen M. *Fathoming the Ocean: The Discovery and Exploration of the Deep Sea*. Cambridge, MA and London: Belknap, 2005.

Rozwadowski, Helen M. 'Ocean's Depths'. *Environmental History* 15, no. 3, 2010, 520–5.

Rozwadowski, Helen M. and David K. Van Keuren. *The Machine in Neptune's Garden: Historical Perspectives on Technology and the Marine Environment*. New York: Science History Publications, 2004.

Runnion, James B. 'The Negro Exodus'. *Atlantic Monthly* 44, no. 262, August 1879, 226–7.

Russell, Edmund. 'Can Organisms Be Technology?'. In *The Illusory Boundary: Environment and Technology in History*, edited by Martin Reuss and Stephen H. Cutcliffe. Charlottesville: University of Virginia Press, 2010, 249–62.

Russell, Edmund, James Allison, Thomas Finger, John K. Brown, Brian Balogh and W. Bernard Carlson. 'The Nature of Power: Synthesizing the History of Technology and Environmental History'. *Technology and Culture* 52, April 2011, 246–59.

Rustan, Agne, Claude Cunningham, William Fourney, Alex Spathis and K.R.Y. Simha, eds. *Mining and Rock Construction Technology Desk Reference: Rock Mechanics, Drilling and Blasting.* London: CRC Press, 2010.

Ruuskanen, Esa. 'Valuing Wetlands and Peatlands: Mires in the Natural Resource and Land Use Policies in the Nordic Countries from the Late Eighteenth Century to the Present Date'. In *Trading Environments. Frontiers, Commercial Knowledge, and Environmental Transformation, 1750–1990*, edited by Gordon M. Winder and Andreas Dix. New York: Routledge, 2016, 118–32.

Ryan, Patrick. 'Meadows under Scrutiny'. *New Scientist*, 20 April 1972, 166.

Ryan , Paul D. 'Caledonides'. In *The Oxford Companion to the Earth*, edited by Paul Hancock and Brian J. Skinner. Oxford: Oxford University Press, 2000.

Sachs, Wolfgang. *For the Love of the Automobile. Looking Back into the History of our Desires.* Berkeley: University of California Press, 1992.

Salaman, Redcliffe N. 'Half a Century of Potato Research', Proceedings of the Second Conference on Potato Virus Diseases Lisse-Wageningen. 25–29 June, 1954.

Sandbach, Francis. 'The Rise and Fall of the Limits to Growth Debate'. *Social Studies of Science* 8, no. 4, 1978, 495–520.

Saraiva, Tiago. *Fascist Pigs: Technoscientific Organisms and the History of Fascism.* Cambridge, MA: MIT Press, 2016.

Schalk, David L. 'La trahison des clercs – 1927 and Later'. *French Historical Studies* 7, no. 2, Autumn 1971, 245–63.

Schatzberg, Eric. 'Technik Comes to America: Changing Meanings of Technology before 1930'. *Technology and Culture* 47, 2006, 486–512.

Schepers, Gerrit W.H. 'Chronology of Asbestos Cancer Discoveries: Experimental Studies of the Saranac Laboratory'. *American Journal of Industrial Medicine* 27, 1995, 602–3.

Schivelbusch, Wolfgang. *The Railway Journey: The Industrialization of Time and Space in the 19th Century.* Berkeley: University of California Press, 1986.

Schmelzer, Matthias. '"Born in the Corridors of the OECD": The Forgotten Origins of the Club of Rome, Transnational Networks, and the 1970s in Global History'. *Journal of Global History* 12, no. 1, 2017, 26–48.

Schmidt, Markus, et al. *Synthetic Biology: The Technoscience and its Societal Consequences.* Dordrecht: Springer, 2009

Schmidt-Gernig, Alexander. 'The Cybernetic Society: Western Future Studies of the 1960s and 1970s and their Predictions for the Year 2000'. In *What the Future Holds: Insights from Social Science*, edited by Richard Cooper and Richard Layard. Cambridge, MA: MIT Press, 2003, 233–60.

Schneider, Daniel. *Hybrid Nature: Sewage Treatment and the Contradictions of the Industrial Ecosystem.* Cambridge, MA: MIT Press, 2011.

Schoijet, Mauricio. 'Limits to Growth and the Rise of Catastrophism'. *Environmental History* 4, 1999, 515–30.

Schrepfer, Susan and Philip Scranton. *Industrializing Organisms: Introducing Evolutionary History.* London: Routledge, 2004.

Schulz, Thorsten. 'Transatlantic Environmental Security in the 1970s? NATO's "Third Dimension" as an Early Environmental and "Human Security" Approach'. *Historical Social Research* 35, 2010, 309–28.

Scott, James C. *Seeing Like a State: How Certain Schemes to Improve the Human Condition Have Failed*. New Haven: Yale University Press, 1998.

Secord, Anne. 'Corresponding Interests: Artisans and Gentlemen in Nineteenth-Century Natural History'. *British Journal for the History of Science* 27, 1994, 383–408.

Secord, Anne. 'Science in the Pub: Artisan Botanists in Early Nineteenth-Century Lancashire'. *History of Science* 32, 1994, 269–315.

Seddon, Quentin. *The Silent Revolution: Farming and the Countryside into the 21st Century*. London: BBC Books, 1989.

Seefried, Elke. 'Towards the Limits to Growth? The Book and its Reception in West Germany and Britain, 1972–1973'. *German Historical Institute London Bulletin* 33, no. 1, 2011, 3–37.

Self, Peter and Herbert Storing. *The State and the Farmer*. London: George Allen & Unwin, 1962.

Seymour, John. 'A New Kind of Man'. *Resurgence* 1, no. 12, March/April 1968, 18–19.

Sheail, John. *Nature in Trust: The History of Nature Conservation in Britain*. Glasgow: Blackie, 1976.

Sheail, John. 'Government and the Perception of Reservoir Development in Britain: An Historical Perspective'. *Planning Perspectives* 1, 1986, 1.

Sheail, John. 'River Regulation in the United Kingdom: An Historical Perspective'. *River Research and Applications* 2, 1988, 3.

Sheail, John. 'Town Wastes, Agricultural Sustainability and Victorian Sewage'. *Urban History* 23, 1996, 189–210

Sheail, John. *Nature Conservation in Britain: The Formative Years*. London: Stationery Office, 1998.

Sheail, John. 'Review of Dale H. Porter, *The Thames Embankment: Environment, Technology and Society in Victorian London*'. *Environment and History* 4, 1998, 371–2.

Sheail, John. *An Environmental History of Twentieth-Century Britain*. London: Palgrave, 2002.

Sheail, John. '*Torrey Canyon*: The Political Dimension'. *Journal of Contemporary History* 42, no. 3, 2007, 485–504.

Sheail, John. 'Pesticides and the British Environment: An Agricultural Perspective'. *Environment and History* 19, 2013, 87–108.

Shennum, R.H. and P. T. Haury. 'A General Description of the Telstar Spacecraft'. *The Bell System Technical Journal* 42, no. 4, July 1963, 820–21.

Shoard, Marion. *The Theft of the Countryside*. London: Temple Smith, 1980.

Short, Brian. *The Battle of the Fields: Rural Community and Authority in Britain during the Second World War*. Woodbridge: Boydell & Brewer, 2014.

Short, Brian, Charles Watkins and John Martin, eds. *The Front Line of Freedom: British Farming in the Second World War*. London: British Agricultural History Society, 2007.

Silvey, Valerie. 'The Contribution of New Varieties to Cereal Yields in England and Wales between 1947 and 1983'. *The Journal of the National Institute of Agricultural Botany* 17, 1986, 155–68.

Simmonds, Norman, W. *Principles of Crop Improvement*. New York: Longman, 1979.

Simmons, I.G. *Environmental History of Great Britain from 10,000 Years Ago to the Present*. Edinburgh: Edinburgh University Press, 2001.

Simms, E. *The Public Life of the Street Pigeon*. London: Hutchinson Radius, 1979.

Simpson, Archibald. *On Compressed Air as a Therapeutic Agent in the Treatment of Consumption. Asthma, Chronic Bronchitis, and Other Diseases*, Edinburgh: Sutherland & Knox, 1857.

Sinclair, Clive. 'Technological Change = Social Change? The Work of the Science Policy Research Unit at Sussex University in the Field of Forecasting'. *Industrial Marketing Management* 2, no. 4, 1973, 375–90.

Sinclair, George. *Useful and Ornamental Planting*. London: Baldwin & Craddock, 1832.

Singer, Charles, ed. *A History of Technology*. Oxford: Clarendon Press, 5 volumes. 1954–8.

Sivaramakrisnan, K. *Modern Forests: Statemaking and Environmental Change in Colonial Eastern India*. Redwood City: Stanford University Press, 1999.

Skelton, Leona. *Tyne after Tyne: An Environmental History of a River's Battle for Protection 1529–2015*. Winwick: White Horse, 2017.

Slotten, Hugh Richard. 'Satellite Communications, Globalization, and the Cold War'. *Technology and Culture* 43, no. 2, 2002, 315–50.

Slotten, Hugh Richard. 'International Governance, Organizational Standards, and the First Global Satellite Communication System'. *Journal of Policy History* 27, no. 3, 2015, 521–49.

Smallman, R.E. and J.F. Knott. 'Sir Alan Cottrell, 17 July 1919 –15 February 2012'. *Biographical Memoir of the Fellows of the Royal Society* 59, 2013, 93–124.

Smil, Vaclav. 'Perils of Long-Range Energy Forecasting: Reflections on Looking Far Ahead'. *Technological Forecasting and Social Change* 65, no. 3, 2000, 251–64.

Smith, Edward H. *Quantock Life and Rambles*. Taunton: The Wessex Press, 1939.

Smith, Kenneth M. 'Redcliffe Nathan Salaman. 1874–1955'. *Biographical Memoirs of Fellows of the Royal Society* 1, 1955.

Smith, Mark J. 'Creating an Industrial Plant: The Biotechnology of Sugar Production in Cuba'. In *Industrializing Organisms: Introducing Evolutionary History*, edited by Susan Schrepfer and Philip Scranton. London: Routledge, 2004.

Smith, Roger. 'BT Bites Back … New Cable More than a Match for "Jaws"'. *BT Journal* 8, no. 2, 1987, 38–41.

Smout, T.C. 'Bogs and People in Scotland since 1600'. In *Exploring Environmental History: Selected Essays*, Edinburgh: Edinburgh University Press, 2009, 104.

Smout, T.C. 'Garrett Hardin, the Tragedy of the Commons and the Firth of Forth'. *Environment and History* 17, 2011, 357–78.

Snoke, L.R. 'Resistance of Organic Materials and Cable Structures to Marine Biological Attack'. *The Bell System Technical Journal* 36, no. 5, September 1957, 1095–27.

Soulé, Michael E. and Bruce A. Wilcox, eds. *Conservation Biology: An Evolutionary Perspective*. Sunderland: Sinauer Associates, 1980.

Spencer, A.J. and J.B. Passmore. *Agricultural Implements and Machinery*. London: HMSO, 1930.

Spicer, Michael. 'Towards a Policy Information and Control System'. *Public Administration* 48, no. 4, 1970, 443–8.

Spicer, Michael. *The Spicer Diaries*. London: Biteback, 2012.

Stamp, Dudley. *Nature Conservation in Britain*. London: Collins New Naturalist, 1969.

Starosielski, Nicole. *The Undersea Network*. Durham, NC: Duke University Press, 2015.

Steers, J.A. 'The East Coast Floods'. *Geographical Journal* 119, no. 3, September 1953, 280–95.

Steffen, Will, Jacques Grinevald, Paul Crutzen and John McNeil. 'The Anthropocene: Conceptual and Historical Perspectives'. *Philosophical Transactions of the Royal Society A* 369, no. 1938, 2011, 842–67.

Stillman, Edmund, James Bellini, William Pfaff, Laurence Schloesing, Jonathan Story, Jean-Jacques De Peretti and Herman Kahn. *L'Envol de la France: Portrait de la France dans les années 80*. Paris: Hachette, 1973.

Stillman, Edmund, James Bellini, William Pfaff, Laurence Schloesing and Michael Barth. *The United Kingdom in 1980: The Hudson Report*. London: Association Business Programmes, 1974.

Stine, Jeffrey K. and Joel A. Tarr. 'At the Intersection of Histories: Technology and the Environment'. *Technology and Culture* 39, 1998, 601–40.

Stirling, Andy. '"Opening Up" and "Closing Down": Power, Participation, and Pluralism in the Social Appraisal of Technology'. *Science, Technology, & Human Values* 33, 2008, 2.

Stoddart, Anna M. *Elizabeth Pease Nichol*. London: J.M. Dent & Co., 1899.

Stoddart, David. 'Our Environment'. *Area* 2, no. 1, 1970, 1–4.

Stone, J. *Morbi Aere Vitali Curantur, or, Philosophy and Application of Condensed Air as a Curative Agent*. Rochester: Rochester Union and Advertising Company's Printing, 1881.

Stone, Richard. 'Review: Decision in Government'. *The Economic Journal* 81, no. 323, 1971, 668–70.

Tarlow, Sarah. *The Archaeology of Improvement in Britain, 1750–1850*. Cambridge: Cambridge University Press, 2007.

Tarr, Joel and Gabriel Dupuy, eds. *Technology and the Rise of the Networked City in Europe and America*. Philadelphia: Temple University Press, 1988.

Taylor, Charles F. 'Dr. Taylor's "Compressed Air-Bath"'. *The Water-Cure Journal and Herald of Reforms* 25, 1857, 51.

Taylor, Peter. *Unruly Complexity: Ecology, Interpretation, Engagement*. Chicago: University of Chicago Press, 2005.

Taylor, Peter and Frederick Buttel. 'How Do We Know We Have Global Environmental Problems? Science and the Globalisation of Environmental Discourse'. *Geoforum* 23, no. 3, 1992, 405–16.

Taylor, Robert. 'The Heath Government and Industrial Relations: Myth and Reality'. in *The Heath Government, 1970–1974: A Reappraisal*, edited by Stuart Ball and Anthony Seldon. New York and London: Longman, 1996, 161–90.

Taylor, Vanessa. 'Watershed Democracy or Ecological Hinterland? London and the Thames River Basin, 1857–1989'. In *Rivers Lost, Rivers Regained: Rethinking City–River Relations*, edited by Martin Knoll, Uwe Lübken and Dieter Schott. Pittsburgh: University of Pittsburgh Press, 2017, 63–81.

Taylor, Vanessa, Heather Chappells, Will Medd and Frank Trentmann. 'Drought is Normal: The Socio-Technical Evolution of Drought and Water Demand in England and Wales, 1893–2006'. *Journal of Historical Geography* 35, 2009, 568–91.

Taylor, William M. *The Vital Landscape: Nature and the Built Environment in Nineteenth-Century Britain*. Burlington: Ashgate, 2004.

Technology Strategy Board. *A Synthetic Biology Roadmap for the UK*. Swindon: Technology Strategy Board, 2012.

Thirsk, Joan. 'The British Agricultural History Society and *The Agrarian History of England and Wales*: New Projects in the 1950s'. *Agricultural History Review* 50, 2002, 155–63.

Thomas, Keith. *Man and the Natural World: Changing Attitudes in England 1500–1800*. Harmondsworth: Penguin, 1984.

Thomas, William. *Rational Action: The Sciences of Policy in Britain and America, 1940–1960*. Cambridge, MA: MIT Press, 2015.

Thomas, William and Lambert Williams. 'The Epistemologies of Non-Forecasting Simulations, Part 1: Industrial Dynamics and Management Pedagogy at MIT'. *Science in Context* 22, no. 2, 2009, 245–70.

Thomson, R. Wodrow. *Ben Rhydding: The Asclepion of England: Its Beauties, Its Ways, and Its Water-Cure*. Second edition. Ilkley: John Shuttleworth, 1862.

Thorarinsson, Sigurdar. 'Present Glacier Shrinkage, and Eustatic Changes of Sea-level'. *Geografiska Annaler* 22, 1940, 131–59.

Thorsheim, Peter. 'The Paradox of Smokeless Fuels: Gas, Coke and the Environment in Britain, 1813–1949'. *Environment and History* 8, 2002, 381–401.

Thorsheim, Peter. *Inventing Pollution: Coal Smoke and Culture in Britain since 1800*. Athens, OH: Ohio University Press, 2006.

Thorsheim, Peter. *Waste into Weapons: Recycling in Britain during the Second World War*. Cambridge: Cambridge University Press, 2015.

Tichelar, Michael. '"Putting Animals into Politics": The Labour Party and Hunting in the First Half of the Twentieth Century'. *Rural History* 17, 2006, 213–34.

Tinker, Jon. 'Britain to Start a Limits to Growth Unit'. *New Scientist*, 15 June 1972, 615.

Tomlinson, John. *Tomlinson's Handy Guide to Ben Rhydding, Bolton Abbey, and the Neighbourhood*. London: Robert Hardwicke, 1864.

Trachtenberg, Marc. *A Constructed Peace: The Making of the European Settlement, 1945–1963*. Princeton: Princeton University Press, 1999.

Train, R.E. 'A New Approach to International Environmental Cooperation: The NATO CCMS'. *Kansas Law Review* 22, 1973–4, 167–91.

Trent, Christopher. *Motoring on English Byways: A Practical Guide for Wayfarers*. London: G.T. Foulis, 1962.

Troup, L.C. 'The Afforestation of Chalk Downland'. *Forestry* 27, no. 2, 1954, 135–44.

Tucker, Jennifer. *Nature Exposed: Photography as Eyewitness in Victorian Science*. Baltimore: Johns Hopkins University Press, 2005.

Tuke, John. *General View of the Agriculture of the North Riding of Yorkshire*. London: McMillan, 1794.

Tully, John. 'A Victorian Ecological Disaster: Imperialism, the Telegraph, and Gutta-Percha'. *Journal of World History* 20, no. 4, 2009, 559–79.

Turchetti, Simone. 'Looking for the Bad Teachers: The Radical Science Movement and its Transnational History'. In *Science Studies during the Cold War and Beyond*, edited by Elena Aranova and Simone Turchetti. New York: Palgrave Macmillan, 2016, 77–102.

Turchetti, Simone. *The Greening Alliance: Science, Environment and the North Atlantic Treaty Organization*. Chicago: University of Chicago Press, forthcoming.

Turner, Michael. *After the Famine: Irish Agriculture, 1850–1914*. Cambridge: Cambridge University Press, 1996.

Tweedale, Geoffrey. *Magic Mineral to Toxic Dust: Turner & Newall and the Asbestos Hazard*. Oxford: Oxford University Press, 2001.

Underwood, Betty. *Ormskirk Workhouse, Two World Wars, NHS. Hospital*. Accrington: Nayler Group, 2007.

Vale, Barbara. *Prefabs: The History of the UK Temporary Housing Programme*. London: Routledge, 2003.

van Harten, A.M. *Mutation Breeding: Theory and Practical Applications*. Cambridge: Cambridge University Press, 1998.

van Horssen, Jessica. '"À faire un peu de poussière": Environmental Health and the Asbestos Strike of 1949'. *Labour/LeTravail* 70, Fall 2012, 101–32.

van Horssen, Jessica. *A Town Called Asbestos: Environmental Health, Contamination, and Resilience in a Resource Community*. Vancouver: University of British Columbia Press, 2016.

Veldman, Meredith. *Fantasy, the Bomb, and the Greening of Britain: Romantic Protest, 1945–1980*. Cambridge: Cambridge University Press, 1994.

Viles, Heather. '"Unswept Stone, Besmeer'd by Sluttish Time": Air Pollution and Building Stone Decay in Oxford, 1790–1960'. *Environment and History* 2, 1996, 359–72.

Vogt, A. 'Aspects of Historical Soil Erosion in Western Europe'. In *The Silent Countdown: Essays in European Environmental History*, edited by Christian Pfister and Peter Brimblecombe. Berlin: Springer Verlag, 1990, 83–91.

von Braun, Wernher. 'Now at Your Service – The World's Most Talkative Satellite'. *Popular Science* 56–7, May 1971, 138.

Waddington, Conrad, ed. *Towards a Theoretical Biology*: *Prolegomena*. Edinburgh: Edinburgh University Press, 1968.

Waddington, Conrad. 'School of the Man-Made Future at Scottish University'. *Futures* 4, no. 4, 1972, 378.

Waddington, Conrad. *Tools for Thought: How to Understand and Apply the Latest Scientific Techniques of Problem Solving*. Basic Books, New York, 1977.

Waff, Craig B. 'Project Echo, Goldstone, and Holmdel: Satellite Communications as Viewed from the Ground Station'. In *Beyond the Ionosphere: Fifty Years of Satellite Communication*, edited by Andrew J. Butrica. Washington, DC: NASA History Office, 1997, 41–50.

Waite, Vincent. *Portrait of the Quantocks*. London: Robert Hale, 1964.

Waldenburg, L. 'On a Portable Pneumatic Apparatus for the Mechanical Treatment of Diseases of the Lungs and Heart'. *British Medical Journal* 1, 11 April 1874.

Warde, Paul. 'Fear of Wood Shortage and the Reality of the Woodland in Europe, c.1450–1850'. *History Workshop Journal* 62, 2006, 28–57.

Warde, Paul. 'The Environment'. In *Local Places, Global Processes: Histories of Environmental Change in Britain and Beyond*, edited by Peter Coates, David Moon and Paul Warde. Oxford: Windgather, 2016, 32–46.

Warde, Paul and Sverker Sörlin. 'Expertise for the Future: The Emergence of Environmental Prediction c.1920–1070'. In *The Struggle for the Long Term in Transnational Science and Politics: Forging the Future*, edited by Jenny Andersson and Egle Rindzevičiūtė. London and New York: Routledge, 2015, 38–62.

Waverley Committee. *Report of the Departmental Committee on Coastal Flooding*. HMSO, 1954.

Weart, Spencer. *The Discovery of Global Warming*. Cambridge, MA: Harvard University Press, 2008.

Wendel, C.H. *Encyclopedia of American Farm Implements and Antiques*. Iola: Krause Publications, 2004.

West, Lesley. 'An Agricultural History Museum'. *Agricultural History* 41, 1967, 269.

White, Richard. *Organic Machine: The Remaking of the Columbia River*. New York: Hill & Wang, 1996.

White, Richard Grant. 'Americanisms'. *The Galaxy* 24, no. 3, September 1877, 383.

Wiener, Martin J. *English Culture and Decline of Industrial Spirit, 1850–1980*. Cambridge: Cambridge University Press, 1981.

Wiener, Norbert. *The Human Use of Human Beings: Cybernetics and Society*. Doubleday, 1954.

Wilkinson, Clive. *The British Navy and the State in the Eighteenth Century*. Martlesham: Boydell & Brewer, 2004.

Williams, C. Theodore. 'Lectures on the Compressed Air Bath and its Uses in the Treatment of Disease'. *British Medical Journal* 1, 25 April 1885, 825.

Williams, C. Theodore. 'The Treatment of Bronchial Asthma'. *The American Journal of the Medical Sciences* 96, 1888.

Williams, Raymond. *Culture and Society 1780–1950*. London: Chatto & Windus, 1958.

Williams, Raymond. *The Country and the City*. London: Chatto & Windus, 1973.

Williams, Watkin. 'Genetics and Plant Breeding'. *Nature* 4485, 1955, 719.

Wilson, Harold. *Labour's Plan for Science: Reprint of Speech by the Rt Hon. Harold Wilson, MP, Leader of the Labour Party at the Annual Conference Scarborough, Tuesday, October 1*. London: The Labour Party, 1963.

Wilson, Norman A.B. *On the Quality of Working Life: A Report Prepared for the Department of Employment*. London: HMSO, 1973.

Winder, Gordon M. and Andreas Dix. 'Commercial Knowledge and Environmental Transformation'. In *Trading Environments. Frontiers, Commercial Knowledge, and Environmental Transformation, 1750–1990*, edited by Gordon M. Winder and Andreas Dix. New York: Routledge, 2016, 14.

Winter, James. *Secure from Rash Assault: Sustaining the Victorian Environment*. Berkeley: University of California Press, 1999.

Withers, William. *A Letter to Sir Henry Steuart, Bart on the Improvement in the Quality of Timber, to Be Effected By the High Cultivation and Quick Growth of Forest-Trees*. London and Holt: Longman, James Shalders, 1829.

Withers, William. *The Acacia Tree, Robinia Pseudo Acacia: Its Growth, Qualities, and Uses*. London and Holt: Longman, James Shalders, 1842.

Wohl, Anthony S. *Endangered Lives: Public Health in Victorian Britain*. London: Dent, 1983.

Wood, Leslie B. *The Restoration of the Tidal Thames*. Bristol: Adam Hilger, 1982.

Woods, Abigail. 'Partnership in Action: Contagious Abortion and the Governance of Livestock Disease in Britain, 1885–1921'. *Minerva* 47, 2009, 2

Woods, Abigail. 'A Historical Synopsis of Farm Animal Disease and Public Policy in Twentieth Century Britain'. *Philosophical Transactions of the Royal Society, B* 366, 2011, 1943–54.

Woods, Abigail. 'Rethinking the History of Modern Agriculture: British Pig Production, c.1910–65'. *Twentieth Century British History* 23, 2012, 165–91.

Woods, Abigail. 'Science, Disease and Dairy Production in Britain, c.1927 to 1980'. *Agricultural History Review* 62, 2014, 294–314.

Woodward, Rachel. *Military Geographies*. Oxford: Blackwell, 2004.

Wright, Nigel. 'The Formulation of British and European Policy toward an International Satellite Telecommunications System: The Role of the British Foreign Office'. In *Beyond the Ionosphere: Fifty Years of Satellite Communication*, edited by Andrew J. Butrica. Washington, DC: NASA History Office, 1997, 157–70.

Wrigley, E.A. 'The Supply of Raw Materials in the Industrial Revolution'. *Economic History Review*, 1962, 15, 1–16.

Young, Arthur. *Arthur Young's Tour in Ireland, 1776–1779*. London: George Bell & Sons, 1892.

Zachmann, Karin. 'Peaceful Atoms in Agriculture and Food: How the Politics of the Cold War Shaped Agricultural Research Using Isotopes and Radiation in Post War Divided Germany'. *Dynamis* 35, 2015, 312.

Zuckerman, Solly. *Monkeys, Men, and Missiles: An Autobiography, 1946–88*. New York: Norton, 1989.

Index